Lecture Notes in Physics

300

Berthold-Georg Englert

Semiclassical Theory of Atoms

Springer-Verlag
Berlin Heidelberg GmbH

Author

Berthold-Georg Englert
Universität München, Sektion Physik
Am Coulombwall 1, D-8046 Garching, F.R.G.

ISBN 978-3-662-13681-2 ISBN 978-3-540-39141-8 (eBook)
DOI 10.1007/978-3-540-39141-8

Library of Congress Cataloging-in-Publication Data. Englert, B.-G. (Berthold-Georg), 1953- Semiclassical theory of atoms / B.-G. Englert. p. cm.–(Lecture notes in physics; 300) Includes bibliographies and indexes. ISBN 0-387-19204-2 (U.S.) 1. Atoms–Models. 2. Thomas-Fermi theory. I. Title. II. Series.
QC173.E56 1988 539'.14–dc 19 88-4949

Originally published by Springer-Verlag Berlin Heidelberg New York in 1988
Softcover reprint of the hardcover 1st edition 1988

2158/3140-543210

Preface

This book grew out of a set of notes that I supplied to the audience of a series of lectures on "The Thomas-Fermi Method in Atomic Physics and Its Refinements" delivered at the University of Munich in 1985. Standard textbook material played a minor role during these lectures; the emphasis was on the novel approach developed by Professor Julian Schwinger and myself, beginning about eight years ago. As a consequence, this book is the first complete, detailed step-by-step presentation of our ideas and their implications.

Naturally, the work of other researchers is not ignored. In particular, I have tried to collect and organize the many pieces of knowledge about the Thomas-Fermi model that are scattered over as many original publications. On the other hand, my intention was not to supply a complete list of every paper on the subject, as this would have been of little value. Thus referencing is selective and I cite only the most relevant papers. On a few occasions honesty demanded critical remarks about someone else's work; I hope that these comments will not be misunderstood as put-downs.

The reader is not expected to have any previous knowledge about the subject. In addition to an open mind, the only prerequisite is a thorough understanding of elementary quantum mechanics, and some familiarity with the phenomenology of atoms is certainly helpful. The text consists of a mixture of general concepts and technical detail. Both need to be absorbed, although some of the latter can be skipped during a first reading. I trust that readers can perform a reasonable selection themselves.

I am grateful for the many insights gained in discussions with a large number of people. Being afraid of forgetting somebody, I shall not even try to list them.

It is a pleasure to thank Mrs. E. Figge, who typed the manuscript with enviable skill.

Garching, February 1988 B.-G. Englert

TABLE OF CONTENTS

Chapter One. Introduction.. 1

Atomic units.. 2

Bohr atoms.. 4

Traces and phase-space integrals.. 10

Bohr atoms with shielding.. 11

The effective potential.. 13

Size of atoms.. 21

Problems.. 24

Chapter Two. Thomas-Fermi Model.. 27

General formalism.. 27

The TF model.. 33

Neutral TF atoms.. 37

Maximum property of the TF potential functional.. 39

An electrostatic analogy.. 41

TF density functional.. 43

Minimum property of the TF density functional.. 45

Upper bounds on B.. 46

Lower bounds on B.. 50

Binding energy of neutral TF atoms.. 54

TF function F(x).. 56

Scaling properties of the TF model.. 66

Highly ionized TF atoms.. 73

Weakly ionized TF atoms.. 80

Arbitrarily ionized TF atoms.. 96

Validity of the TF model.. 99

Density and potential functionals.. 104

Relation between the TF approximation and Hartree's method.... 119

Problems.. 124

Chapter Three. Strongly Bound Electrons.. 130

Qualitative argument.. 130

First quantitative derivation of Scott's correction.. 131

Scott's original argument.. 138

TFS energy functional.. 139

TFS density.. 146

Consistency.......... 148
Scaling properties of the TFS model.......... 155
Second quantitative derivation of Scott's correction.......... 158
Some implications concerning energy.......... 162
Electron density at the site of the nucleus.......... 164
Numerical procedure.......... 168
Numerical results for neutral mercury.......... 170
Problems.......... 173

Chapter Four. Quantum Corrections and Exchange.......... 175
Qualitative arguments.......... 176
Quantum corrections I (time transformation function).......... 177
Quantum corrections II (leading energy correction).......... 190
The von Weizsäcker term.......... 194
Quantum corrections III (energy).......... 196
Airy averages.......... 199
Validity of the TF approximation.......... 207
Quantum corrected $E_1(V+\zeta)$.......... 210
Quantum corrected density.......... 213
Exchange I (general).......... 221
Self energy.......... 225
Exchange II (leading correction).......... 225
History.......... 230
Energy corrections for ions.......... 232
Ionization energies.......... 236
Minimal binding energies (chemical potentials).......... 239
Shielding of the nuclear magnetic moment.......... 241
Simplified new differential equation (ES model).......... 244
An application of the ES model. Diamagnetic susceptibilities.. 255
Improved (?) ES model. Electric polarizabilities.......... 260
Exchange III. (Exchange potential).......... 275
New differential equation.......... 278
Small distances.......... 282
Large distances.......... 283
Numerical results.......... 286
Problems.......... 293

Chapter Five. Shell Structure.......... 295
Experimental evidence.......... 297
Qualitative arguments.......... 299
Bohr atoms.......... 300

TF quantization 305
Fourier formulation 312
Isolating the TF contribution 314
Lines of degeneracy 315
Classical orbits 320
Degeneracy in the TF potential 323
TF degeneracy and the systematics of the Periodic Table 327
General features of N_{qu} 329
General features of E_{qu} 333
Linear degeneracy. Scott correction 335
Perturbative approach to E_{osc} 336
ℓ-quantized TF model 338
$j\neq 0$ terms. Leading energy oscillation 342
Fresnel integrals 347
λ oscillations 351
ν oscillations 358
Semiclassical prediction for E_{osc} 360
Other manifestations of shell structure 363
Problems 367

Chapter Six. Miscellanea 370
Relativistic corrections 370
Kohn-Sham equation 377
Wigner's phase-space functions 379
Problems 381

Footnotes 383
Chapter One 383
Chapter Two 384
Chapter Three 387
Chapter Four 389
Chapter Five 394
Chapter Six 395

Index 397
Names 397
Subjects 398

Chapter One

INTRODUCTION

Atoms that contain many electrons possess a degree of complexity so high that it is impossible to give an exact answer even when we are asking simple questions. We are therefore compelled to resort to approximate descriptions. Two main approaches have been pursued in theoretical atomic physics. One is the Hartree-Fock(HF) method and its refinements; it can be viewed as a generalization of Schrödinger's description of the hydrogen atom to many-electron systems; it is, by construction, the more reliable the smaller the number of electrons. The other one is the Thomas-Fermi (TF) treatment and its improvements; this one uses the picture of an electronic atmosphere surrounding the nucleus; it is the better the larger the number of electrons. For this reason, the TF method is frequently called the "statistical theory of the atoms."

Throughout these lectures we shall be concerned with the TF approach, thereby concentrating on more recent developments. The repetition of material that has been presented in textbooks[1] already will be limited to the minimal amount necessary to make the lectures self-contained. The derivation of known results will, wherever feasible, be done differently, and - I believe - more elegantly, than in standard texts on the subject.

It should be realized that the methods of the TF approach are in no way limited to atomic physics. Besides the immediate modifications for applying the formalism to molecules or solids, there exists the possibility of employing the technics in astrophysics and in nuclear physics. The latter application naturally requires appropriate changes reflecting the transition from the Coulomb interaction of the electrons to the much more complicated nucleon-nucleon forces.

In these lectures we shall confine the discussion to atoms, however. This has the advantage of keeping the complexity of most calculations at a rather low level, so that we can fully focus on the properties of the TF method without being distracted by the technical complications that arise from the considerations of molecular structure or from our incomplete knowledge of the nuclear forces, for instance. Restricting ourselves to atoms is further advantageous because it enables us to compare predictions of TF theory with those of other methods, like HF calculations. The ultimate test of a theoretical description is, of course, the comparison of its implications with experimental data. Whenever possible, we shall therefore measure the accuracy of the TF predictions by confronting it with experimental results.

Lack of experimental data sometimes forces us into relying upon HF results for comparison. The same situation occurs when quantities of a more theoretical nature are discussed (as, e.g., the nonrelativistic binding energy, which is not available from experiments). Such a procedure must not be misunderstood as an attempt of reproducing HF predictions by TF theory. The TF method is not an approximation to the HF description, but an independent approach to theoretical atomic physics. [Incidentally, it is the historically older one: TF theory originated in the years 1926 (Thomas) and 1927 (Fermi), whereas the HF model did not exist prior to 1928 (Hartree) and 1930 (Fock).][2] The two approaches should not be regarded as competing with each other, but as supplementing one another. Each of the two methods is well suited for studying certain properties of atoms. For example, if one is interested in the ionization energy of oxygen, a HF calculation will produce a reliable result; but if you want to know how the total binding energy varies over the entire Periodic Table, the TF model will tell you. Tersely: the HF method for specific information about a particular atom, the TF method for the systematics of all atoms. There is, of course, a certain overlap of the two approaches, and they are not completely unrelated. We shall discuss their connection to some extent in Chapter Two.

Atomic units. All future algebraic manipulations are eased significantly when atomic units are used for measuring distances, energies, etc. Let us briefly consider the many-particle Hamilton operator

$$H_{mp} = \sum_{j=1}^{N} \frac{1}{2m} \vec{p}_j^{\,2} - \sum_{j=1}^{N} \frac{Ze^2}{r_j} + \frac{1}{2} {\sum_{j,k=1}^{N}}' \frac{e^2}{r_{jk}} \tag{1-1}$$

of an atom with nuclear charge Ze and N electrons, each of mass m and carrying charge -e. The third sum is primed to denote the omission of the term with j = k. Obviously, r_j stands for the distance between the nucleus and the j-th electron, whereas r_{jk} is the distance from the j-th to the k-th electron, and $\vec{p}_j$ the momentum of the j-th electron. This H_{mp} is accompanied by the commutation relations

$$[\vec{r}_j, \vec{p}_k] = i \hbar \overleftrightarrow{1} \delta_{jk} \tag{1-2}$$

and the injunctions caused by the Fermi statistics that the electrons obey. Equations (1) and (2) contain three dimensional parameters: m, e, $\hbar$. But none of them can possibly be used as expansion variable of a perturbation series because together they do no more than set the atomic scale. To see this in detail, let us rewrite (1) and (2) with the aid of the Bohr radius

$$a_0 = \frac{\hbar^2}{me^2} = 0.5292\ \text{Å} \tag{1-3}$$

and twice the Rydberg energy

$$E_0 = \frac{e^2}{a_0} = \frac{me^4}{\hbar^2} = 27.21\text{eV}. \tag{1-4}$$

Equations (1) and (2) now appear as

$$H_{mp}/E_0 = \sum_j \frac{1}{2}(p_j/\frac{\hbar}{a_0})^2 - \sum_j \frac{Z}{(r_j/a_0)} + \frac{1}{2}\sum_{j,k}{}' \frac{1}{(r_{jk}/a_0)} \tag{1-5}$$

and

$$[(\vec{r}_j/a_0)\ ,\ (\vec{p}_k/\frac{\hbar}{a_0})] = i\,\overleftrightarrow{1}\,\delta_{jk}\ . \tag{1-6}$$

If we then introduce the dimensionless quantities $\vec{r}_j/a_0$, $\vec{p}_j/\frac{\hbar}{a_0}$, and H_{mp}/E_0 as relevant objects, all reference to m, e, and $\hbar$ disappears. Using the same letters as for the dimensional quantities, we now have

$$H_{mp} = \sum_j \frac{1}{2}\,p_j^2 - \sum_j \frac{Z}{r_j} + \frac{1}{2}\sum_{j,k}{}' \frac{1}{r_{jk}} \tag{1-7}$$

and

$$[\vec{r}_j,\vec{p}_k] = i\,\overleftrightarrow{1}\,\delta_{jk}\ . \tag{1-8}$$

Equations (7) and (8) are identical with Eqs. (1) and (2) except that instead of the macroscopics units (cm, erg, etc.) atomic untis are used. Formally, the transition from (1) and (2) to (7) and (8) can be done by "setting e = $\hbar$ = m = 1," but the meaning of this colloquial procedure is made precise by the argument presented above.

Besides simplifying the algebra, the use of atomic units also prevents us from trying such foolish things like "expanding the energy in powers of $\hbar$," a phrase that one meets surprisingly frequently in the literature. The energy is nothing but E_0 times a dimensionless function of Z and N, it depends on $\hbar$ only through $E_0 \sim 1/\hbar^2$. We shall see later, what is really meant when the foregoing phrase is used.

The many particle problem defined by Eqs. (7) and (8) cannot be solved exactly. It is much too complicated. This is true even when the number of electrons is only two, the situation of helium-like atoms. There is a branch of research[3] in which rigorous theorems about the system (7) and (8) are proved, such as (disappointingly rough) limits on the total binding energy. One can show for example, that for $N=Z\to\infty$ the many particle problem reduces to the original TF model, which we shall describe in the next Chapter. In these lectures, we shall not follow those highly mathematized lines. I prefer rather simple physical arguments instead of employing the machinery of functional analysis. Also, it is my impression that those "rigorous" methods are of little help when it comes to improving the description by going beyond the original TF model. Finally, let us not forget that <u>mathematical</u> theorems about (7) and (8) are not absolute knowledge about real atoms, because in putting down the Hamilton operator (7) we have already made <u>physical</u> approximations: the finite size and mass of the nucleus is disregarded; so are all relativistic effects including magnetic interactions and quantum electrodynamical corrections; other than electric interactions are neglected - no reference is made to gravitational and weak forces. Of course, both attitudes, the highly mathematical one and the more physical one, are valuable, but there is danger in judging one by the standards of the other.

<u>Bohr atoms.</u> We continue the introductory remarks by studying a very simple model in order to illustrate a few basic concepts. This primitive theoretical model neglects the inter-electronic interaction, thus treating the electrons as independently bound by the nucleus. But even if fermions do not interact they are aware of each other through the Pauli principle. Therefore, such noninteracting electrons (NIE) will fill the successive Bohr shells of the Coulomb potential with two electrons in each occupied orbital state.

For the present purpose it would be sufficient to consider the situation of m full Bohr shells. But with an eye on a later discussion of shell effects, in Chapter Five, let us additionally suppose that the (m+1)th shell is filled by a fraction μ, $0\le\mu<1$. Since the multiplicity of the shell with principal quantum number m' is $2m'^2$-fold, the total number, N, of electrons then is (see Problem 1)

$$N = \sum_{m'=1}^{m} 2m'^2 + \mu 2(m+1)^2 \tag{1-9}$$

$$= \frac{2}{3}(m+\tfrac{1}{2})^3 - \frac{1}{6}(m+\tfrac{1}{2}) + 2\mu(m+1)^2 \quad .$$

The total binding energy for a nucleus of charge Z is even simpler,

$$-E = \sum_{m'=1}^{m} 2m'^2 \frac{Z^2}{2m'^2} + \mu 2(m+1)^2 \frac{Z^2}{2(m+1)^2} \tag{1-10}$$

$$= Z^2(m+\mu) \quad ,$$

which uses the single particle binding energy $Z^2/(2m'^2)$. If we understand Eq. (9) as defining m and μ as functions of N, then Eq. (10) displays $-E(Z,N)$. Towards the objective of making this functional dependence explicit we proceed from noting that

$$(m+\tfrac{1}{2})^3 - \frac{1}{4}(m+\tfrac{1}{2}) \le \frac{3}{2}N < (m+\tfrac{3}{2})^3 - \frac{1}{4}(m+\tfrac{3}{2}) \quad . \tag{1-11}$$

Consequently, if y solves the equation

$$y^3 - \frac{1}{4}y = \frac{3}{2}N \quad , \tag{1-12}$$

then m is the integer part of $y - \frac{1}{2}$. (For $N>0$, there is just one solution larger than 1/2.) We use the standard Gaussian notation,

$$m = [y - 1/2] \quad . \tag{1-13}$$

For the sequel the introduction of $\langle y\rangle$, defined by

$$\langle y\rangle = y - [y+1/2] \quad , \tag{1-14}$$

that is

$$y - \langle y\rangle = \text{integer},$$

$$-\frac{1}{2} \le \langle y\rangle < \frac{1}{2} \quad , \tag{1-15}$$

will prove useful. We employ it in writing

$$m = y - 1 - \langle y - 1\rangle = y - 1 - \langle y\rangle \quad . \tag{1-16}$$

The latter equality is based upon the obvious periodicity of $\langle y\rangle$,

$$\langle y + 1\rangle = \langle y\rangle \quad . \tag{1-17}$$

We can now insert both Eq.(12) and Eq.(15) into Eq.(9),

$$\frac{2}{3}(y^3-\frac{1}{4}y) = \frac{2}{3}(y-\frac{1}{2}-\langle y\rangle)^3 - \frac{1}{6}(y-\frac{1}{2}-\langle y\rangle) + 2\mu(y-\langle y\rangle)^2 \quad , \tag{1-18}$$

and solve for μ. The result is

$$\mu = \frac{1}{2} + \langle y\rangle + (\langle y\rangle^2 - \frac{1}{4}) \frac{y-\frac{2}{3}\langle y\rangle}{(y-\langle y\rangle)^2} \quad . \tag{1-19}$$

As a consequence of $y>\frac{1}{2}$, the denominator here is nonzero. Also, one easily checks that, as y increases from $m+\frac{1}{2}$ to $m+\frac{3}{2}$, μ grows monotonically from zero to one, as it should.

The combination of Eqs.(10),(16),and (19) now produces

$$-E = Z^2\{y - \frac{1}{2} + (\langle y\rangle^2-\frac{1}{4}) \frac{y-\frac{2}{3}\langle y\rangle}{(y-\langle y\rangle)^2}\} \quad , \tag{1-20}$$

with y(N) from Eq.(12). Let us first observe that this binding energy is a continuous function of y - and therefore of N - although $\langle y\rangle$ occasionally jumps from $+\frac{1}{2}$ to $-\frac{1}{2}$. Next, we note that for large N, Eq.(12) is solved by

$$y(N) = (\frac{3}{2}N)^{1/3} + \frac{1}{12}(\frac{3}{2}N)^{-1/3} + \ldots \quad , \tag{1-21}$$

so that the oscillatory contribution in (20) is of order $N^{-1/3}$. Consequently, the binding energy of NIE is

$$-E = Z^2 \{(\frac{3}{2}N)^{1/3} - \frac{1}{2} + \ldots.\} \quad , \tag{1-22}$$

where the ellipsis indicates oscillatory terms of order $N^{-1/3}$ and smaller. The physical origin of these terms is the process of the filling of shells. We shall disregard them here with the promise of returning later when we shall engage in a more detailed discussion of shell effects.

Expansion (21) is expected to be good for large N. However, just the two terms displayed explicitly form a practically perfect formula even for small N. An impressive way of demonstrating the high quality of this two-term approximation is to look at the values predicted for N, at which closed shells occur. For $y=\frac{3}{2},\frac{5}{2},\frac{7}{2},\ldots$, the exact answer of Eq. (12) is N=2,10,28, ..., whereas Eq. (21) produces N=1.99987, 9.999974, 27.999991, ... ; even for the first shell the agreement is better than one hundredth of a percent.

We have just learned an important lesson: a few terms of an asymptotic expansion like Eq. (21) may be, and frequently are, a highly accurate approximation even for very moderate values of N. Such considerations based upon large numbers are the origin of the label "statistical" that is attached to TF theory. The fundamental physical approximation is, however, rather a semiclassical one.

This will become clearer when we now answer the question how one can find the leading term in (22) somewhat more directly, without utilizing our detailed knowledge of the energy and degeneracy of bound states in the Coulomb potential.

The count of electrons is evaluated in Eq. (9) as the sum of the multiplicities of all occupied <u>shells</u>. Equivalently, we could have summed over all occupied <u>states</u>,

$$N = \sum_{\text{all states}} \begin{Bmatrix} 1 \text{ if the state is occupied} \\ 0 \text{ if the state is not occupied} \end{Bmatrix} . \tag{1-23}$$

Since a given state is occupied (or not) if its binding energy, $-E_{\text{state}}$, is larger than a certain amount, ζ , (or less), we can employ Heaviside's unit step function,

$$\eta(x) = \begin{cases} 1 \text{ for } x > 0 \\ 0 \text{ for } x < 0 \end{cases} , \tag{1-24}$$

in writing

$$N(\zeta) = \sum_{\text{all states}} \eta(-E_{\text{state}} - \zeta) . \tag{1-25}$$

Such a sum over all eigenstates of an operator, here the single-particle Hamilton operator for NIE,

$$H_{NIE} = \frac{1}{2}p^2 - \frac{Z}{r} \quad , \tag{1-26}$$

is more concisely expressed as a trace. We then have

$$N(\zeta) = \text{tr} \;\; \eta(-H_{NIE} - \zeta) \quad . \tag{1-27}$$

[Do not worry about the possibility of partly filled shells. Then ζ equals the binding energy of the respective shell, and the freedom of assigning any value between 0 and 1 to $\eta(x=o)$ enables us to describe the situation of a fractionally filled shell. More about this in Chapter Five.] Analogously, we can express the energy of Eq. (10) as the sum over the single-particle energies of all occupied states,

$$E(\zeta) = \text{tr}\; H_{NIE}\; \eta(-H_{NIE} - \zeta) \quad . \tag{1-28}$$

The identity

$$\frac{d}{dx}(x\; \eta(x)) = \eta(x) \quad , \tag{1-29}$$

used in the form

$$-\int_x^\infty dx' \;\; \eta(-x') = x\; \eta(-x) \quad , \tag{1-30}$$

can be used to relate $E(\zeta)$ to $N(\zeta)$:

$$\begin{aligned} E(\zeta) &= \text{tr}\; (H_{NIE} + \zeta)\; \eta\; (-H_{NIE} - \zeta) - \zeta\; \text{tr}\; \eta(-H_{NIE} - \zeta) \\ &= -\;\text{tr} \int_\zeta^\infty d\zeta' \;\; \eta(-H_{NIE} - \zeta') - \zeta\; N(\zeta) \quad , \end{aligned} \tag{1-31}$$

or,

$$E(\zeta) = -\;\zeta\; N(\zeta) - \int_\zeta^\infty d\zeta' \;\; N(\zeta') \quad . \tag{1-32}$$

We see that $E(\zeta)$ is immediately available as soon as we know $N(\zeta)$. This is no surprise. Recall that $N(\zeta)$ signifies the number of states with binding energy larger than ζ. Consequently, $N(\zeta)$ is discontinuous at all values of ζ equal to the binding energy of a (Bohr) shell, and the size of the jump of $N(\zeta)$ at such a discontinuity is the multiplicity of the respective shell. So $N(\zeta)$, regarded as a function of ζ, tells us both the energy and the multiplicity of all shells.

The problem is now reduced to evaluating the trace in Eq.(27) in an appropriate, approximate way. [Remember, it is the leading term of Eq.(22) only, that we want to derive simply.] First an intuitive argument. The count of states is the spin multiplicity of two, times the count of orbital states. There is roughly one orbital state per phase-space volume $(2\pi\hbar)^3$ [$=(2\pi)^3$ in atomic units], so that

$$N(\zeta) \cong 2\int\frac{(d\vec{r}\,')(d\vec{p}\,')}{(2\pi)^3}\,\eta\left(-\left(\tfrac{1}{2}p'^2-\frac{Z}{r'}\right)-\zeta\right) \quad , \qquad (1\text{-}33)$$

where primes have been used to distinguish numbers from operators. The step function equals unity in the classically allowed domain of the phase-space and vanishes outside. Therefore, Eq.(33) represents the extreme semiclassical (or should we say: semiquantal?) limit, in which the possibility of finding the quantum-mechanical system outside the classical allowed region is ignored.

Some support for the approximation (33) is supplied by its implications. After performing the momentum integration, we have

$$N(\zeta) \cong \int(d\vec{r}\,')\,\frac{1}{3\pi^2}\left[2\left(\frac{Z}{r'}-\zeta\right)\right]^{3/2} \quad , \qquad (1\text{-}34)$$

where the square root is understood to vanish for negative arguments. Then the r' integration produces, with $x = \zeta r'/Z$,

$$\begin{aligned} N(\zeta) &\cong (2Z^2/\zeta)^{3/2}\,\frac{4}{3\pi}\int_0^1 dx\; x^{1/2}(1-x)^{3/2} \\ &= \frac{2}{3}\left(\frac{Z^2}{2\zeta}\right)^{3/2} \quad . \qquad (1\text{-}35) \end{aligned}$$

(the integral has the value $\pi/16$.) Insertion into Eq.(32) results in

$$-E(\zeta) \cong Z^2\left(\frac{Z^2}{2\zeta}\right)^{1/2} \quad , \qquad (1\text{-}36)$$

which combined with (35) is

$$-E \cong Z^2\left(\tfrac{3}{2}N\right)^{1/3} \quad . \qquad (1\text{-}37)$$

Indeed, here is the leading term of Eq.(22), now very simply reproduced by the semiclassical counting of states. Please note that the steps from Eq.(32) to Eq.(37) did not require any knowledge about the energy and multiplicity of bound states in the Coulomb potential.

Traces and phase-space integrals. One does not have to rely upon intuition alone when writing down Eq.(33). A general way of evaluating the trace of a function of position operator $\vec{r}$ and the conjugate momentum operator $\vec{p}$ is

$$\begin{aligned} \operatorname{tr} F(\vec{r},\vec{p}) &= \int (d\vec{r}\,')\, \langle \vec{r}\,' \mid F(\vec{r},\vec{p}) \mid \vec{r}\,' \rangle \\ &= \int (d\vec{r}\,')(d\vec{p}\,')\, \langle \vec{r}\,' \mid F(\vec{r},\vec{p}) \mid \vec{p}\,' \rangle \langle \vec{p}\,' \mid \vec{r}\,' \rangle \quad . \end{aligned} \tag{1-38}$$

We have left out the factor of two for the spin multiplicity here, because it is irrelevant for the present discussion. If now $F(\vec{r},\vec{p})$ is ordered such that all $\vec{r}$'s stand to the left of all $\vec{p}$'s, then

$$\langle \vec{r}\,' \mid F(\vec{r},\vec{p}) \mid \vec{p}\,' \rangle = F(\vec{r}\,',\vec{p}\,')\, \langle \vec{r}\,' \mid \vec{p}\,' \rangle \quad . \tag{1-39}$$

This, combined with the position-momentum transformation functions

$$\begin{aligned} \langle \vec{r}\,' \mid \vec{p}\,' \rangle &= \frac{1}{(2\pi)^{3/2}}\, e^{i\,\vec{r}\,'\cdot\vec{p}\,'} \quad , \\ \langle \vec{p}\,' \mid \vec{r}\,' \rangle &= \frac{1}{(2\pi)^{3/2}}\, e^{-i\,\vec{r}\,'\cdot\vec{p}\,'} \quad , \end{aligned} \tag{1-40}$$

and inserted into (38) produces

$$\operatorname{tr} F(\vec{r},\vec{p}) = \int \frac{(d\vec{r}\,')(d\vec{p}\,')}{(2\pi)^3}\, F(\vec{r}\,',\vec{p}\,') \tag{1-41}$$

Equation (41) is an <u>exact</u> statement for an <u>ordered</u> operator $F(\vec{r},\vec{p})$. If the operator, of which the trace is desired, is not ordered, one can sometimes do the ordering explicitly. An example is (see Problem 3)

$$e^{-i(\frac{1}{2}\vec{p}^{\,2} - \vec{F}\cdot\vec{r})t} =$$

$$= e^{i\vec{F}\cdot\vec{r}\,t}\, e^{-i\frac{1}{2}(\vec{p}+\frac{1}{2}\vec{F}t)^2 t}\, e^{-iF^2t^3/24} , \qquad (1\text{-}42)$$

where $\vec{F}$ is a constant vector. With the aid of (42), all other functions of $\frac{1}{2}\vec{p}^2 - \vec{F}\cdot\vec{r}$ can be ordered if expressed as the appropriate Fourier integral. We shall have a use for Eq. (42) later, in Chapter Four.

With the exception of a few relatively simple instances, the ordering of an operator $F(\vec{r},\vec{p})$ is practically impossible. However, even then Eq. (41) is not useless. Inasmuch as the ordering process involves the evaluation of commutators of functions of $\vec{r}$ with functions of $\vec{p}$, $\langle\vec{r}'|F(\vec{r},\vec{p})|\,\vec{p}'\rangle$ differs from $F(\vec{r}',\vec{p}')\langle\vec{r}'|\vec{p}'\rangle$ by commutator terms. Under circumstances when these commutators are small,

$$\mathrm{tr}\, F(\vec{r},\vec{p}) \cong \int \frac{(d\vec{r}')(d\vec{p}')}{(2\pi)^3} F(\vec{r}',\vec{p}') \qquad (1\text{-}43)$$

can be used as the basis for approximations. Since the noncommutativity of $\vec{r}$ and $\vec{p}$ becomes insignificant in the semiclassical limit, Eq. (43) manifests a highly semiclassical approximation.

All refinements of (43) are due to the noncommutativity of position and momentum. This is at the heart of quantum mechanics, and we shall therefore call these improvements "quantum corrections," notwithstanding the fact that (41) is already a quantum mechanical result. The starting point is clearly the semiclassical (or, semiquantal) picture, not the classical theory of the atom, which does not exist in the first place.

For the trace of Eq. (27) all this means that the semiclassical evaluation of Eq. (33) will be a reliable approximation, if the deBroglie wavelength of an individual electron is small compared to the typical distance over which the Coulomb potential varies significantly. This condition is satisfied if both the number of electrons and the nuclear charge are large, because in this situation the electronic cloud is very dense. In this sense the semiclassical approximation is equivalent to a large-N, a statistical one.

Bohr atoms with shielding. The primitive model of NIE constitutes a sensible approximation for highly ionized atoms only, in which the dynamics is governed by the Coulomb potential of the nucleus, and the electron-electron interaction is negligible. Consequently, the results obtained above should not be taken seriously, unless $N \ll Z$. For instance, the binding energy of a neutral atom is expected to differ significantly from the prediction of Eq. (37),

$$-E \cong \left(\frac{3}{2}\right)^{1/3} Z^{7/3} = 1.145\ Z^{7/3} \quad , \tag{1-44}$$

because of the screening of the nuclear Coulomb potential by the inner shells. Let us, therefore, try to get a feeling for the importance of the electron-electron interactions by refining the NIE picture.

Without inner-shell screening, each full Bohr shell contributes the amount of Z^2 to the binding energy [see Eq. (10)]. We now suppose that this remains true for the first shell, whereas the effective Z-value for the second shell is Z-2, since the total charge of the nucleus together with the first shell is (-Z+2)e. Similarily, the third shell sees Z-2-8=Z-10, and so on. In this picture,[4] the screening of the inner shells is so effective that the m'-th shell is exposed to a Coulomb potential $-Z_{m'}/r$ with

$$Z_{m'} = Z - \sum_{m''=1}^{m'-1} 2m''^2 \quad . \tag{1-45}$$

Its contribution to the binding energy is $Z^2_{m'}$, so that we have

$$-E = \sum_{m'=1}^{m} Z^2_{m'} + \mu\ Z^2_{m+1} \quad . \tag{1-46}$$

We are interested in the leading term only and can, therefore, evaluate the various sums over m' by means of

$$\sum_{m'=1}^{m} m'^{\nu} \cong \frac{1}{\nu+1} m^{\nu+1} \quad . \tag{1-47}$$

We can then also disregard all effects of the filling of shells, since the terms proportional to μ are of a lower order, both in N [Eq. (9)] and in E. To leading order, we have

$$N \cong \frac{2}{3} m^3 \quad , \tag{1-48}$$

$$Z_{m'} \cong Z - \frac{2}{3}(m'-1)^3 \quad ,$$

as well as

$$-E = \sum_{m'=1}^{m} Z_{m'}^2 \cong \sum_{m'=1}^{m} [Z - \frac{2}{3}(m'-1)^3]^2$$
$$\cong Z^2 m - \frac{1}{3} Z m^4 + \frac{4}{63} m^7 \quad , \tag{1-49}$$

or,

$$-E = Z^2(\tfrac{3}{2}N)^{1/3} - \tfrac{1}{3}Z(\tfrac{3}{2}N)^{4/3} + \tfrac{4}{63}(\tfrac{3}{2}N)^{7/3}$$
$$= Z^2(\tfrac{3}{2}N)^{1/3}[1 - \frac{1}{2}\frac{N}{Z} + \frac{1}{7}(\frac{N}{Z})^2] \quad . \tag{1-50}$$

For neutral atoms, $N = Z$, the prediction for the total binding energy is now

$$-E \cong (1 - \frac{1}{2} + \frac{1}{7})\ Z^2(\tfrac{3}{2}Z)^{1/3}$$
$$= \frac{9}{14}(\tfrac{3}{2})^{1/3}\ Z^{7/3} = 0.736\ Z^{7/3} \quad . \tag{1-51}$$

The comparison with (44) shows that the inner-shell screening reduces the total binding energy by roughly one third. It certainly is a substantial effect in a neutral atom.

Incidentally, it is remarkable that the numerical coefficient in (51) differs from the correct answer (see the next Chapter) by less than 5%. In view of the crude way, in which the electron-electron interaction has been taken into account, this is much better than one could possibly expect.

The effective potential. The model that we just studied possesses one particularly unsatisfactory feature: the inner shells influence the outer ones, but not vice versa. There is action but no reaction - hardly a good way of describing interaction.

In our present, preliminary attempt of resolving this insufficiency, in the framework of a modified Bohr model, we shall continue to assume that the various Bohr shells are geometrically separated. The m'-th shell is supposed to be spherical of a certain radius, $R_{m'}$. Then the potential energy of an electron with this shell is

$$U_{m'}(r) = \begin{cases} 2m'^2/R_{m'} & \text{for} \quad r<R_{m'} \\ 2m'^2/r & \text{for} \quad r>R_{m'} \end{cases} \quad , \tag{1-52}$$

if the electron is situated at a distance r from the nucleus. The picture is no longer asymmetric now, since the energy of the m"-th shell in the electrostatic field of the m'-th shell,

$$2m''^2\, U_{m'}(R_{m''}) = (2m''^2)(2m'^2)/\mathrm{Max}(R_{m'},R_{m''}) \quad , \tag{1-53}$$

remains unaltered if m' and m" are interchanged; action and reaction are equal.

The total potential energy of the electrons in the m'-th Bohr shell is the sum of the potential energy with the nucleus and with all other shells,

$$E_{pot,m'} = 2m'^2 [- \frac{Z}{R_{m'}} + \sum_{m''=1}^{m}{}' U_{m''}(R_{m'}) + \mu U_{m+1}(R_{m'})] \quad . \tag{1-54}$$

In this sum, the prime is a reminder to delete the term with m"=m'. Including this term would mean to include the self energy of the m'-th shell. This is not undesirable, though, because the self energy of the shell consists mostly of the interaction energy of the individual electrons in the shell. The unphysical electron self-energy can be expected to be a relatively small fraction of the shell self-energy. Thus we feel justified in dropping the prime on the sum in (54), implying

$$E_{pot,m'} = 2m'^2\, V(R_{m'}) \quad , \tag{1-55}$$

which introduces the effective potential

$$V(r) = -\frac{Z}{r} + \sum_{m'=1}^{m} U_{m'}(r) + \mu\, U_{m+1}(r) \quad . \tag{1-56}$$

It is the same for all shells, i.e. for all electrons.

Before going on, let us supply additional evidence in favor of the introduction of the effective potential. It comes from evaluating the total self energy of all shells. This is

$$E_{sse} = \frac{1}{2}\sum_{m'=1}^{m} 2m'^2\, U_{m'}(R_{m'}) + \frac{1}{2}\mu(2(m+1)^2)\mu\, U_{m+1}(R_{m+1})$$

$$= \frac{1}{2}\sum_{m'=1}^{m} \frac{(2m'^2)^2}{R_{m'}} + \frac{1}{2}\,\frac{\mu^2(2(m+1)^2)^2}{R_{m+1}} \quad ; \tag{1-57}$$

the subscript sse stands for same-shell electrons. It now becomes necessary to specify the radii of the shells, $R_{m'}$. An electron of the m'-th shell moves in a potential of the form $-Z_{m'}/r$ + const. This is, besides the here irrelevant additive constant, a Coulomb potential. Consequently, the expectation values of the kinetic and potential energy of this electron are $Z_{m'}^2/(2m'^2)$ and $-Z_{m'}^2/m'^2$+ const, respectively. It is natural to define $R_{m'}$ by equating this potential-energy expectation-value to $-Z_{m'}/R_{m'}$ + const. This means

$$\frac{Z_{m'}^2}{m'^2} = \frac{Z_{m'}}{R_{m'}} \quad , \tag{1-58}$$

or,

$$\frac{m'^2}{R_{m'}} = Z_{m'} = Z - \sum_{m''=1}^{m'-1} 2m''^2 \quad ; \tag{1-59}$$

the latter equality is Eq.(45). (Of course, no claims are made that this represents the one and only way of defining $R_{m'}$. The electrons of a Bohr shell are not geometrically confined to a small range of r, so that there cannot be a unique, physical value ascribed to $R_{m'}$.)

Upon inserting (59) into (57), we have

$$\begin{aligned} E_{sse} &= \sum_{m'=1}^{m} 2m'^2 \, Z_{m'} + \mu^2 \, 2(m+1)^2 \, Z_{m+1} \\ &= \sum_{m'=1}^{m} 2m'^2 \, [Z - \tfrac{2}{3}(m' - \tfrac{1}{2})^3 + \tfrac{1}{6}(m' - \tfrac{1}{2})] \\ &\qquad + \mu^2 \, 2(m+1)^2 \, [Z - \tfrac{2}{3}(m + \tfrac{1}{2})^3 + \tfrac{1}{6}(m + \tfrac{1}{2})] \quad , \end{aligned} \tag{1-60}$$

of which the leading terms are [Eqs.(47) and (48)]

$$\begin{aligned} E_{sse} &\cong \frac{2}{3} Z m^3 - \frac{2}{9} m^6 \\ &\cong Z^2 \left[\frac{N}{Z} - \frac{1}{2}\left(\frac{N}{Z}\right)^2 \right] \quad . \end{aligned} \tag{1-61}$$

For a neutral atom (N/Z = 1), this is

$$E_{sse} \cong \frac{1}{2} Z^2 \quad , \tag{1-62}$$

and does not contribute to the leading term of the binding energy, which is $\sim Z^{7/3}$.

Since the self energy of the m'-th shell is proportional to $(2m'^2)^2$, i.e., to the square of the number of electrons it contains, whereas the sum of all electron self-energies is proportional to their number ($2m'^2$ for those of the m'-th shell, N for the whole atom), the error made in the total binding energy by the inclusion of the electron self-energy is very small on the scale set by the leading term (proportional to $Z^{7/3}$). Moreover, as soon as we shall have included the exchange interaction into the description, the electronic self energy will be exactly cancelled by the equally unphysical self-exchange energy. In other words: there is no reason at all to worry about the self energy; at the present stage is does not contribute significantly, and later it is going to be taken care of automatically.

In the effective potential, the use of which now being justified, the electrons move independently. As the main consequence, the complicated many-particle problem is reduced to an effective single-particle one. In our present model the total kinetic energy is

$$E_{kin} = \sum_{m'=1}^{m} Z_{m'}^2 + \mu Z_{m+1}^2 \tag{1-63}$$

[recall, once more, that the kinetic energy of an electron in a Coulomb potential $-Z_{m'}/r$ + const , as is the situation in the m'-th shell, is given by $Z_{m'}^2/(2m'^2)$]. Further, the independent-particle (IP) potential energy can be expressed with the aid of the effective potential V,

$$E_{IP,pot} = \sum_{m'=1}^{m} 2m'^2 V(R_{m'}) + \mu 2(m+1)^2 V(R_{m+1}) \tag{1-64}$$

Together they constitute an approximation to the independent-particle energy E_{IP} :

$$E_{IP} = E_{kin} + E_{IP,pot} \tag{1-65}$$

$$= \sum_{m'=1}^{m} Z_{m'}^2 + \sum_{m'=1}^{m} 2m'^2 V(R_{m'}) + \mu[Z_{m+1}^2 + 2(m+1)^2 V(R_{m+1})] \quad .$$

This is, however, not the energy of the system. Because of the use of the effective potential, the electron-electron interaction is counted twice in (65). In addition to the energy of the m'-th shell in the elec-

trostatic field of the m"-th shell, $2m'^2\, U_{m''}(R_{m'})$, E_{IP} also contains $2m''^2\, U_{m'}(R_{m''})$, for any pair m',m"; and the two are equal, as we have seen earlier, in Eq.(53). Consequently, we have to remove the electron-electron interaction energy once. This is conveniently achieved by expressing this energy in terms of the electric field made by the electrons,

$$\vec{E} = -\vec{\nabla}(V - (-\frac{Z}{r})) \quad . \tag{1-66}$$

We have carefully substracted the contribution to V that stems from the nuclear charge. What we have to add to E_{IP} in order to remove the doubly counted interaction energy, is then

$$\begin{aligned} E_2 &= -\frac{1}{8\pi}\int(d\vec{r})\ \vec{E}^2 \\ &= -\frac{1}{8\pi}\int(d\vec{r})\,[\vec{\nabla}(V + \frac{Z}{r})]^2 \quad . \end{aligned} \tag{1-67}$$

The evaluation of this integral is facilitated by the observation that in our model

$$V(r) + \frac{Z}{r} = \frac{Z-Z_{m'+1}}{r} + \text{const} \quad \text{for } R_{m'}<r<R_{m'+1} \quad . \tag{1-68}$$

Accordingly,

$$[\vec{\nabla}(V+\frac{Z}{r})]^2 = \frac{(Z-Z_{m'+1})^2}{r^4} \quad \text{for} \quad R_{m'}<r<R_{m'+1} \quad , \tag{1-69}$$

which holds for $m' = 1,2,...,m$. Additionally, we need

$$[\vec{\nabla}(V+\frac{Z}{r})]^2 = \begin{cases} 0 & \text{for} \quad r<R_1 \\ \frac{N^2}{r^4} & \text{for} \quad r>R_{m+1} \end{cases} \quad . \tag{1-70}$$

At this stage, we have

$$E_2 = -\frac{1}{8\pi}\left[\int_0^{R_1} + \int_{R_1}^{R_2} + \ldots + \int_{R_m}^{R_{m+1}} + \int_{R_{m+1}}^{\infty}\right] 4\pi r^2 dr\ [\vec{\nabla}(V+\frac{Z}{r})]^2 =$$

$$= -\frac{1}{2}\sum_{m'=1}^{m}\int_{R_{m'}}^{R_{m'+1}} dr\,\frac{(Z-Z_{m'+1})^2}{r^2} - \frac{1}{2}\int_{R_{m+1}}^{\infty} dr\,\frac{N^2}{r^2} \tag{1-71}$$

$$= -\frac{1}{2}\sum_{m'=1}^{m}(Z-Z_{m'+1})^2\left(\frac{1}{R_{m'}} - \frac{1}{R_{m'+1}}\right) - \frac{1}{2}\frac{N^2}{R_{m+1}} \quad .$$

A unit shift of the summation index transforms the sum with $R_{m'+1}$ into an equivalent one with $R_{m'}$:

$$E_2 = -\frac{1}{2}\sum_{m'=1}^{m}\frac{(Z-Z_{m'+1})^2}{R_{m'}} + \frac{1}{2}\sum_{m'=2}^{m+1}\frac{(Z-Z_{m'})^2}{R_{m'}} - \frac{1}{2}\frac{N^2}{R_{m+1}} \quad . \tag{1-72}$$

The recognition that $Z - Z_1 = 0$, combined with

$$(Z-Z_{m'+1})^2 - (Z-Z_{m'})^2$$

$$= (Z_{m'}-Z_{m'+1})[(Z-Z_{m'}) + (Z-Z_{m'+1})]$$

$$= 2m'^2(2\sum_{m''=1}^{m'-1}2m''^2 + 2m'^2) \tag{1-73}$$

$$= 2m'^2\,R_{m'}\,(2\sum_{m''=1}^{m'-1}U_{m''}(R_{m'}) + U_{m'}(R_{m'})) \quad ,$$

which uses Eqs.(45) and (52), as well as [Eqs.(45),(52) and (9)]

$$N^2 - (Z-Z_{m+1})^2 = \mu 2(m+1)^2[2\sum_{m''=1}^{m}2m''^2 + \mu 2(m+1)^2]$$

$$= \mu 2(m+1)^2 R_{m+1}(2\sum_{m''=1}^{m}U_{m''}(R_{m+1}) + \mu U_{m+1}(R_{m+1})) \quad , \tag{1-74}$$

turns (72) into

$$E_2 = -\left\{\sum_{m'=1}^{m}2m'^2\sum_{m''=1}^{m'-1}U_{m''}(R_{m'}) + \mu 2(m+1)^2\sum_{m''=1}^{m}U_{m''}(R_{m+1})\right\} - \tag{1-75}$$

$$-\left\{ \frac{1}{2} \sum_{m'=1}^{m} 2m'^2 U_{m'}(R_{m'}) + \frac{1}{2}[\mu 2(m+1)^2][\mu U_{m+1}(R_{m+1})] \right\} .$$

The contents of the two curly brackets are immediately recognized as the interaction energy of the pairs of shells and the self energy E_{sse} of the individual shells, respectively. Indeed, E_2 is the negative of the electron-electron interation energy, as it should be.

Before adding E_{IP} of (65) and E_2 of (75) to get the total energy itself, it is useful to rewrite $E_{IP,pot}$. From (64) we get

$$\begin{aligned}
E_{IP,pot} &= \sum_{m'=1}^{m} 2m'^2 [-\frac{Z}{R_{m'}} + \sum_{m''=1}^{m} U_{m''}(R_{m'}) + \mu U_{m+1}(R_{m'})] \\
&\quad + \mu 2(m+1)^2 [-\frac{Z}{R_{m+1}} + \sum_{m'=1}^{m} U_{m'}(R_{m+1}) + \mu U_{m+1}(R_{m+1})] \\
&= \sum_{m'=1}^{m} \frac{2m'^2}{R_{m'}} (-Z + \sum_{m''=1}^{m'-1} 2m''^2) + \sum_{m'=1}^{m} 2m'^2 U_{m'}(R_{m'}) \\
&\quad + \sum_{m'=1}^{m} \sum_{m''=m'+1}^{m} 2m'^2 U_{m''}(R_{m'}) \qquad (1\text{-}76) \\
&\quad + \mu \sum_{m'=1}^{m} 2m'^2 U_{m+1}(R_{m'}) \\
&\quad + \mu \frac{2(m+1)^2}{R_{m+1}} (-Z + \sum_{m'=1}^{m} 2m'^2) \\
&\quad + [\mu 2(m+1)^2] \times [\mu U_{m+1}(R_{m+1})] .
\end{aligned}$$

After using Eq. (59) and the m'-m" symmetry of $2m'^2 U_{m''}(R_{m'})$ [the action-reaction symmetry that we observed in Eq. (53)], this reads

$$\begin{aligned}
E_{IP,pot} &= -2 \sum_{m'=1}^{m} Z_{m'}^2 - 2\mu Z_{m+1}^2 \\
&\quad + \left\{ \sum_{m''=1}^{m} \sum_{m'=1}^{m''-1} 2m''^2 U_{m'}(R_{m''}) + \mu (2m+1)^2 \sum_{m'=1}^{m} U_{m'}(R_{m+1}) \right\} +
\end{aligned}$$

$$+\ 2\left\{\frac{1}{2}\sum_{m'=1}^{m} 2m'^2 U_{m'}(R_{m'}) + \frac{1}{2}[\mu 2(m+1)^2][\mu U_{m+1}(R_{m+1})]\right\}$$

$$= -\ 2\sum_{m'=1}^{m} Z_{m'}^2 - 2\mu\ Z_{m+1}^2 - E_2 + E_{sse} \quad . \tag{1-77}$$

Finally, we obtain the total binding energy

$$-E = -\ (E_{kin} + E_{IP,pot} + E_2) \tag{1-78}$$

$$= \sum_{m'=1}^{m} Z_{m'}^2 + \mu\ Z_{m+1}^2 - E_{sse} \quad .$$

We compare this with Eq.(46) and notice that the more symmetrical treatment of the electrons leads to an additional term, E_{sse}, in the energy. This is very satisfactory because E_{sse} is the interaction energy of electrons in the same shell (plus the innocuous electron self-energy), which was left out when (46) was derived.

Equation (78) can be simplified. First, we use

$$2m'^2 = Z_{m'} - Z_{m'+1} \tag{1-79}$$

and

$$\mu 2(m+1)^2 = Z_{m+1} - (Z-N) \quad , \tag{1-80}$$

which are consequences of Eqs.(45) and (9), to rewrite E_{sse} of (60) as

$$E_{sse} = \sum_{m'=1}^{m} (Z_{m'} - Z_{m'+1})Z_{m'} + \mu\ Z_{m+1}^2 - \mu(Z-N)Z_{m+1} \quad . \tag{1-81}$$

Then we insert this into (78). The outcome is

$$-E = \sum_{m'=1}^{m} Z_{m'}\ Z_{m'+1} + \mu(Z-N)Z_{m+1} \quad . \tag{1-82}$$

We can now evaluate the sum over m', express m and μ in terms of y, as given in Eqs.(16) and (19), and pick out the two leading contributions to −E. They are

$$-E = (Z^2y - \frac{1}{3}Zy^4 + \frac{4}{63}y^7) - \frac{1}{2}Z^2 + O(Z^{5/3}\sim Zy^2\sim y^5) \qquad (1\text{-}83)$$

To this order, y is simply given by $(\frac{3}{2}N)^{1/3}$ [Eq.(21)], so that

$$-E = Z^2(\frac{3}{2}N)^{1/3}(1 - \frac{1}{2}\frac{N}{Z} + \frac{1}{7}(\frac{N}{Z})^2) - \frac{1}{2}Z^2$$
$$+ O(ZN^{2/3}\sim N^{5/3}) \quad . \qquad (1\text{-}84)$$

The neutral-atom binding energy predicted by our improved model of Bohr atoms with shielding is, consequently,

$$-E = 0.736\ Z^{7/3} - \frac{1}{2}Z^2 + O(Z^{5/3}) \quad . \qquad (1\text{-}85)$$

Without shielding, that is: without accounting for the electron-electron interaction, the result was [Eq.(22) for N = Z]

$$-E = 1.145\ Z^{7/3} - \frac{1}{2}Z^2 + O(Z^{5/3}) \quad . \qquad (1\text{-}86)$$

Whereas the screening of the nuclear potential by the inner electrons reduces the leading term by 5/14 ≅ 1/3, it does not affect the Z^2 term at all. We shall see later, in Chapter Three, that this next-to-leading term is a consequence of the Coulomb shape of the effective potential for small r. It is the same for all potentials with $V \cong -Z/r$ for $r \to o$, for which reason it is independent of N [Eqs.(22) and (84) confirm this]. The two examples that we looked at so far, the Coulomb potential of (26) and the V(r) of (56), both have this property.

<u>Size of atoms.</u> A last application of our model of Bohr atoms with shielding consists in studying the Z dependence of the size of neutral atoms. The individual Bohr shells shrink proportional to $1/Z_m$, as Z increases [see Eq.(59)]. This would mean that the size of an atom is roughly given by 1/Z, if there were not the necessity of filling additional shells to compensate for the growth of nuclear charge. Clearly, a qualified statement about atomic size requires the evaluation of some average of r over the atom.

According to Eq.(59), it is the inverse of R_m, that is easy to handle. We shall therefore measure the size R of an atom as

$$\frac{1}{R} = \frac{1}{N} \langle\frac{1}{r}\rangle \quad , \tag{1-87}$$

where $\langle 1/r\rangle$ denotes the expectation value of $1/r$. In our model it is given by

$$\langle\frac{1}{r}\rangle = \sum_{m'=1}^{m} \frac{2m'^2}{R_{m'}} + \mu \frac{2(m+1)^2}{R_{m+1}} \quad . \tag{1-88}$$

It has a simple physical significance: $\langle 1/r\rangle$ is the electrostatic energy of the electrons in the field of a unit charge situated at the location of the nucleus, $r=o$; or, equivalently, the electrostatic energy of this unit charge in the field of the electrons. As such it can be evaluated in terms of the effective potential V:

$$\langle\frac{1}{r}\rangle = (V + \frac{Z}{r})(r=o) = \sum_{m'=1}^{m} U_{m'}(o) + \mu U_{m+1}(o) \quad . \tag{1-89}$$

Indeed, Eq.(52) assures us of the equivalence of (88) and (89).

After employing Eq.(59) to rewrite (88),

$$\begin{aligned}\langle\frac{1}{r}\rangle &= 2\sum_{m'=1}^{m} Z_{m'} + 2\mu\, Z_{m+1} \\ &= 2\sum_{m'=1}^{m} (Z - \tfrac{2}{3}m'(m'-1)(m'-\tfrac{1}{2})) + 2\mu(Z-N) + 4\mu^2(m+1)^2\end{aligned} \tag{1-90}$$

[the latter equality also uses Eq.(80)], we can sum over m' and then identify the leading contributions with the aid of Eqs.(16), (19), and (21). The result is

$$\langle\frac{1}{r}\rangle = 2Z(\tfrac{3}{2}N)^{1/3}(1 - \tfrac{1}{4}\tfrac{N}{Z}) - Z(1 - \tfrac{N}{Z}) + O(N^{2/3} \sim ZN^{-1/3}) \quad , \tag{1-91}$$

which for neutral atoms reads

$$\langle\frac{1}{r}\rangle = (\tfrac{3}{2}Z)^{4/3} + O(Z^{2/3}) \quad . \tag{1-92}$$

Consequently, the atomic size is [Eq.(87)]

$$R \sim Z^{-1/3} \quad . \tag{1-93}$$

Heavier atoms are geometrically smaller. This prediction of our rather simple model will remain valid in more realistic treatments.

A remarkable observation is the agreement of Eq.(92) with the Z derivative of Eq.(84) to leading order,

$$\langle\frac{1}{r}\rangle \cong \frac{\partial}{\partial Z}(-E) \quad . \tag{1-94}$$

Its physical significance becomes transparent when we exhibit the change in the binding energy that is caused by increasing the nuclear charge Z by the infinitesimal amount δZ:

$$\delta(-E) = \frac{\partial}{\partial Z}(-E)\delta Z \cong \langle\frac{\delta Z}{r}\rangle \quad . \tag{1-95}$$

This says that the change in the binding energy is <u>mainly</u> given by the electrostatic energy of the extra nuclear charge; the induced alterations of the shell radii R_m, do not contribute to $\delta(-E)$ to leading order. This result of the model must be contrasted with the corresponding implication of the exact treatment based upon the many-particle Hamilton operator (7). In general, an infinitesimal change of a parameter in a Hamilton operator causes a change in the energy, which is equal to the expectation value of the respective change of the Hamilton operator[5]. In the present discussion, this statement reads

$$\delta(-E) = \langle\delta(-H_{mp})\rangle = \langle\sum_j \frac{\delta Z}{r_j}\rangle = \langle\frac{\delta Z}{r}\rangle \quad , \tag{1-96}$$

which says that the change in the binding energy is <u>entirely</u> given by the electrostatic energy of the extra nuclear charge δZ.

As we see, in our model of Bohr atoms with shielding, Eq.(96) is not obeyed exactly, but approximately. This minor deficiency could possibly be removed by a slightly different definition of the R_m, [Eq. (59)]. It is not worth the trouble, though.

The models studied in this Introduction not only provided a first insight into the general characteristics of complex atoms, but also made us somewhat familiar with a few important ideas: the concept of the effective potential, the semiclassical evaluation of traces through phase-space integrals, and relations of the kind illustrated by Eq.(32) are the central ones. The next Chapter, devoted to the Thomas-Fermi model, will use them for a first self-consistent description.

Problems

1-1. Sums of powers of m', as, e.g., in Eq. (9), can be conveniently evaluated following the pattern of this example:

$$\sum_{m'=1}^{m} m' = \sum_{m'=1}^{m} \frac{1}{2}[(m'+\frac{1}{2})^2-(m'-\frac{1}{2})^2]$$

$$= \frac{1}{2}\sum_{m'=1}^{m}(m'+\frac{1}{2})^2 - \frac{1}{2}\sum_{m'=o}^{m-1}(m'+\frac{1}{2})^2$$

$$= \frac{1}{2}(m+\frac{1}{2})^2 - \frac{1}{2}(0+\frac{1}{2})^2 = \frac{1}{2}(m+\frac{1}{2})^2 - \frac{1}{8} \quad .$$

Show that the other sums, that occur in this Chapter, are given by

$$\sum_{m'=1}^{m} m'^2 = \frac{1}{3}(m+\frac{1}{2})^3 - \frac{1}{12}(m+\frac{1}{2}) \quad ,$$

$$\sum_{m'=1}^{m} m'^3 = \frac{1}{4}(m+\frac{1}{2})^4 - \frac{1}{8}(m+\frac{1}{2})^2 + \frac{1}{64} \quad ,$$

$$\sum_{m'=1}^{m} m'^4 = \frac{1}{5}(m+\frac{1}{2})^5 - \frac{1}{6}(m+\frac{1}{2})^3 + \frac{7}{240}(m+\frac{1}{2}) \quad ,$$

$$\sum_{m'=1}^{m} m'^5 = \frac{1}{6}(m+\frac{1}{2})^6 - \frac{5}{24}(m+\frac{1}{2})^4 + \frac{7}{96}(m+\frac{1}{2})^2 - \frac{1}{128} \quad ,$$

and

$$\sum_{m'=1}^{m} m'^6 = \frac{1}{7}(m+\frac{1}{2})^7 - \frac{1}{4}(m+\frac{1}{2})^5 + \frac{7}{48}(m+\frac{1}{2})^3 - \frac{31}{1344}(m+\frac{1}{2}) \quad .$$

1-2. Use the periodicity of $\langle y\rangle$ [see Eq. (17)] to write it as a Fourier series,

$$\sum_{m=1}^{\infty} \frac{(-1)^m}{\pi m} \sin(2\pi m y) = - \langle y\rangle \quad .$$

Integrate this repeatedly to evaluate

$$\sum_{m=1}^{\infty} \frac{(-1)^m}{(\pi m)^2} \cos(2\pi m y) \;, \sum_{m=1}^{\infty} \frac{(-1)^m}{(\pi m)^3} \sin(2\pi m y) \quad ,$$

$$\sum_{m=1}^{\infty} \frac{(-1)^m}{(\pi m)^4} \cos(2\pi m y) \;.$$

<u>1-3.</u> In order to establish Eq.(43), first use the one-dimensional statements

$$\frac{1}{2}p^2 - F x = e^{-i\frac{p^3}{6F}} (-F x) e^{i\frac{p^3}{6F}} \quad ,$$

$$e^{-iFxt} p\, e^{iFxt} = p + F x t \quad ,$$

which are illustrations of

$$e^{-i f(p)} x\, e^{i f(p)} = x - \frac{d}{dp} f(p)$$

and

$$e^{-i f(x)} p\, e^{i f(x)} = p + \frac{d}{dx} f(x) \quad ,$$

to show that

$$e^{-i(\frac{1}{2}p^2 - F x)t} = e^{-i\frac{p^3}{6F}} e^{iFxt} e^{i\frac{p^3}{6F}}$$

$$= e^{iFxt} e^{-i\frac{1}{6F}(p+Ft)^3} e^{i\frac{p^3}{6F}}$$

$$= e^{iFxt} e^{-i\frac{1}{2}(p+\frac{1}{2}Ft)^2 t} e^{-iF^2t^3/24} \quad .$$

Generalize to three dimensions and arrive at Eq.(42).

<u>1-4.</u> The average value of r^2 for an orbital state in the Bohr atom is

$$(\overline{r^2})_{m',\ell'} = \frac{m'^2}{2Z^2}[5m'^2 - 3\ell'(\ell'+1)] \quad ,$$

where m' is the principal quantum number and $\ell'=o,\ldots,m'-1$ the angular momentum quantum number. Average this over the ℓ' values to find

$$\overline{(r^2)}_{m'} = \frac{m'^2}{4Z^2}(7m'^2+5) \quad .$$

A measure for the size R of the atom is the average of r^2,

$$N R^2 = \overline{(r^2)}_{\text{atom}} = \sum_{m'=1}^{m} 2m'^2\,\overline{(r^2)}_{m'} + \mu\, 2(m+1)^2\,\overline{(r^2)}_{m+1} \quad .$$

Show that

$$R \cong \frac{\sqrt{3/4}}{Z}\left[\left(\frac{3}{2}N\right)^{2/3} + \frac{1}{2}\right]$$

for large N. Compare with Eqs. (87) and (92).

1-5. The contribution of a full Bohr shell, with principal quantum number m', to n_o, the electron density at the site of the nucleus, is given by

$$\frac{(2Z)^3}{4\pi}\left(\frac{1}{m'}\right)^3 \quad .$$

Show that

$$n_o = \frac{(2Z)^3}{4\pi}\left[\sum_{m'=1}^{\infty}\left(\frac{1}{m'}\right)^3 - \frac{1}{2}\left(\frac{3}{2}N\right)^{-2/3} + O(N^{-4/3})\right] \quad ,$$

for a Bohr atom (without shielding) that contains N electrons.

1-6. Derive the identity

$$\sum_{m'=1}^{m} f(m') - \int_1^m dy f(y) = \int_1^m dy \,\langle y - \tfrac{1}{2}\rangle \frac{df(y)}{dy} + \frac{1}{2}[f(m)+f(1)]$$

(which, incidentally, was first proven by Euler) and use it to confirm Eq. (47).

Chapter Two

THOMAS - FERMI MODEL

The crude models of the preceding Chapter taught us that it may be useful to treat the electrons in an atom (or ion) as if they were moving independently in an effective potential. We shall now take this idea very seriously, without, however, making explicit assumptions about the effective potential, V. It is clear that V possesses the general structure[1,2]

$$V = -\frac{Z}{r} + [\text{electron-electron part}] \quad , \tag{2-1}$$

and the challenge consists in finding the electron-electron part in a consistent way. The fundamental tool for achieving this aim is the electrostatic Poisson equation

$$-\frac{1}{4\pi}\nabla^2 V_{es} = n \quad , \tag{2-2}$$

which relates the electron density, n, to the electrostatic potential, V_{es}, due to the electrons. As soon as we shall have managed to express both V_{es} and n in terms of V, Eq.(2) will determine the effective potential.

General formalism. The dynamics of the electrons is controlled by the independent-particle Hamilton operator

$$H = \frac{1}{2}p^2 + V(\vec{r}) \quad . \tag{2-3}$$

The electrons fill the eigenstates of H successively in such a way that all states with binding energy larger than a certain value, ζ, are occupied, whereas those with less binding energy are not. The parameter ζ is thus determined by the requirement that the count of occupied states equals the number of electrons N. Just as in Eq.(1-27) this is expressed as

$$N = \operatorname{tr} \eta(-H-\zeta) \quad , \tag{2-4}$$

where we remember that the spin mulitplicity of two is included in the trace.

The sum of independent-particle energies is, analogously,

$$E_{IP} = \mathrm{tr}\; H\eta(-H-\zeta) \quad . \tag{2-5}$$

The combination $H+\zeta$, that appears in the argument of Heaviside's step function η, invites rewriting E_{IP} as

$$E_{IP} = \mathrm{tr}(H+\zeta)\eta(-H-\zeta) - \zeta \mathrm{tr}\;\eta(-H-\zeta) \quad , \tag{2-6}$$

which, with the aid of (4) and the definition

$$E_1 \equiv \mathrm{tr}(H+\zeta)\eta(-H-\zeta) \quad , \tag{2-7}$$

reads

$$E_{IP} = E_1 - \zeta N \quad . \tag{2-8}$$

In this equation, N is the given number of electrons, and both E_{IP} and E_1 are function(al)s of the effective potential V and the minimum binding energy ζ.

Let us make contact with Eqs.(1-27) and (1-32), in that we write

$$E_1(\zeta) = -\int_{\zeta}^{\infty} d\zeta'\; N(\zeta') \quad , \tag{2-9}$$

where

$$N(\zeta') = \mathrm{tr}\;\eta(-H-\zeta') \tag{2-10}$$

is the count of states with binding energy exceeding ζ'. Equation (4) appears now as

$$N = N(\zeta) \quad . \tag{2-11}$$

Equation (9) can be equivalently presented as a differential statement. If ζ deviates from its correct value [which is determined by Eq.(11)] by the amount $\delta\zeta$, then E_1 is off by

$$\delta_\zeta E_1 = \frac{\partial E_1}{\partial \zeta}\,\delta\zeta = N(\zeta)\delta\zeta = N\delta\zeta \quad . \tag{2-12}$$

This has the important implication that E_{IP} of Eq.(8) is stationary under variations of ζ (around its correct value, of course):

$$\delta_\zeta E_{IP} = \delta_\zeta E_1 - N\delta\zeta = o \quad . \tag{2-13}$$

In addition to ζ, E_1 and E_{IP} also depend on V. The local response of both energies to variations of the potential exhibits the electron density n:

$$\delta_V E_{IP} = \delta_V E_1 = \int (d\vec{r}\,')\delta V(\vec{r}\,')\; n(\vec{r}\,') \quad . \tag{2-14}$$

Although this is intuitively obvious, let us supply a formal proof. The first equality follows immediately from (8), because N is the given number of electrons and ζ is a parameter that we regard as independent of V. For the second equality, we need the following identity:

$$\delta_H \operatorname{tr} f(H) = \operatorname{tr} \delta H\, f'(H) \quad , \tag{2-15}$$

which expresses the change in the trace of a function of an operator H as the trace of the product of the change in the operator, δH, and the derivative of that function. [Note that (15) is not true without the trace operation, unless δH commutes with H:

$$\delta_H f(H) = \delta H\, f'(H) \quad \text{only if} \quad [\delta H, H] = o. \tag{2-16}$$

Under the trace the possible noncommutativity does not matter.] In our application,

$$\begin{aligned} f(H) &= (H+\zeta)\eta(-H-\zeta) \quad , \\ f'(H) &= \eta(-H-\zeta) \end{aligned} \tag{2-17}$$

[compare with Eq. (1-29)], and $\delta H = \delta V$. Accordingly,

$$\begin{aligned} \delta_V E_1 &= \operatorname{tr} \delta V\, \eta(-H-\zeta) \\ &= 2 \int (d\vec{r}\,')\; \langle \vec{r}\,' | \delta V(\vec{r}) \eta(-H(\vec{p},\vec{r}) - \zeta) | \vec{r}\,' \rangle \quad . \end{aligned} \tag{2-18}$$

We use, again, primes to distinguish numbers from operators; the factor of two is, once more, the spin multiplicity. Now, since

$$\langle \vec{r}\,' | \delta V(\vec{r}) = \delta V(\vec{r}\,')\; \langle \vec{r}\,' | \quad , \tag{2-19}$$

and, anticipating that

$$2\langle\vec{r}'|\eta(-H-\zeta)|\vec{r}'\rangle = n(\vec{r}') \quad , \tag{2-20}$$

Eq.(18) implies Eq.(14). Indeed, equation (20) is nothing but the representation of the density as the sum of squared wavefunctions over all occupied states. Upon labelling these wavefunctions by their energies E' and additional quantum numbers, α, the left-hand side of (20) is

$$\begin{aligned} &2\sum_{E',\alpha} \psi^*_{E',\alpha}(\vec{r}')\ \eta(-E'-\zeta)\psi_{E',\alpha}(\vec{r}') \\ &= 2\sum_{E',\alpha}|\psi_{E',\alpha}(\vec{r}')|^2\eta(-E'-\zeta) \quad , \end{aligned} \tag{2-21}$$

which is recognized as the usual definition of the density.

For consistency, the integrated density must equal the number of electrons,

$$N = \int(d\vec{r}')n(\vec{r}') \quad . \tag{2-22}$$

This follows immediately from Eq.(20):

$$\begin{aligned} \int(d\vec{r}')n(\vec{r}') &= 2\int(d\vec{r}')\langle\vec{r}'|\eta(-H-\zeta)|\vec{r}'\rangle \\ &= \mathrm{tr}\ \eta(-H-\zeta) = N(\zeta) = N \quad . \end{aligned} \tag{2-23}$$

Another, and more instructive, proof makes use of (i) the definition of n in Eq.(14); (ii) the circumstance that E_1 does not depend on V and ζ individually, but only on the sum $V+\zeta$; (iii) Equation (12). Consider infinitesimal changes in ζ and V such that $\delta V(\vec{r}) = -\delta\zeta$.[3] Then $\delta(V+\zeta)=o$, implying $\delta E_1=o$. In view of Eqs.(12) and (14) this means

$$\begin{aligned} o = \delta_\zeta E_1 + \delta_V E_1 &= N\delta\zeta + \int(d\vec{r}')(-\delta\zeta)n(\vec{r}') \\ &= \delta\zeta\left(N-\int(d\vec{r})n(\vec{r})\right) \quad , \end{aligned} \tag{2-24}$$

which is equivalent to (22). This second proof has the advantage of remaining valid when the trace in E_1 is evaluated approximately. There is no assurance that the densities derived from (14) and (20) are identical in a certain approximation. If they are not, Eq.(14) is the preferable definition. (We shall, indeed, be confronted with this possibility later, in Chapter Four.)

Equation (14) relates the density to the effective potential, so that we have taken care of the right-hand side of Eq.(2). We are left with the problem of expressing the electrostatic potential of the electrons, V_{es}, in terms of V.

We proceed from noting that E_{IP} is not the energy of the system. Just as in the preceding Chapter [recall the remark after Eq.(1-65)], the use of the effective potential causes a double counting of the electron-electron interaction energy, E_{ee}. The interaction potential V_{ee} which is the electron-electron part of V in Eq. (1), is naturally given as the response of E_{ee} to variations of the density,

$$\delta E_{ee} = \int (d\vec{r}\,') \; \delta n(\vec{r}\,') \; V_{ee}(\vec{r}\,') \quad . \tag{2-25}$$

[Please do not miss the analogy to Eq.(14).] Since V and ζ are the fundamental quantities in our "potential-functional formalism," $\delta n(\vec{r})$ must be regarded as the change in the density induced by variations of V and ζ.

Some evidence in favor of (25) is supplied by considering the electrostatic interaction energy

$$E_{es} = \frac{1}{2} \int (d\vec{r}\,')(d\vec{r}\,'') \; \frac{n(\vec{r}\,')n(\vec{r}\,'')}{|\vec{r}\,'-\vec{r}\,''|} \quad , \tag{2-26}$$

for which

$$\delta E_{es} = \int (d\vec{r}\,')\delta n(\vec{r}\,') \int (d\vec{r}\,'') \; \frac{n(\vec{r}\,'')}{|\vec{r}\,'-\vec{r}\,''|} \quad . \tag{2-27}$$

Thus, Eq.(25) implies the familiar expression

$$V_{es}(\vec{r}\,') = \int (d\vec{r}\,'') \; \frac{n(\vec{r}\,'')}{|\vec{r}\,'-\vec{r}\,''|} \quad , \tag{2-28}$$

which is equivalent to the Poisson equation (2).

The electron-electron interaction energy, as it is incorrectly contained in E_{IP} ("double counting of pairs"), is

$$\begin{aligned} &\mathrm{tr}\, V_{ee}\, \eta(-H-\zeta) \\ &= 2 \int (d\vec{r}\,') \; V_{ee}(\vec{r}\,')\langle \vec{r}\,'|\eta(-H-\zeta)|\vec{r}\,'\rangle \\ &= \int (d\vec{r}\,') V_{ee}(\vec{r}\,') n(\vec{r}\,') \quad ; \end{aligned} \tag{2-29}$$

the last step uses Eq.(20). Consequently, the correct energy expression is

$$E = E_{IP} - \int (d\vec{r}) V_{ee} n + E_{ee} \quad . \tag{2-30}$$

The second term removes the incorrect account for the electron-electron interaction contained in E_{IP}, and the last term adds the correct amount.

The energy of Eq.(30) is endowed with the important property of being stationary under variations of both V and ζ,

$$\delta_\zeta \, E = \delta_V \, E = 0. \tag{2-31}$$

In order to see this, first appreciate

$$\begin{aligned} &\delta\left(- \int (d\vec{r}) V_{ee} n + E_{ee}\right) \\ &= - \int (d\vec{r}) (\delta V_{ee} n + V_{ee} \delta n) + \int (d\vec{r}) \delta n \, V_{ee} \\ &= - \int (d\vec{r}) n \delta V_{ee} \quad , \end{aligned} \tag{2-32}$$

which is an implication of Eq.(25). Further, a consequence of Eqs.(13) and (14) is

$$\begin{aligned} \delta E_{IP} &= \delta_\zeta E_{IP} + \delta_V \, E_{IP} \\ &= \int (d\vec{r}) n \delta V \quad . \end{aligned} \tag{2-33}$$

Then, the change in E is

$$\delta E = \int (d\vec{r}) n (\delta V - \delta V_{ee}) = \int (d\vec{r}) n \delta (V - V_{ee}) \tag{2-34}$$

In view of [Eq.(1)]

$$V = - \frac{Z}{r} + V_{ee} \quad , \tag{2-35}$$

the variation $\delta(V-V_{ee})$ vanishes, and Eq.(34) implies Eq.(31), indeed.

It is useful to separate E_{ee} into the classical electrostatic part, E_{es}, of Eq.(26), and the remainder E'_{ee}, which consists of the exchange interaction and possibly other effects. Accordingly, we write

$$E_{ee} = E_{es} + E'_{ee} \quad , \tag{2-36}$$

and likewise

$$V_{ee} = V_{es} + V'_{ee} \quad . \tag{2-37}$$

The electrostatic contribution to the energy (30) can be rewritten, with the aid of the Poisson equation (2), in terms of the electrostatic field $-\vec{\nabla}V_{es}$:

$$\begin{aligned} -\int(d\vec{r})n\,V_{es} + E_{es} &= -\frac{1}{2}\int(d\vec{r})n\,V_{es} \\ &= \frac{1}{8\pi}\int(d\vec{r})\,(\nabla^2 V_{es})V_{es} = -\frac{1}{8\pi}\int(d\vec{r})\,(\vec{\nabla}V_{es})^2 \ . \end{aligned} \tag{2-38}$$

[The surface term of the partial integration is zero, because $V_{es} \cong N/r$ for large r.] Further, we combine Eqs.(35) and (37) into

$$V_{es} = V + \frac{Z}{r} - V'_{ee} \quad , \tag{2-39}$$

thereby expressing V_{es} in terms of V, as needed in (2). The energy now reads

$$\begin{aligned} E = E_{IP} &- \frac{1}{8\pi}\int(d\vec{r})\,[\vec{\nabla}(V+\frac{Z}{r}-V'_{ee})]^2 \\ &- \int(d\vec{r})n\,V'_{ee} + E'_{ee} \quad . \end{aligned} \tag{2-40}$$

This expression for the energy is our basis for approximations. Various models emerge depending upon the accuracy to which the trace in E_{IP} [Eqs.(7) and (8)] is evaluated, and upon the extent to which E'_{ee} is taken into account. Of course, a consistent description requires a balanced treatment of both.

The TF model. The simplest model based upon Eq.(40) is the TF model. It neglects E'_{ee} entirely [then V'_{ee} also disappears from (40)], and evaluates the trace of Eq.(7) in the highly semiclassical approximation of Eq.(1-43). The TF energy expression is therefore

$$\begin{aligned} E_{TF} = 2\int\frac{(d\vec{r})\,(d\vec{p})}{(2\pi)^3}\,(\tfrac{1}{2}p^2+V+\zeta)\,\eta(-\tfrac{1}{2}p^2-V-\zeta) - \zeta N \\ - \frac{1}{8\pi}\int(d\vec{r})\,[\vec{\nabla}(V+\frac{Z}{r})]^2 \quad . \end{aligned} \tag{2-41}$$

We recognize the last term as the quantity E_2 of Eq.(1-67), which was there introduced to remove the doubly counted (electrostatic) interaction energy; the term plays the same role here. The phase-space integral is the TF version of E_1, properly denoted by $(E_1)_{TF}$. We shall, however, suppress the subscript TF until it will become a necessary distinction from other models.

The step function cuts off the momentum integral at the (r-dependent) maximal momentum (the so-called "Fermi momentum")

$$P = \sqrt{-2(V+\zeta)} \quad , \tag{2-42}$$

so that

$$\begin{aligned} E_1 &= \int(d\vec{r})\,\frac{2}{(2\pi)^3}\,4\pi\int_0^P dp\;p^2\left(\tfrac{1}{2}p^2-\tfrac{1}{2}P^2\right) \\ &= \int(d\vec{r})\,\frac{1}{\pi^2}\left(\frac{1}{10}-\frac{1}{6}\right)P^5 \quad , \end{aligned} \tag{2-43}$$

or, square roots of negative arguments being zero,

$$E_1 = \int(d\vec{r})\left(-\frac{1}{15\pi^2}\right)[-2(V+\zeta)]^{5/2} \quad . \tag{2-44}$$

This is the <u>Thomas-Fermi result for E_1</u>. The entire <u>energy functional in the TF model</u> is then

$$\begin{aligned} E_{TF} &= E_1 + E_2 - \zeta N \\ &= \int(d\vec{r})\left(-\frac{1}{15\pi^2}\right)[-2(V+\zeta)]^{5/2} - \frac{1}{8\pi}\int(d\vec{r})\left[\vec{\nabla}\left(V+\frac{Z}{r}\right)\right]^2 - \zeta N \quad . \end{aligned} \tag{2-45}$$

Is there any reality to it? Yes. Look back to Chapter One, where (45) has been used unconsciously for the Coulomb potential $V=-Z/r$. In this situation, E_2 equals zero, and E_{TF} gives the leading term of Eq.(1-22) [see Eqs.(1-26) through (1-37)]. Since V is essentially equal to the Coulomb potential in a highly ionized atom, we conclude

$$E_{TF} \cong -Z^2\left(\tfrac{3}{2}N\right)^{1/3} \quad \text{for} \quad N \ll Z \quad . \tag{2-46}$$

We shall return to highly ionized systems in a while and find the modification of (46) when accounting for the electron-electron repulsion. Before doing so, we have to study some implications of Eq.(45).

The stationary property of E_{TF} with respect to variations of V and ζ reads

$$o = \delta E_{TF} = \int(d\vec{r})\delta V\{\frac{1}{3\pi^2}[-2(V+\zeta)]^{3/2} + \frac{1}{4\pi}\nabla^2(V+\frac{Z}{r})\}$$

$$+ \delta\zeta\{\int(d\vec{r})\frac{1}{3\pi^2}[-2(V+\zeta)]^{3/2} - N\}$$

$$- \frac{1}{4\pi}\int(d\vec{r})\vec{\nabla}\cdot(\delta V\,\vec{\nabla}(V+\frac{Z}{r})) \quad . \tag{2-47}$$

The value of the last integral is zero, because the equivalent integration over a remote surface vanishes in view of $\delta V = o$ for $r\to\infty$. The variations of V and ζ are independent, so that the two curly brackets equal zero individually. Accordingly,

$$-\frac{1}{4\pi}\nabla^2(V+\frac{Z}{r}) = \frac{1}{3\pi^2}[-2(V+\zeta)]^{3/2} \tag{2-48}$$

and

$$\int(d\vec{r})\frac{1}{3\pi^2}[-2(V+\zeta)]^{3/2} = N \quad , \tag{2-49}$$

of which the first is the Poisson equation, and the second the normalization of the density to N. Obviously, Eq.(49) is the TF version of (11), as we notice that Eq.(10) is realized as

$$N(\zeta') = 2\int\frac{(d\vec{r})(d\vec{p})}{(2\pi)^3}\eta(-\frac{1}{2}p^2-V-\zeta') = \int(d\vec{r})\frac{1}{3\pi^2}[-2(V+\zeta')]^{3/2}. \tag{2-50}$$

This, inserted into Eq.(9), reproduces (44), as it should.

On the right-hand side of (48) as well as under the integral of (49) we have the TF density

$$n = \frac{1}{3\pi^2}[-2(V+\zeta)]^{3/2} \quad . \tag{2-51}$$

In the classically forbidden domain, characterized by $V>-\zeta$, this density vanishes. There is a sharp boundary assigned to atoms in the TF model. In contrast, in an exact quantum mechanical description the transition from the classically allowed to the classically forbidden region is smooth. We have just learned about one of the deficiencies of the TF model. It is going to be removed later when we shall incorporate quantum corrections of the sort discussed briefly after Eq.(1-43).

The differential equation (48) for V, known as the TF equation for V, is supplemented by the constraint (49) and the short distance behavior of V,

$$r\,V \to -\,Z \quad \text{for} \quad r \to o \quad . \tag{2-52}$$

It signifies the physical requirement that for $r \to o$, the effective potential is mainly given by the electrostatic potential energy of an electron with the nucleus; formally, (52) is necessary to ensure the finiteness of E_2. Consequently, we have the following situation: for small r, the potential is large negative, and the density is large; as r increases the potential becomes less and less negative; finally, at the edge of the classically allowed region, it equals $-\zeta$, and the argument of the square root in (51) turns negative; beyond this distance, r_o, the density is zero, so that (48) is the homogeneous Poisson equation. Gauss's law, combined with Eqs.(49) and (52), then implies

$$V = -\,\frac{Z-N}{r} \quad \text{for} \quad r \geq r_o \quad , \tag{2-53}$$

and the radius r_o of the atom is determined by

$$V(r=r_o) = -\,\zeta \quad , \tag{2-54}$$

or,

$$\zeta = \frac{Z-N}{r_o} \quad . \tag{2-55}$$

The electric field $-\vec{\nabla}V$ is continuous (there are no charged surfaces in an atom); in particular, at the edge we have

$$\frac{d}{dr}\,V(r)\Big|_{r_o} = \frac{Z-N}{r_o^2} = \frac{\zeta}{r_o} \quad . \tag{2-56}$$

Neutral systems, N=Z, have $\zeta = o$, so that both V and dV/dr vanish at $r=r_o$. Consequently, the TF equation for V, Eq.(48), requires $r_o = \infty$, since for a finite r_o it cannot have a solution satisfying these boundary conditions. We have just learned that neutral TF atoms are infinitely large, they do not have an "outside", only an "inside".

It is useful to measure $V+\zeta$ as as multiple of the potential of the nucleus by introducing a function f(x),

$$V + \zeta = -\,\frac{Z}{r}\,f(x) \quad , \tag{2-57}$$

the argument of which is related to the physical distance r by

$$x = Z^{1/3}r/a \quad , \quad a = \frac{1}{2}\left(\frac{3\pi}{4}\right)^{2/3} = 0.8853\ldots \tag{2-58}$$

The constant a is chosen such that the differential equation for f(x),

$$\frac{d^2}{dx^2} f(x) = \frac{[f(x)]^{3/2}}{x^{1/2}} \quad , \tag{2-59}$$

called the TF equation for f(x), is free of numerical factors. The boundary conditions (52),(54), and (56) translate into

$$f(o) = 1 \ , \quad f(x_o) = o \ , \quad -x_o \frac{d}{dx_o} f(x_o) = 1 - \frac{N}{Z} \equiv q \quad , \tag{2-60}$$

which introduces q, the degree of ionization. Of course, x_o is related to r_o through (58). Equation (53) now appears as

$$f(x) = q(1-x/x_o) \quad \text{for} \quad x \geq x_o \quad . \tag{2-61}$$

Please notice that Z and N do not appear individually in Eqs.(59) and (60). Consequently, f(x) is solely determined by the degree of ionization, q, so that all ions with the same q possess a common shape of the potential and of the density. The potential V itself does, of course, depend on Z; first through the factor Z/r, but then also because of the Z dependence of the TF variable x of Eq. (58). The factor $Z^{1/3}$ there implies the same shrinking of heavier atoms that we have already observed in Chapter One, when considering Bohr atoms with shielding, see Eq.(1-93).

For illustration, Fig.1 shows a sketch of f(x) for q = 1/2, for which $x_o \cong 3$. The geometrical significance of the third equation in (60) is indicated.

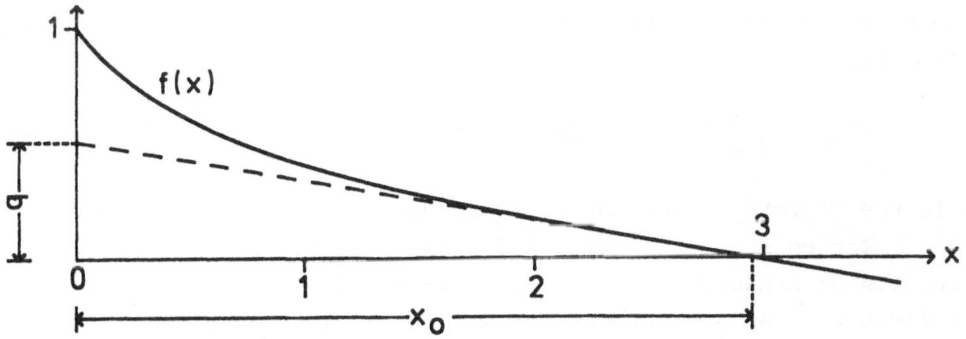

Fig. 2-1. Sketch of f(x) *for* q = 1/2.

<u>Neutral TF atoms.</u> For the solution of Eqs.(59) and (60) that belongs to

q=o, we write F(x) and call it the <u>TF function</u>. It obeys

$$\frac{d^2}{dx^2} F(x) = \frac{[F(x)]^{3/2}}{x^{1/2}} \quad , \tag{2-62}$$

and is subject to

$$F(o) = 1 \quad , \quad F(\infty) = o \quad . \tag{2-63}$$

Its initial slope B,

$$F(x) = 1 - Bx + \ldots \qquad \text{for} \quad x \ll 1 \quad , \tag{2-64}$$

has an important physical significance. We insert (64) into (57), use (58), and arrive at

$$V(r) \cong -\frac{Z}{r} + \frac{B}{a} Z^{4/3} \qquad \text{for} \quad r \to o \quad . \tag{2-65}$$

The additive constant is the interaction energy of an electron, near the nucleus, with the main body of electrons. We can use it to immediately write down the change in energy caused by an infinitesimal change of the nuclear charge Z to Z + δZ. It is the analogous electrostatic energy of that additional charge, where a minus sign is needed to connect with the known energy, which is that of an electron:

$$\delta E_{TF} = -\frac{B}{a} Z^{4/3} \delta Z \quad . \tag{2-66}$$

The simultaneous increase of the number of electrons from N=Z to N=Z+δZ has no effect on the energy since $\partial E/\partial N = -\zeta = o$ for N=Z, see Eq.(55). Consequently,

$$-E_{TF} = \frac{3}{7} \frac{B}{a} Z^{7/3} \qquad \text{for} \qquad N = Z \quad . \tag{2-67}$$

This is the TF formula for the total binding energy of neutral atoms.

The constant B is well known numerically. But before quoting the results of a numerical integration of Eqs.(62) and (63), let us use our insight to find an estimate for B. Indeed, in view of the physical approximations that led to the TF model, there is no need, at this stage of the development, of knowing B better than within a few percent. A first crude estimate is given by the comparison of (67) with (1-51), the result obtained in the model of Bohr atoms with shielding:

$$B \cong \frac{7a}{3}\,\frac{9}{14}\left(\frac{3}{2}\right)^{1/3} = \frac{9}{8}\left(\frac{\pi}{2}\right)^{2/3} = 1.52 \quad . \tag{2-68}$$

We have no way of judging, how accurate this number may be, but shall see later that it deviates by less than 5% from the correct value.

The stationary property of the energy functional (45) provides a tool for obtaining good estimates for B. If we evaluate $E_{TF}(V,\zeta)$ for a trial potential V and $\zeta = o$ (this much we know for sure when N=Z), the deviation of $E_{TF}(V,\zeta=o)$ from $-\frac{3}{7}\frac{B}{a}Z^{7/3}$ will be of second order in the error of V. As we shall see in the following section, the energy functional has a maximum for the correct potential. Consequently, any trial V gives an upper bound for the constant B:

$$B \leq -\frac{7}{3}a\,Z^{-7/3}\,E_{TF}(V,\zeta) \quad , \quad (\text{for } N = Z) \quad , \tag{2-69}$$

where the equal sign holds only for ζ=o and $V = -(Z/r)F(x)$.

<u>Maximum property of the TF potential functional.</u> Let us consider finite deviations from the correct potential V and the correct value for ζ, denoted by ΔV and $\Delta\zeta$, respectively, as distinguished from the infinitesimal variations δV and $\delta\zeta$. Whereas $\Delta\zeta$ is quite arbitrary, ΔV is subject to

$$\begin{aligned} r\,\Delta V &\to o \quad \text{for} \quad r \to o \quad , \\ \Delta V &\to o \quad \text{for} \quad r \to \infty \quad , \end{aligned} \tag{2-70}$$

which are consequences of (52) and the normalization $V(r\to\infty)$=o. The deviations of the three terms of E_{TF} in (45) are then

$$\Delta E_1 = \int(d\vec{r})\left(-\frac{1}{15\pi^2}\right)\left([-2(V+\Delta V+\zeta+\Delta\zeta)]^{5/2} - [-2(V+\zeta)]^{5/2}\right) \quad , \tag{2-71}$$

and

$$\begin{aligned} \Delta E_2 &= -\frac{1}{8\pi}\int(d\vec{r})\left([\vec{\nabla}(V+\Delta V+\frac{Z}{r})]^2 - [\vec{\nabla}(V+\frac{Z}{r})]^2\right) \\ &= -\frac{1}{8\pi}\int(d\vec{r})\,[\vec{\nabla}(\Delta V)]^2 - \frac{1}{4\pi}\int(d\vec{r})\,\vec{\nabla}(\Delta V)\cdot\vec{\nabla}(V+\frac{Z}{r}) \quad , \end{aligned} \tag{2-72}$$

which after a partial integration and the use of Eq. (48) reads

$$\Delta E_2 = -\frac{1}{8\pi}\int(d\vec{r})\,[\vec{\nabla}(\Delta V)]^2 - \int(d\vec{r})\,\Delta V\,\frac{1}{3\pi^2}[-2(V+\zeta)]^{3/2} \quad , \tag{2-73}$$

as well as

$$\Delta(-\zeta N) = -(\Delta\zeta)N = -\int(d\vec{r})\,\Delta\zeta\,\frac{1}{3\pi^2}[-2(V+\zeta)]^{3/2} \quad , \tag{2-74}$$

where Eq. (49) has been employed. Accordingly,

$$\begin{aligned}\Delta E_{TF} &= \Delta E_1 + \Delta E_2 + \Delta(-\zeta N)\\ &= \int(d\vec{r})\left(-\frac{1}{15\pi^2}\right)\{[-2(V+\zeta)-2(\Delta V+\Delta\zeta)]^{5/2} - [-2(V+\zeta)]^{5/2}\\ &\qquad + 5(\Delta V+\Delta\zeta)[-2(V+\zeta)]^{3/2}\}\\ &\quad - \frac{1}{8\pi}\int(d\vec{r})\,[\vec{\nabla}(\Delta V)]^2 \quad .\end{aligned} \tag{2-75}$$

The contents of the curly brackets is of the structure

$$\begin{aligned}&[u+v]^{5/2} - u^{5/2} - \frac{5}{2}\,v\,u^{3/2}\\ &= \frac{15}{4}\int_o^v dv'\,(v-v')\,[u+v']^{1/2} \geq o \quad ,\end{aligned} \tag{2-76}$$

where $u = -2(V+\zeta)$ and $v = -2(\Delta V+\Delta\zeta)$. The equal sign in (76) holds only if $v = o$, or, if $u+v' \leq o$ over the whole range of integration (under which circumstance the square root vanishes). This implies

$$\Delta E_{TF} \leq o \ ; \ = o \ \text{ only for } \ \Delta V = o \ \text{ and } \ \Delta\zeta = o \quad . \tag{2-77}$$

In words: the TF potential functional of Eq. (45) has an absolute maximum at the correct V and ζ.

This maximum property might come as a surprise, as one naively expects the electrons to arrange themselves such that the energy achieves a minimum. True, but it is not different electron distributions that we compare; the competition is among different potentials. In the same sense, in which it is natural for the right density to minimize the energy, it is common for the right potential to maximize it. Let us illustrate this point by the analogous (and closely related) situation in electrostatics.

An electrostatic analogy. Consider the problem of finding the electrostatic potential, Φ, to a given charge density, ρ, in the vacuum.[4] They are related to each other by the Poisson equation

$$-\frac{1}{4\pi}\nabla^2\Phi = \rho \quad . \tag{2-78}$$

The electrostatic energy can be expressed in various ways:

$$\begin{aligned} E &= \frac{1}{2}\int(d\vec{r})\,\rho\Phi = \frac{1}{8\pi}\int(d\vec{r})\,(\vec{\nabla}\Phi)^2 \\ &= \int(d\vec{r})\,[\rho\Phi - \frac{1}{8\pi}(\vec{\nabla}\Phi)^2] \quad . \end{aligned} \tag{2-79}$$

If we insert the Φ that obeys (78) into any of these expressions, they all give the same answer. Suppose, however, that we do not know the correct Φ and have to resort to using an approximate one. In this situation, it is advisable to employ the third version of (79) in calculating the energy, because, unlike the other ones, this expression is stationary at the correct Φ:

$$\delta\int(d\vec{r})\,[\rho\Phi - \frac{1}{8\pi}(\vec{\nabla}\Phi)^2] = \int(d\vec{r})\,\delta\Phi[\rho + \frac{1}{4\pi}\nabla^2\Phi] = 0 \quad . \tag{2-80}$$

A finite deviation $\Delta\Phi$ from the right electrostatic potential results in the second order error in E that is given by

$$\Delta E = -\frac{1}{8\pi}\int(d\vec{r})\,[\vec{\nabla}(\Delta\Phi)]^2 < 0 \quad ; \tag{2-81}$$

the energy is maximal for the right Φ. The analogy to the TF functional is, indeed, close, since the same term occurs also in (75).

Here is a little application of the stationary property of the electrostatic "potential functional."[5] Instead of inserting $\Phi(\vec{r})$, we evaluate the energy for $\Phi(\lambda\vec{r})$:

$$\begin{aligned} E(\lambda) &= \int(d\vec{r})\ \rho\Phi(\lambda\vec{r}) - \frac{1}{8\pi}\int(d\vec{r})\,[\vec{\nabla}\Phi(\lambda\vec{r})]^2 \\ &= \int(d\vec{r})\ \rho\Phi(\lambda\vec{r}) - \frac{1}{\lambda}\frac{1}{8\pi}\int(d\vec{r})\,[\vec{\nabla}\Phi(\vec{r})]^2 \quad . \end{aligned} \tag{2-82}$$

For $\lambda=1$, it is the correct energy. Consequently,

$$\frac{d}{d\lambda} E(\lambda) = 0 \quad \text{for} \quad \lambda=1 \quad , \tag{2-83}$$

which implies

$$\frac{1}{8\pi} \int (d\vec{r}) (\vec{\nabla}\Phi)^2 = \int (d\vec{r}) \; \rho\vec{r}\cdot(-\vec{\nabla}\Phi) \quad . \tag{2-84}$$

We have thus found an unusual expression for the electrostatic energy: the integral of the scalar product of the dipole density $\rho\vec{r}$ with the electric field $-\vec{\nabla}\Phi$. Note, in particular, that there is no factor of 1/2. Since a translated charge distribution $\rho(\vec{r}+\vec{R})$ has the same electrostatic energy,

$$\int (d\vec{r}) \rho(\vec{r}+\vec{R}) \vec{r} \cdot (-\vec{\nabla}\Phi(\vec{r}+\vec{R})) = \int (d\vec{r}) \rho(\vec{r}) \vec{r}\cdot[-\vec{\nabla}\Phi(\vec{r})] \quad , \tag{2-85}$$

we find, after substituting $\vec{r}\rightarrow\vec{r}-\vec{R}$ on the left hand side, that the self force of any charge density vanishes:

$$\int (d\vec{r}) \rho(-\vec{\nabla}\Phi) = 0 \quad . \tag{2-86}$$

(The stresses, of course, do not.)

A different problem is that of finding the correct charge density on the surface, S, of a conductor carrying a given total charge, Q. In this situation, the relevant equations are

$$\int dS' \frac{\sigma(\vec{r}\,')}{|\vec{r}-\vec{r}\,'|} = \text{const. for } \vec{r} \text{ on S} \quad , \tag{2-87}$$

and

$$\int dS \; \sigma(\vec{r}) = Q \quad , \tag{2-88}$$

where σ denotes the surface charge density. Here the stationary energy expression is

$$E = \frac{1}{2}\int dS dS' \frac{\sigma(\vec{r})\sigma(\vec{r}\,')}{|\vec{r}-\vec{r}\,'|} + \Phi_o(Q-\int dS\sigma(\vec{r})) \quad . \tag{2-89}$$

The last term incorporates the constraint (88). Infinitesimal variations of both σ and Φ_o imply Eqs.(87) and (88), thereby identifying Φ_o as the (constant) electrostatic potential on S. This energy is a minimum if only σ's obeying (88) are allowed in the competition, i.e., if various distributions of the same, given, amount of charge are compared.

We get

$$\Delta E = \frac{1}{2}\int dSdS' \; \frac{\Delta\sigma(\vec{r})\;\Delta\sigma(\vec{r}\,')}{|\vec{r} - \vec{r}\,'|} \quad , \tag{2-90}$$

where $\Delta\sigma$ is the deviation from the optimal density σ. Since this is the electrostatic energy of some charge distribution, it is, indeed, positive.

TF density functional. This digression into the realm of electrostatics raises the question if it is possible to write down a functional of the density, in addition to the potential functional of (45), thus getting upper bounds on the energy, lower ones on the constant B. This can be done, indeed. It requires appropriate rewriting of (45), whereby the potential is replaced in terms of the density. Both Eq.(51) and the electrostatic relation

$$V(\vec{r}) = -\frac{Z}{r} + \int (d\vec{r}\,') \; \frac{n(\vec{r}\,')}{|\vec{r}-\vec{r}\,'|} \tag{2-91}$$

can and must be used in this process.

We start by undoing the step from Eq.(43) to Eq.(44), so that E_1 is split into the kinetic energy, E_{kin}, and a potential energy part:

$$\begin{aligned} E_1 &= \int (d\vec{r}) \; \frac{1}{10\pi^2}[-2(V+\zeta)]^{5/2} - \int (d\vec{r}) \; \frac{1}{6\pi^2}[-2(V+\zeta)]^{1+3/2} \\ &= \int (d\vec{r}) \; \frac{1}{10\pi^2}(3\pi^2 n)^{5/3} + \int (d\vec{r})\,(V+\zeta)n \\ &= E_{kin} + \int (d\vec{r})\,(V+\frac{Z}{r})n - \int (d\vec{r}) \; \frac{Z}{r}n + \zeta\int (d\vec{r})n \quad . \end{aligned} \tag{2-92}$$

E_2 is rewritten by first performing a partial integration, then making use of the Poisson equation, followed by employing Eq. (91):

$$\begin{aligned} E_2 &= -\frac{1}{8\pi}\int (d\vec{r})\,[\vec{\nabla}(V+\frac{Z}{r})]^2 = \frac{1}{2}\int (d\vec{r})\,(V+\frac{Z}{r}) \; \frac{1}{4\pi}\nabla^2(V+\frac{Z}{r}) \\ &= -\frac{1}{2}\int (d\vec{r})\,(V+\frac{Z}{r})n(\vec{r}) \quad = -\frac{1}{2}\int (d\vec{r})\,(d\vec{r}\,')\frac{n(\vec{r})n(\vec{r}\,')}{|\vec{r}-\vec{r}\,'|} \end{aligned} \tag{2-93}$$

Combining the two last versions into

$$E_2 = \frac{1}{2} \int (d\vec{r})(d\vec{r}\,') \frac{n(\vec{r})n(\vec{r}\,')}{|\vec{r}-\vec{r}\,'|} - \int (d\vec{r})(V+\zeta)n \tag{2-94}$$

makes the potential disappear from the sum of E_1 and E_2. The resulting <u>TF density functional</u> is

$$\begin{aligned} E &= E_1 + E_2 - \zeta N \\ &= \int (d\vec{r}) \frac{1}{10\pi^2}(3\pi^2 n)^{5/3} - \int (d\vec{r}) \frac{Z}{r}n + \frac{1}{2}\int (d\vec{r})(d\vec{r}\,') \frac{n(\vec{r})n(\vec{r}\,')}{|\vec{r}-\vec{r}\,'|} \\ &\quad - \zeta(N - \int (d\vec{r})n) \quad . \end{aligned} \tag{2-95}$$

All we know at this stage is that Eq.(95) gives the correct value of the energy, provided we insert the correct density. To be useful this functional has to be stationary about the right density. Not surprisingly, it is:

$$\begin{aligned} \delta E &= \int (d\vec{r})\ \delta n(\vec{r}) [\tfrac{1}{2}(3\pi^2 n(\vec{r}))^{2/3} - \frac{Z}{r} + \int (d\vec{r}\,') \frac{n(\vec{r}\,')}{|\vec{r}-\vec{r}\,'|} + \zeta] \\ &\quad - \delta\zeta\ (N - \int (d\vec{r})n) = 0 \quad , \end{aligned} \tag{2-96}$$

which uses Eqs.(51) and (91) in the combination[6]

$$V(\vec{r}) = -\frac{1}{2}(3\pi^2 n(\vec{r}))^{2/3} - \zeta = -\frac{Z}{r} + \int (d\vec{r}) \frac{n(\vec{r}\,')}{|\vec{r}-\vec{r}\,'|} \quad , \tag{2-97}$$

and the constraint (49), now reading

$$\int (d\vec{r})n = N \quad . \tag{2-98}$$

The successive terms in Eq.(95) have the physical significance of the kinetic energy, the potential energy between the nucleus and the electrons, and the electron-electron potential energy. The last term incorporates the constraint (98), thereby identifying ζ as the corresponding Lagrangian multiplier. In contrast, the potential functional of Eq.(45) consists of the sum of independent particle energies, $E_1 - \zeta N$, plus the removal of the doubly counted electron-electron-interaction energy, E_2. It is important to appreciate this difference in structure.

Let us now check if the density functional does have the expected property of being minimal for the correct n and ζ.

Minimum property of the TF density functional. In analogy to the previous discussion of the maximum property of the TF potential functional, we consider finite deviations Δn and $\Delta\zeta$ from the correct n and ζ. Again, $\Delta\zeta$ is quite arbitrary, whereas Δn is restricted by the requirement that the density be non-negative,

$$n + \Delta n \geq 0 \quad \text{for all } r \quad . \tag{2-99}$$

The derivation of (95) made use of (51) so that negative densities had been implicitly excluded.

The various contributions to ΔE are then

$$\Delta E_{kin} = \int (d\vec{r}) \frac{(3\pi^2)^{5/3}}{10\pi^2} [(n+\Delta n)^{5/3} - n^{5/3}] \quad , \tag{2-100}$$

and

$$\begin{aligned} &\Delta\left(- \int (d\vec{r}) \frac{Z}{r} n + \frac{1}{2}\int (d\vec{r}) (d\vec{r}\,') \frac{n(\vec{r})n(\vec{r}\,')}{|\vec{r}-\vec{r}\,'|}\right) \\ &= \int (d\vec{r}) \Delta n(\vec{r}) \left[- \frac{Z}{r} + \int (d\vec{r}\,') \frac{n(\vec{r}\,')}{|\vec{r}-\vec{r}\,'|}\right] \\ &\qquad + \frac{1}{2} \int (d\vec{r}) (d\vec{r}\,') \frac{\Delta n(\vec{r})\ \Delta n(\vec{r}\,')}{|\vec{r}-\vec{r}\,'|} \\ &= \int (d\vec{r}) \frac{(3\pi^2)^{5/3}}{10\pi^2} \left[- \frac{5}{3} n^{2/3} \Delta n\right] - \zeta \int (d\vec{r}) \Delta n \\ &\qquad + \frac{1}{2} \int (d\vec{r}) (d\vec{r}\,') \frac{\Delta n(\vec{r})\ \Delta n(\vec{r}\,')}{|\vec{r}-\vec{r}\,'|} \quad , \end{aligned} \tag{2-101}$$

which uses Eq. (97), as well as

$$\Delta\left[- \zeta \left(N - \int (d\vec{r}) n\right)\right] = (\zeta + \Delta\zeta) \int (d\vec{r}) \Delta n \quad . \tag{2-102}$$

Consequently,

$$\begin{aligned} \Delta E = &\int (d\vec{r}) \frac{(3\pi^2)^{5/3}}{10\pi^2} \left[(n+\Delta n)^{5/3} - n^{5/3} - \frac{5}{3} n^{2/3} \Delta n\right] \\ &+ \frac{1}{2} \int (d\vec{r}) (d\vec{r}\,') \frac{\Delta n(\vec{r})\ \Delta n(\vec{r}\,')}{|\vec{r}-\vec{r}\,'|} + \end{aligned}$$

$$+ \Delta\zeta \int (d\vec{r})\, \Delta n \quad . \tag{2-103}$$

The first term here is positive definite, which becomes obvious when we write it [compare Eq. (76)] in the form

$$\int (d\vec{r})\, \frac{(3\pi^2)^{5/3}}{10\pi^2}\, \frac{10}{9} \int_0^{\Delta n} d\nu\, (\Delta n - \nu)\,(n+\nu)^{-1/3} \tag{2-104}$$

$$\geq o \;;\quad = o \quad \text{only if} \quad \Delta n(\vec{r}) = o \quad \text{for all } \vec{r} \quad .$$

The second term in (103) is the electrostatic energy of the charge density $\Delta n(\vec{r})$, thus it is also positive, unless $\Delta n = o$ everywhere. The third term does not have a definite sign. Therefore we restrict the class of trial densities n and trial ζ's such that

$$\Delta\zeta \int (d\vec{r})\, \Delta n = o \quad . \tag{2-105}$$

Then

$$\Delta E \;\geq o \;;\quad = o \quad \text{only for} \quad \Delta n(\vec{r}) = o \quad \text{for all } \vec{r} \;; \tag{2-106}$$

the TF density functional of Eq. (95) has an absolute minimum at the correct density, provided Eq. (105) holds.

In general, satisfying (105) will mean to consider only such trial densities that obey the constraint (98), since then

$$\int (d\vec{r})\, \Delta n = o \quad . \tag{2-107}$$

The main exception are neutral atoms, about which we know that $\zeta = o$. Consequently, trial values for ζ need not be chosen, so that $\Delta\zeta = o$. Then Eq. (105) is satisfied without restricting the density according to (107). This observation will prove useful, when seeking lower bounds on the constant B.

Upper bounds on B. We pick up the story at Eq. (69). The calculation is considerably simplified by employing the TF variables x, x_o, and f(x), which have been introduced in Eqs. (57) through (60). In these, the TF potential functional appears as $[f'(x) \equiv \frac{d}{dx} f(x)]$

$$E = -\,(Z^{7/3}/a)\{\frac{2}{5} \int_0^{\infty} dx\, \frac{[f(x)]^{5/2}}{x^{1/2}} + \frac{1}{2} \int_0^{\infty} dx\, [f'(x) + \frac{q}{x_o}]^2 +$$

$$+ \frac{q(1-q)}{x_o}\} \quad , \tag{2-108}$$

where, replacing V and ζ, it is now f(x) and x_o that have to be found.[7] Whereas arbitrary variations of x_o may be considered, f(x) is subject to

$$f(o) = 1 \tag{2-109}$$

and

$$f'(x) = -\frac{q}{x_o} \quad \text{for} \quad x \to \infty \quad . \tag{2-110}$$

The first of these is Eq.(52), the second comes from the inclusion of ζ, in Eq.(57), into the definition of f(x). In (108) it is needed to ensure the finiteness of the second integral. Note, in particular, that the trial functions do not have to obey

$$f(x_o) = o \quad , \quad -x_o f'(x_o) = q \tag{2-111}$$

[see (60)]. This, and the differential equation

$$f''(x) = \frac{[f(x)]^{3/2}}{x^{1/2}} \tag{2-112}$$

[see (59)] are implications of the stationary property of (108). Here is how it works: infinitesimal variations of f(x) cause a change in E,

$$\begin{aligned} o &= \delta[-(a/Z^{7/3})E] \\ &= \int_o^\infty dx\, \delta f(x)\{\frac{[f(x)]^{3/2}}{x^{1/2}} - f''(x)\} + \int_o^\infty dx\, \frac{d}{dx}\{\delta f(x)[f'(x)+\frac{q}{x_o}]\} \\ &= \int_o^\infty dx\, \delta f(x)\{\frac{[f(x)]^{3/2}}{x^{1/2}} - f''(x)\} \quad . \end{aligned} \tag{2-113}$$

where the first equality is the stationary property and the last one uses (109) and (110) in finding the null value of the integrated total differential. Thus (112) is implied. We combine it with (110) to conclude that beyond a certain (yet unspecified) $\bar{x}$, f(x) is negative and linear:

$$f(x) = q\,\frac{\bar{x}-x}{x_o} \quad \text{for} \quad x \geq \bar{x} \quad . \tag{2-114}$$

Next, we consider variations of x_o. They produce

$$o = \delta[-(a/Z^{7/3})E]$$

$$= \delta(\frac{1}{x_o})\ q\ \{\ \int_o^\infty dx[f'(x)+\frac{q}{x_o}]\ +\ (1-q)\}\quad , \tag{2-115}$$

implying the vanishing of the contents of the curly brackets. In view of (114), the integration stops at $\bar{x}$:

$$\begin{aligned} o &= \int_o^{\bar{x}} dx[f'(x)+\frac{q}{x_o}]\ +\ (1-q) \\ &= f(\bar{x})\ -\ f(o)\ +\ q\frac{\bar{x}}{x_o}\ +\ 1\ -\ q\ =\ q(\frac{\bar{x}}{x_o}\ -1)\quad ; \end{aligned} \tag{2-116}$$

the last step makes use of (109) and (114). Now we see that $\bar{x}=x_o$, so that (114) becomes (61) and implies (111).

Let us now turn to neutral atoms, q=o. The maximum property of the functional (108), combined with the known form of the neutral atom binding energy , Eq.(67), reads

$$\frac{3}{7}\ B\ \leq\ \frac{2}{5}\int_o^\infty dx\ \frac{[f(x)]^{5/2}}{x^{1/2}}\ +\ \frac{1}{2}\int_o^\infty dx[f'(x)]^2\quad , \tag{2-117}$$

where the equal sign holds only for f(x)=F(x). Note that x_o disappeared together with q, so that we do not need to use explicitly our knowledge of $x_o=\infty$ for q=o. According to (109) and (110), the competition in (117) is among trial functions that are subject to

$$f(o)\ =\ 1\qquad ,\qquad f'(x\rightarrow\infty)\ =\ o. \tag{2-118}$$

For any trial f(x), we can always change the scale,

$$f(x)\ \rightarrow\ f(\mu x)\qquad ,\ (\mu>o)\ , \tag{2-119}$$

and obtain another trial function. The optimal choice for μ minimizes the right hand side of

$$\frac{3}{7}\ B\ \leq\ \frac{2}{5}\int_o^\infty dx\ \frac{[f(\mu x)]^{5/2}}{x^{1/2}}\ +\ \frac{1}{2}\int_o^\infty dx[\frac{d}{dx}f(\mu x)]^2\ =$$

$$= \mu^{-1/2}\,\frac{2}{5}\int_0^\infty dx\,\frac{[f(x)]^{5/2}}{x^{1/2}} + \mu\,\frac{1}{2}\int_0^\infty dx\,[f'(x)]^2 \tag{2-120}$$

It is given by

$$\mu^{3/2} = \frac{2}{5}\int_0^\infty dx\,\frac{[f(x)]^{5/2}}{x^{1/2}} \bigg/ \int_0^\infty dx\,[f'(x)]^2 \quad . \tag{2-121}$$

We insert it into (210) and arrive at the scale invariant version of (117):

$$B \le \frac{7}{2}\left[\frac{2}{5}\int_0^\infty dx\,\frac{[f(x)]^{5/2}}{x^{1/2}}\right]^{2/3}\left[\int_0^\infty dx\,[f'(x)]^2\right]^{1/3} \quad , \tag{2-122}$$

where now the equal sign holds for $f(x)=F(\mu x)$ with arbitrary $\mu>o$.

We are now ready to invent trial functions and produce upper bounds on B. Before doing so, let us make a little observation. If f equals F, the optimal μ in (120) is unity, since the equal sign in (117) holds only for $f(x)=F(x)$. Consequently, the numerator and denominator in (121) are equal for $f=F$. In this situation the related sum in (117) is (3/7)B. We conclude

$$\frac{2}{5}\int_0^\infty dx\,\frac{[F(x)]^{5/2}}{x^{1/2}} = \frac{2}{7}\,B \quad , \tag{2-123}$$

and

$$\frac{1}{2}\int_0^\infty dx\,[F'(x)]^2 = \frac{1}{7}\,B \quad . \tag{2-124}$$

An independent (and rather clumsy) derivation of these equations uses the differential equation obeyed by $F(x)$ [Eq.(62)], combined with some partial integrations. Equations (123) and (124) can be employed for an immediate check of the equality in (122) for $f(x)=F(\mu x)$:

$$B = \frac{7}{2}\,(\mu^{-1/2}\,\frac{2}{7}\,B)^{2/3}\,(\mu\,\frac{2}{7}\,B)^{1/3} \quad . \tag{2-125}$$

More about relations like (123) and (124) will be said in the section on the scaling properties of the TF model.

A very simple trial function is

$$f(x) = \frac{1}{1+x} \quad , \tag{2-126}$$

for which

$$\frac{2}{5}\int_0^\infty dx\,\frac{[f(x)]^{5/2}}{x^{1/2}} = \frac{2}{5}\int_0^\infty dx\; x^{-1/2}(1+x)^{-5/2} = \frac{8}{15} \tag{2-127}$$

[the integral, in terms of Euler's Beta function, is $B(\frac{1}{2},2) = \frac{4}{3}$], and

$$\int_0^\infty dx[f'(x)]^2 = \int_0^\infty \frac{dx}{(1+x)^4} = \frac{1}{3} \quad . \tag{2-128}$$

Accordingly,

$$B < \frac{14}{3}\,5^{-2/3} = 1.596 \quad . \tag{2-129}$$

A better value is obtained for

$$f(x) = \left(\frac{1}{1+x}\right)^{4/3} \quad , \tag{2-130}$$

when

$$B < 2^{-19/9}\left(\frac{77}{3}\right)^{1/3}\left[\pi\left(\frac{2}{3}\right)!\right]^{2/3}\left[\left(\frac{1}{3}\right)!\right]^{-4/3} = 1.5909 \quad . \tag{2-131}$$

This number is, as we shall see, very close to the actual one; so there is no point in considering more complicated trial functions.

Lower bounds on B. In order to express the density functional of Eq. (95) in terms of TF variables, we write

$$\begin{aligned} n(\vec{r}) &= -\frac{1}{4\pi}\nabla^2\left(V+\frac{Z}{r}\right) \\ &= -\frac{1}{4\pi}\frac{1}{r}\frac{d^2}{dr^2}\left[r\left(V+\frac{Z}{r}\right)\right] \\ &= \frac{1}{4\pi}\frac{1}{r}\frac{d^2}{dr^2}(Zf(x)) \\ &= \frac{1}{4\pi}\frac{Z^2}{a^3}\frac{f''(x)}{x} \quad , \end{aligned} \tag{2-132}$$

or, more conveniently here,

$$n(\vec{r}) = \frac{1}{4\pi}\frac{Z^2}{a^3}\frac{g'(x)}{x} \quad . \tag{2-133}$$

The function $g(x)$, thus introduced, differs from $f'(x)$, at most, by a constant. We choose this constant to be q/x_o,

$$g(x) = f'(x) + \frac{q}{x_o} \quad , \tag{2-134}$$

which in view of (110) is equivalent to requiring

$$g(x \to \infty) = o \quad . \tag{2-135}$$

With (133) and (135) we have for the interaction energy between the nucleus and the electrons

$$-\int(d\vec{r})\,\frac{Z}{r}\,n = -\frac{Z^{7/3}}{a}\int_o^\infty dx\; g'(x) = \frac{Z^{7/3}}{a}\,g(o) \quad , \tag{2-136}$$

whereas the electron-electron energy is

$$\frac{1}{2}\int(d\vec{r})(d\vec{r}\,')\,\frac{n(\vec{r})n(\vec{r}\,')}{|\vec{r}-\vec{r}\,'|} = \frac{Z^{7/3}}{a}\,\frac{1}{2}\int_o^\infty dx[g(x)]^2 \quad . \tag{2-137}$$

[This quantity equals $-E_2$, so that Eq.(134), used in the second integral of (108), gives this result.] The remaining contributions to the density functional can be expressed in terms of $g(x)$ immediately. We arrive at

$$E = \frac{Z^{7/3}}{a}\Big\{\frac{3}{5}\int_o^\infty dx\; x^{1/3}[g'(x)]^{5/3} + g(o) + \frac{1}{2}\int_o^\infty dx[g(x)]^2$$
$$- \frac{q}{x_o}\Big[1-q-\int_o^\infty dx\; x\; g'(x)\Big]\Big\} \quad . \tag{2-138}$$

Again arbitrary variations of x_o may be considered, whereas g is restricted by the requirement of non-negative densities,

$$g'(x) \geq o \quad , \tag{2-139}$$

and by Eq.(135). Together, they imply

$$g(x) \leq o \quad . \tag{2-140}$$

According to the discussion of Eq.(103), Eq.(138) supplies upper bounds on the energy, provided that Eq.(105) is obeyed. Expressed in terms of x_o and g, it reads

$$\Delta(\frac{1}{x_o}) \int_o^\infty dx\ x\ \Delta g'(x) = o \quad . \tag{2-141}$$

We did notice already [see the remark following Eq.(107)], that in the situation of neutral atoms, our knowledge of $x_o=\infty$ results in $\Delta(\frac{1}{x_o})=o$, so that (141) is satisfied without further ado. In particular, $g(x)$ need not be subject to

$$\int_o^\infty dx\ x\ \Delta g'(x) = o \quad , \tag{2-142}$$

or [this is Eq.(98)], more precisely,

$$\int_o^\infty dx\ x\ g'(x) = 1-q\ ;\ = 1 \quad \text{for} \quad q=o \quad . \tag{2-143}$$

The minimum property of the functional (138), together with the known form of the neutral atom (q=o) binding energy, Eq.(67), implies

$$\frac{3}{7}\, B \geq -\{\frac{3}{5}\int_o^\infty dx\ x^{1/3}[g'(x)]^{5/3} + g(o) + \frac{1}{2}\int_o^\infty dx[g(x)]^2\} \ , \tag{2-144}$$

where the equal sign holds only for $g(x) = F'(x)$. For this $g(x)$, the value of the two integrals is $\frac{3}{7}B$ and $\frac{1}{7}B$, respectively, as follows from Eqs.(123) and (124), and the differential equation (62) obeyed by $F(x)$. Consequently,

$$F'(o) = -B \quad , \tag{2-145}$$

which is nothing more than the original definition of B in (64).

As in the preceding section, we can consider changes of the scale. Here the possible scalings are even more general,

$$g(x) \rightarrow \mu_1 g(\mu_2 x) \quad , (\mu_1,\mu_2 > o) \quad , \tag{2-146}$$

because there is no analog to the restriction $f(o)=1$, that we had to watch before. The optimal choices for μ_1 and μ_2 maximize the right hand side of

$$\frac{3}{7}B \geq -\left\{\frac{3}{5}\int_o^\infty dx\, x^{1/3}\left[\mu_1\frac{d}{dx}g(\mu_2 x)\right]^{5/3} + \mu_1 g(o) + \frac{1}{2}\int_o^\infty dx\left[\mu_1 g(\mu_2 x)\right]^2\right\}$$

$$= -\left\{\mu_1^{5/3}\,\mu_2^{1/3}\,\frac{3}{5}\int_o^\infty dx\, x^{1/3}[g'(x)]^{5/3} + \mu_1 g(o) + \frac{\mu_1^2}{\mu_2}\,\frac{1}{2}\int_o^\infty dx[g(x)]^2\right\}$$

(2-147)

They are

$$\mu_1 = \left(\frac{1}{7}\right)^{4/3} \frac{[-g(o)]^{4/3}}{\frac{1}{5}\int_o^\infty dx\, x^{1/3}[g'(x)]^{5/3}} \left(\frac{1}{2}\int_o^\infty dx[g(x)]^2\right)^{-1/3} \qquad (2\text{-}148)$$

and

$$\mu_2 = \left(\frac{1}{7}\right)^{4/3} \frac{[-g(o)]^{4/3}}{\frac{1}{5}\int_o^\infty dx\, x^{1/3}[g'(x)]^{5/3}} \left(\frac{1}{2}\int_o^\infty dx[g(x)]^2\right)^{2/3} . \qquad (2\text{-}149)$$

Inserted into (147) they produce the scale invariant version of (144):

$$B \geq \left(\frac{1}{7}\right)^{4/3} \frac{[-g(o)]^{7/3}}{\left(\frac{1}{5}\int_o^\infty dx\, x^{1/3}[g'(x)]^{5/3}\right)\left(\frac{1}{2}\int_o^\infty dx[g(x)]^2\right)^{1/3}} , \qquad (2\text{-}150)$$

where the equal sign holds only for $g(x)=\mu_1 F'(\mu_2 x)$ with arbitrary $\mu_1,\mu_2>o$. Indeed, for such a $g(x)$, we get

$$B = \left(\frac{1}{7}\right)^{4/3} \frac{[\mu_1 B]^{7/3}}{\left(\mu_1^{5/3}\mu_2^{1/3}\,\frac{1}{7}B\right)\left(\frac{\mu_1^2}{\mu_2}\,\frac{1}{7}B\right)^{1/3}} . \qquad (2\text{-}151)$$

The main contribution to the energy of an atom comes from the vicinity of the nucleus. Now, Eqs.(62) and (63) imply,

$$F''(x) \cong \frac{1}{\sqrt{x}} \text{ for } x \to o . \qquad (2\text{-}152)$$

Consequently, a good trial g(x) has to be such that

$$g'(x) \sim \frac{1}{\sqrt{x}} \quad \text{for} \quad x \to o \quad . \qquad (2\text{-}153)$$

An example is

$$g(x) = -(1+\sqrt{x})^{-\alpha} \quad , \quad \alpha > o \quad . \qquad (2\text{-}154)$$

It turns out, that the right hand side of (150) increases with α, so that we may immediately consider the limit $\alpha \to \infty$. The scaling invariance helps in this limit, since it allows to evaluate $g(x/\alpha^2)$ for $\alpha \to \infty$, instead of $g(x)$ for $\alpha \to \infty$. The limiting trial function is a simple exponential:

$$g(x) = \lim_{\alpha \to \infty}[-(1+\frac{1}{\alpha}\sqrt{x})^{-\alpha}] = -e^{-\sqrt{x}} \quad . \qquad (2\text{-}155)$$

For this g(x), we have in (150)

$$B > (\tfrac{1}{7})^{4/3} \frac{[-(-1)]^{7/3}}{(\frac{3}{25}\,2^{-2/3})(\frac{1}{4})^{1/3}} = \frac{25}{3}(\tfrac{2}{7})^{4/3} \qquad (2\text{-}156)$$

$$= 1.5682 \quad .$$

Binding energy of neutral TF atoms. We have found an upper bound on B in (131) and a lower one in (156). Now we combine the two and state

$$1.5682 < B < 1.5909 \quad , \qquad (2\text{-}157)$$

or

$$B = 1.580 \pm 0.012 \quad . \qquad (2\text{-}158)$$

The margin in (158) is about 1.5% of the average value, so that we know B with a precision of 0.75%. Please appreciate how little numerical effort was needed in obtaining this result. In view of the crude physical picture that we are still using, the value for B in (158) is entirely sufficient. A higher precision is not called for at this stage of the development.

Inserted into (67), this B value produces

$$-E_{TF} = 0.765\ Z^{7/3} \quad , \qquad (2\text{-}159)$$

which is the TF prediction for the total binding energy of neutral

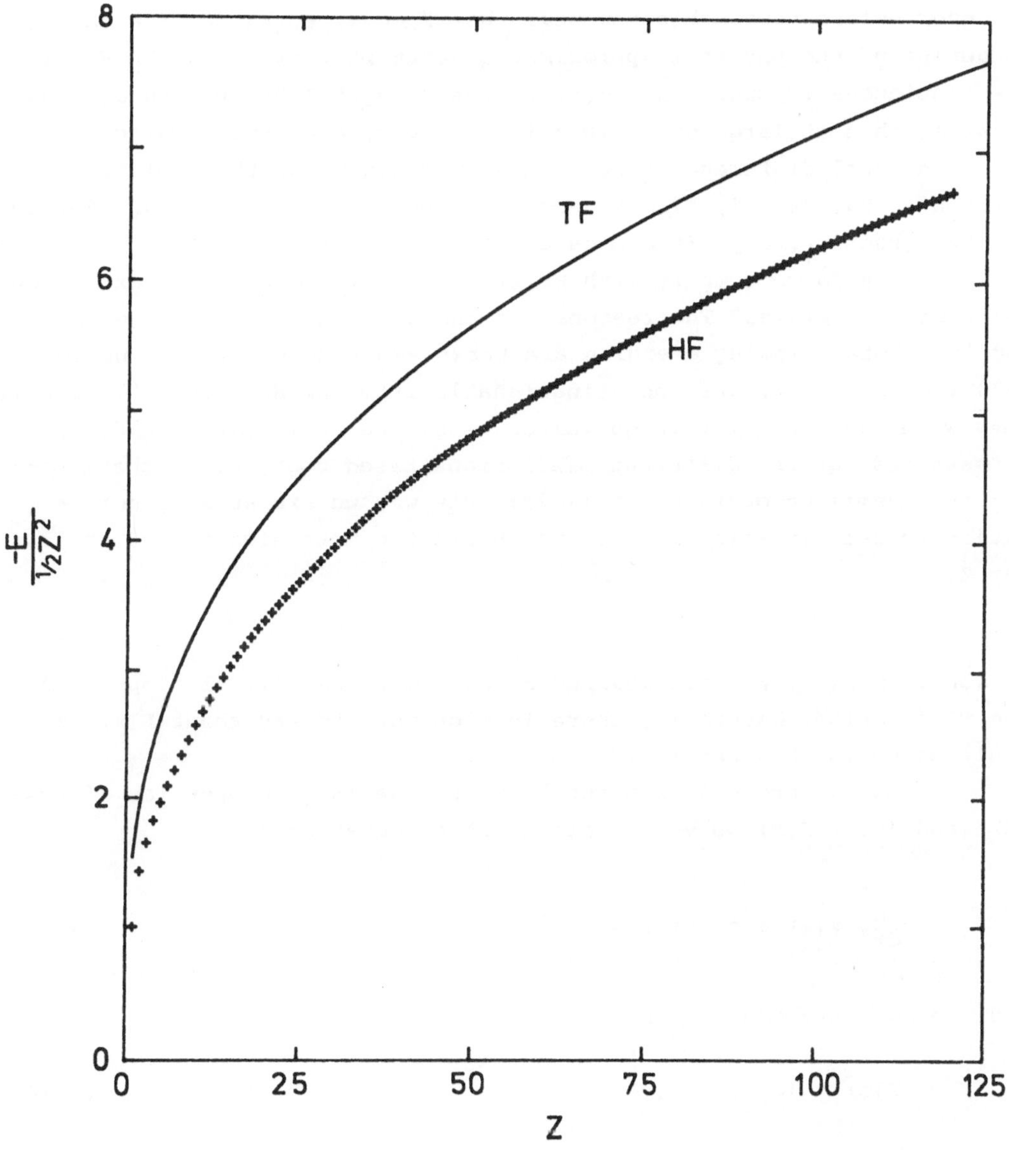

Fig. 2-2. Comparison of the TF prediction (160) with HF binding energies (crosses).

atoms. In Fig.2 the quantity

$$\frac{-E_{TF}}{\frac{1}{2}Z^2} = 1.53\ Z^{1/3} \tag{2-160}$$

is compared to the corresponding HF predictions[8] for integer values of Z. This plot shows that (160) does reproduce the general trend of the atomic binding energies. Altough the need for refinements is clear, it is nevertheless remarkable how well the TF model works despite the crudeness of the physical approximation that it represents. In Fig.2 the continuous TF curve is closer to the integer-Z HF crosses at small Z values than at large ones. This is, however, a deception since it is the fractional difference that counts. The amount of this relative deviation is 29, 24, 21, 17, 15, and 13 percent for Z=10, 20, 30, 60, 90, and 120, respectively. It decreases with increasing Z.

Why do we compare with HF predictions, and not with experimental binding energies? The reasons are the ones mentioned in the Introduction. Total binding energies are known experimentally only up to $Z \cong 20$ (in Fig.2 they are indistinguishable from the HF crosses). Even if they were available for large values of Z, the TF result should still be measured against different predictions based upon, e.g., the many-particle Hamilton operator of (1-7); this way we are sure to not be misled by relativistic effects, which are the more significant the larger Z.

TF function F(x). We have learned a lot about the initial slope B of the TF function. Naturally, there is much more to say about F(x). We shall do so in this section.

Let us proceed from recalling the defining properties of Eqs. (62) and (63). F(x) obeys the differential equation

$$\frac{d^2}{dx^2} F(x) = F''(x) = \frac{[F(x)]^{3/2}}{x^{1/2}} \tag{2-161}$$

and the boundary conditions

$$F(o) = 1 \quad , \quad F(\infty) = o \quad . \tag{2-162}$$

Upon using $\sqrt{x}$ as the main variable, the differential equation (161) appears as

$$\frac{d}{d\sqrt{x}} F(x) = 2\sqrt{x}\, F'(x) \quad ,$$

$$\frac{d}{d\sqrt{x}} F'(x) = 2[F(x)]^{3/2} \quad . \tag{2-163}$$

Whereas (161) is singular at x=o , this system of differential equations is well behaved at $\sqrt{x}$=o . We conclude that around x=o , F(x) can be expanded in powers of $\sqrt{x}$:

$$F(x) = \sum_{k=o}^{\infty} s_k\, x^{k/2} \quad , \tag{2-164}$$

which has become known as Baker's series.[9] The comparison with

$$F(x) = 1 - Bx + O(x^{3/2}) \tag{2-165}$$

[this is Eq.(64)] shows

$$s_o = 1 \quad , \quad s_1 = o \quad , \quad s_2 = -B \quad . \tag{2-166}$$

For the successive calculation of the s_k's for k>2, we need a recurrence relation. We gain it by inserting (164) into the differential equation (161). The left hand side is simple:

$$F''(x) = \sum_{k=o}^{\infty} s_k \frac{k}{2}(\frac{k}{2} - 1)x^{k/2-2}$$

$$= \frac{3}{4} s_3\, x^{-1/2} + \sum_{\ell=1}^{\infty} x^{(\ell-1)/2} \frac{(\ell+1)(\ell+3)}{4} s_{\ell+3} \quad , \tag{2-167}$$

where s_1=o has been used, and the summation index shifted ($k=\ell+3$). The right hand side of Eq.(161) is nonlinear in F(x), so that the power series becomes more complicated. We have

$$[F(x)]^{3/2}/x^{1/2} = x^{-1/2}[1 + \sum_{k=2}^{\infty} s_k\, x^{k/2}]^{3/2}$$

$$= x^{-1/2}[1 + \sum_{j=1}^{\infty} \binom{3/2}{j} (\sum_{k=2}^{\infty} s_k\, x^{k/2})^j] \quad , \tag{2-168}$$

where the binomial theorem is employed. Next, we make the j-th power of the sum over k explicit by writing it as the product of j sums over k_1 , k_2 , ... , k_j ; then the Kronecker Delta symbol ,

$$\delta_{\ell,k} = \begin{cases} 1 & \text{for } k = \ell \\ o & \text{for } k \neq \ell \end{cases} \quad , \tag{2-169}$$

is used to collect all terms of order $x^{-1/2}\, x^{\ell/2} = x^{(\ell-1)/2}$:

$$[F(x)]^{3/2}/x^{1/2} = x^{-1/2}[1+\sum_{j=1}^{\infty}\binom{3/2}{j}\sum_{k_1=2}^{\infty}\sum_{k_2=2}^{\infty}\cdots\sum_{k_j=2}^{\infty} s_{k_1}\, s_{k_2}\cdots$$

$$\times\cdots s_{k_j}\, x^{(k_1+k_2+\ldots+k_j)/2}]$$

$$= x^{-1/2} + \sum_{\ell=1}^{\infty} x^{(\ell-1)/2} \sum_{j=1}^{\infty}\binom{3/2}{j}\sum_{k_1=2}^{\infty}\cdots\sum_{k_j=2}^{\infty} s_{k_1}\cdots s_{k_j}$$

$$\times\, \delta_{\ell,k_1+k_2+\ldots+k_j} \; . \tag{2-170}$$

Since each k is at least two, we have

$$\ell = k_1 + k_2 + \ldots + k_j \geq 2j \geq 2 \quad , \tag{2-171}$$

so that the summation over ℓ starts really at $\ell=2$, and the largest j does not exceed $\ell/2$: $j \leq [\ell/2]$, which makes use of the Gaussian notation for the largest integer contained in $\ell/2$. The individual k-summations stop, at the latest, at $\ell-2(j-1)$, since, again, the other k's, which are j-1 in number, are not less than two each. Accordingly,

$$[F(x)]^{3/2}/x^{1/2} = x^{-1/2} + \sum_{\ell=2}^{\infty} x^{(\ell-1)/2} \sum_{j=1}^{[\ell/2]}\binom{3/2}{j}\sum_{k_1=2}^{\ell-2j+2}\cdots$$

$$\times\cdots\sum_{k_j=2}^{\ell-2j+2} s_{k_1}\cdots s_{k_j}\,\delta_{\ell,k_1+\ldots+k_j} \; . \tag{2-172}$$

This must equal (167), implying

$$s_3 = 4/3 \quad , \quad s_4 = 0 \quad , \tag{2-173}$$

and

$$s_{\ell+3} = \frac{4}{(\ell+1)(\ell+3)} \sum_{j=1}^{[\ell/2]} \binom{3/2}{j} \sum_{k_1=2}^{\ell-2j+2} \cdots \sum_{k_j=2}^{\ell-2j+2} s_{k_1} \cdots s_{k_j} \delta_{\ell, k_1+\ldots+k_j} \tag{2-174}$$

$$\text{for} \quad \ell = 2,3,\ldots \quad .$$

The s_k with the largest index occurs on the right hand side for j=1; it is s_ℓ. Thus $s_{\ell+3}$ is here expressed in terms of s_k's with $k \leq \ell$, so that this is a recurrence relation, indeed. For illustration, consider ℓ=2,3,4, and 5. There is only the j=1 term for ℓ=2 and ℓ=3:

$$s_5 = \frac{4}{15}\binom{3/2}{1} s_2 = \frac{4}{15}\frac{3}{2}(-B) = -\frac{2}{5}B \quad ,$$

$$s_6 = \frac{1}{6}\binom{3/2}{1} s_3 = \frac{1}{6}\frac{3}{2}\frac{4}{3} = \frac{1}{3} \quad . \tag{2-175}$$

For ℓ=4 and ℓ=5 , there are both the j=1 and the j=2 contribution:

$$s_7 = \frac{4}{35}[\binom{3/2}{1} s_4 + \binom{3/2}{2} s_2^2] = \frac{4}{35}[\frac{3}{2}0 + \frac{3}{8}(-B)^2] = \frac{3}{70}B^2 \quad , \tag{2-176}$$

$$s_8 = \frac{1}{12}[\binom{3/2}{1} s_5 + \binom{3/2}{2}(s_2 s_3 + s_3 s_2)]$$

$$= \frac{1}{12}[\frac{3}{2}(-\frac{2}{5}B) + \frac{3}{8}2(-B)\frac{4}{3}] = -\frac{2}{15}B \quad .$$

It is not difficult (only boring) to compute more s_k's. Let us be content with what we have so far:

$$F(x) = 1 - Bx + \frac{4}{3}x^{3/2} - \frac{2}{5}Bx^{5/2} + \frac{1}{3}x^3 + \frac{3}{70}B^2x^{7/2} - \frac{2}{15}Bx^4 + O(x^{9/2}) \tag{2-177}$$

The B dependence of the coefficients and their complicated recurrence relation (174) prohibit asking for the range of convergence of the expansion (164). We can, however, test the quality of the approximation to F(x) obtained by terminating the summation at, say, k=8. This is done by inserting the truncated series into the differential equation obeyed by F(x), and comparing both sides:

$$\sum_{k=3}^{8} \frac{k(k-2)}{4} s_k x^{(k-4)/2} \cong [\sum_{k=o}^{8} s_k x^{k/2}]^{3/2}/x^{1/2} \quad . \tag{2-178}$$

For B=1.580 , our estimate in (158), the comparison is made in Table 1. It shows us, that this approximation to F(x) solves the differential equation with an accuracy of 1% for $\sqrt{x} \lesssim 0.4$; of $\frac{1}{10}$% for $\sqrt{x} \lesssim 0.25$; of $\frac{1}{100}$% for $\sqrt{x} \lesssim 0.20$. This kind of analysis can be repeated for sums truncated at a value of k much larger. One observes that a highly accurate solution to the differential equation (161) is given by these sums for $\sqrt{x} \lesssim 0.4$ only. This is, therefore, the (numerical) range of convergence of the series in Eq.(164); as a consequence, this expansion in powers of $\sqrt{x}$ is utterly useless.[10]

Table 2-1. Left hand side (LHS) and right hand side (RHS) of Eq.(178), and their relative deviation (DEV) for $\sqrt{x}$ = 0.05, ..., 0.50 . For B the value of Eq.(158) is used.

$\sqrt{x}$	LHS	RHS	DEV
0.05	19.8866012	19.8866017	2.3×10^{-8}
0.10	9.783683	9.783698	1.5×10^{-6}
0.15	6.35805	6.35816	1.7×10^{-5}
0.20	4.60944	4.60991	1.0×10^{-4}
0.25	3.5373	3.5387	4.0×10^{-4}
0.30	2.8071	2.8106	1.2×10^{-3}
0.35	2.275	2.282	3.3×10^{-3}
0.40	1.867	1.882	7.8×10^{-3}
0.45	1.542	1.568	1.7×10^{-2}
0.50	1.27	1.32	3.4×10^{-2}

For a precise knowledge of F(x), we cannot rely upon (164) because of its small range of convergence. The differential equation (161) itself has to be integrated numerically. It is not advisable to attempt doing this by starting from x=o with F(o)=1 and F'(o)=-B , using a suitable guess for B [as in Eq.(158)]. If the chosen value for B is too large, the numerical F(x) will turn negative eventually; if B is too small, it will start growing instead of decreasing steadily. One could imagine pinning down the correct value of B by an iteration based on this distinction between trial B's that are too large or too small. This is not going to work, unfortunately, because rounding-off errors cause a wrong large-x behaviour of the numerical F(x), even if one would start with the correct value of B. This difficulty can be circumvented, however, by integrating inwards from $x=\infty$ towards $x=o$ instead of outwards. Let us, therefore, turn our attention to the large-x properties of F(x).

We start by noting that $144/x^3$ is a particular solution of the differential equation (161).[11] Of the two boundary conditions in (162) it satisfies the one at $x=\infty$, whereas it is infinite at $x=o$. It is clear, that F(x) approaches $144/x^3$ for $x\to\infty$ from "below":

$$F(x) \lesssim \frac{144}{x^3} \qquad \text{for} \qquad x \to \infty \quad . \tag{2-179}$$

This invites the ansatz

$$F(x) = \frac{144}{x^3} G(y(x)) \tag{2-180}$$

with

$$y(x) \to o \qquad \text{for} \qquad x \to \infty \tag{2-181}$$

and

$$G(y) = 1 \qquad \text{for} \qquad y = o \quad . \tag{2-182}$$

The best choice for the function y(x) must be found from inserting (180) into the differential equation obeyed by F(x), Eq.(161). This leads to

$$\frac{1}{12}[xy'(x)]^2 G''(y(x)) + \frac{1}{12}[x^2y''(x) - 6xy'(x)]G'(y(x)) + G(y(x))$$
$$= [G(y(x))]^{3/2} \quad , \tag{2-183}$$

which takes on a scale invariant form if we choose $xy'(x)$ to be proportional to y(x):

$$xy'(x) = -\gamma y(x) \quad . \tag{2-184}$$

The optimal value for $\gamma > o$ has to be determined. Equation (184) and its immediate consequence

$$x^2y''(x) = (x\frac{d}{dx} - 1)\, xy'(x) = \gamma(\gamma+1)y(x) \quad , \tag{2-185}$$

used in (183), produce

$$\frac{\gamma^2}{12}\, y^2 G''(y) + \frac{\gamma(\gamma+7)}{12}\, y\, G'(y) + G(y) = [G(y)]^{3/2} \quad . \tag{2-186}$$

The aforementioned scale invariance is obvious here: with G(y) also G(μy) is a solution to (186), for arbitrary μ. A unique value for γ is now implied by the requirement that G(y) be regular at y=o,

$$G(y) = 1 - y + O(y^2) \quad \text{for} \quad y \to o \quad . \tag{2-187}$$

Note that because of the scale invariance of (186), the coefficient in front of the term linear in y can be chosen to be minus one [it has to be negative to not be in conflict with Eq.(179)]. With (187), Eq.(186) reads

$$1 - [1 + \frac{1}{12}\gamma(\gamma+7)]y + O(y^2) = 1 - \frac{3}{2}y + O(y^2) \quad , \tag{2-188}$$

whence

$$\gamma(\gamma+7) = 6 \quad , \tag{2-189}$$

or,

$$\gamma = \frac{1}{2}(-7+\sqrt{73}) = 0.77200... \quad . \tag{2-190}$$

The second solution to (189) is $-(\gamma+7) = -7.772...$ and of no use to us in the present context.

The differential equation (186) is simplified a little bit by making use of (189):

$$\frac{\gamma^2}{12}\, y^2 G''(y) + \frac{1}{2}\, y\, G'(y) + G(y) = [G(y)]^{3/2} \quad . \tag{2-191}$$

G(y) is thereby subject to (187), which determines the solution to (191) entirely. This does not mean that we know F(x) after finding G(y), since the implication of Eq.(184)

$$y(x) = \beta\, x^{-\gamma} \tag{2-192}$$

contains an undetermined constant, β. Its value is fixed by the requirement F(x=o)=1. This is, of course, analogous to the previous situation when F'(o)=-B was determined by F(x→∞)=o.

Since G(y) is, by construction, regular at y=o, we can expand it in powers of y:[12]

$$G(y) = \sum_{k=o}^{\infty} c_k\, y^k \quad , \tag{2-193}$$

where

$$c_o = 1 \quad , \quad c_1 = -1 \quad . \tag{2-194}$$

The steps that led us to (174) can be repeated here for (191) and (193) with the appropriate changes. Comparing powers of y^ℓ on both sides of Eq.(191) gives

$$\begin{aligned}&[\tfrac{\gamma^2}{12}\,\ell(\ell-1) + \tfrac{1}{2}\,\ell + 1]c_\ell \\ &\quad = \sum_{j=1}^{\ell}\binom{3/2}{j}\sum_{k_1=1}^{\ell-j+1}\cdots\sum_{k_j=1}^{\ell-j+1} c_{k_1}\cdots c_{k_j}\delta_{\ell,k_1+k_2+\ldots+k_j}\end{aligned} \tag{2-195}$$

for $\ell\geq 1$. The j=1 term equals $\frac{3}{2}c_\ell$ and has to be brought over to the left hand side. We then arrive at the recurrence relation

$$c_\ell = \frac{12}{(\gamma^2\ell+6)(\ell-1)}\sum_{j=2}^{\ell}\binom{3/2}{j}\sum_{k_1=1}^{\ell-j+1}\cdots\sum_{k_j=1}^{\ell-j+1}c_{k_1}\cdots c_{k_j}\delta_{\ell,k_1+\ldots+k_j}, \tag{2-196}$$

$$\text{for} \quad \ell = 2,\ 3,\ \ldots \quad .$$

For example,

$$\begin{aligned} c_2 &= \frac{12}{2\gamma^2+6}\binom{3/2}{2}c_1^2 = \frac{9}{4\gamma^2+12} = \frac{201+21\sqrt{73}}{608} \\ &= 0.625697\ldots \quad , \\ c_3 &= \frac{12}{(3\gamma^2+6)2}\left\{\binom{3/2}{2}(c_1c_2+c_2c_1)+\binom{3/2}{3}c_1^3\right\} \\ &= -\frac{3-\gamma^2/8}{(\gamma^2+2)(\gamma^2+3)} = -\frac{15377+1813\sqrt{73}}{98496} \\ &= -0.313386\ldots \quad . \end{aligned} \tag{2-197}$$

As we did before, in Eq.(178) and Table 1, we can again insert truncated versions of (193) into the differential equation (191) in order to find the numerical range of convergence of this series. The outcome is: the expansion (193) represents a highly accurate solution to (191) for $y \lesssim 1$, or, $x \gtrsim \beta^{1/\gamma}$. Anticipating that the actual value of β is about 13, this is $x \gtrsim 30$.[13] Does this mean that the expansion (193) is as useless as the one of Eq. (164)?

No. The power series of $G(y)$, for some $o<y<1$, is needed to get away from $x=\infty$, i.e., $y=o$, when integrating the differential equation for $F(x)$, Eq. (161), inwards. Here is a brief description of the essential ingredients of a computer program calculating $F(x)$ for the whole range of x, $o \leq x < \infty$: (i) find $G(y)$ and $G'(y)$ for a suitably chosen y_1 ($\cong 0.3$ is a good choice) by employing (193), truncated at a sufficiently large k (depends on the chosen y_1 and the accuracy of the machine); (ii) integrate numerically the differential equation (191) up to a certain y_2 ($\cong 5$ is a good choice), so that we now know $G(y_2)$ and $G'(y_2)$ within the accuracy of the computer (the standard Runge-Kutta scheme is well suited for the numerical integration); (iii) choose a trial value, $\tilde{\beta}$ ($\cong 13$), for β, and use Eqs. (192) and (180) to find $\tilde{x}_2 = (\tilde{\beta}/y_2)^{1/\gamma}$ together with $\tilde{F}(\tilde{x}_2)$ and $\tilde{F}'(\tilde{x}_2)$; (iv) now integrate the differential equation for $F(x)$, in the form (163) with $\sqrt{x}$ as the relevant argument, down to $x=o$. At this stage, we have <u>a</u> solution to (161), the one corresponding to $\beta=\tilde{\beta}$. This $\tilde{F}(x)$ obeys $\tilde{F}(x=\infty)=o$, but not $\tilde{F}(x=o)=1$. Fortunately, one does not have to vary $\tilde{\beta}$ until $\tilde{F}(x=o)=1$ in order to find $F(x)$. Instead, the observation that, if $\tilde{F}(x)$ obeys (161), so does $\mu^3\tilde{F}(\mu x)$ for arbitrary $\mu>o$, enables us to simply rescale $\tilde{F}(x)$. The last step in the procedure is therefore: (v) set

$$F(x) = \tilde{F}(\tilde{x})/\tilde{F}(o) \quad , \tag{2-198}$$

where

$$\tilde{x} = x/[\tilde{F}(o)]^{1/3} \quad . \tag{2-199}$$

Accordingly, we have B given by

$$B = -F'(o) = -\tilde{F}'(o)/[\tilde{F}(o)]^{4/3} \quad , \tag{2-200}$$

and, as a consequence of

$$y_2 = \beta\, x_2^{-\gamma} = \tilde{\beta}\, \tilde{x}_2^{-\gamma} \quad , \tag{2-201}$$

β is related to $\tilde{\beta}$ through

$$\beta = \beta(x_2/\tilde{x}_2)^\gamma = \tilde{\beta}[\tilde{F}(o)]^{\gamma/3} \quad . \tag{2-202}$$

The sensitivity of the numerical results for B, β, and F(x), to the rounding-off errors of the computer can be tested by varying the parameters y_1, y_2, and $\tilde{\beta}$. Ideally, the outcome should be independent of them, numerically it is not. The little dependence that one observes shows how many decimals of the results can be trusted. For example, the realization of the procedure just described gave

$$B = 1.58807102261 \tag{2-203}$$

and

$$\beta = 13.270973848 \tag{2-204}$$

on a computer with a 15-decimal arithmetic.[14] For illustration, in Fig.3 we give a plot of F(x) for $o \leq x \leq 10$.

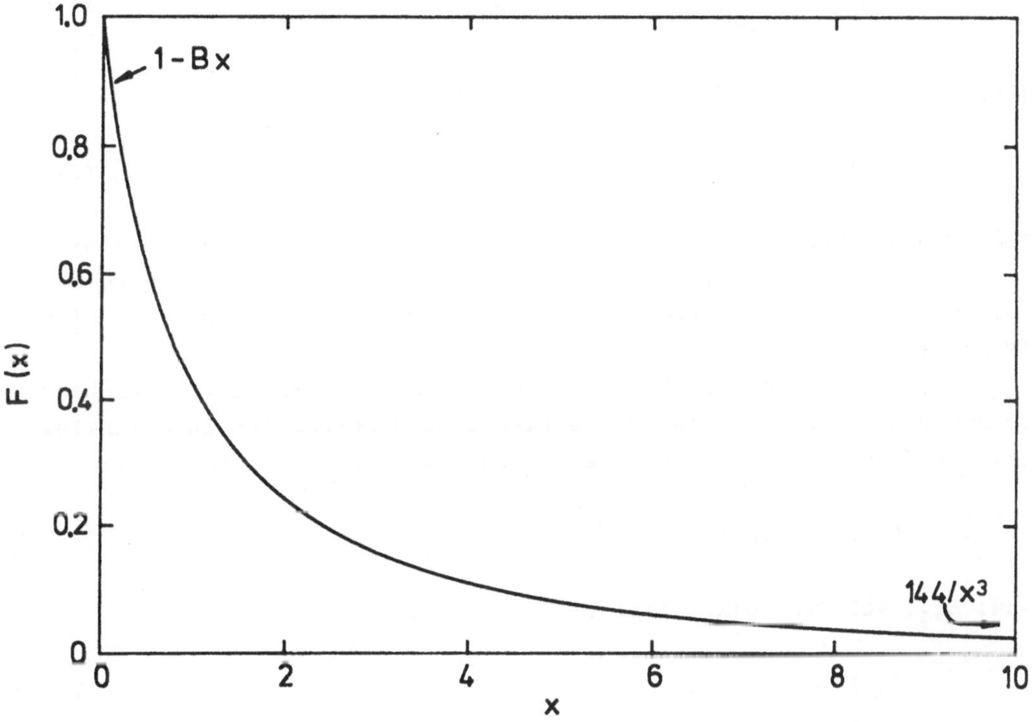

Fig.2-3. The TF function F(x).

Now that we know the actual value of B, let us look back at the bounds that we found earlier, Eq.(157). The upper bound is extremely good: it is too large by less than 0.18%. On the other hand, the lower bound is significantly worse: it is too small by 1.25%. This is a first sign of the superiority of the potential functional over the density functional.

With (203) we can give more significant decimals in the TF binding energy formula. Inserting B into Eq. (67) gives

$$- E_{TF} = 0.768745\ Z^{7/3} \quad . \tag{2-205}$$

There is no point in displaying more than six decimals.

Scaling properties of the TF model. In step (v) of our "computer program" for $F(x)$ we made, in Eqs. (198) and (199), use of the invariance of the TF equation for $f(x)$,

$$f''(x) = \frac{[f(x)]^{3/2}}{x^{1/2}} \tag{2-206}$$

[this is Eq. (59)], under the transformation

$$f(x) \rightarrow \mu^3 f(\mu x) \quad , \quad (\mu > o) \quad . \tag{2-207}$$

The TF model itself is not invariant under such a scaling, because the boundary condition $f(o)=1$ fixes the scale. Therefore, we have to be somewhat more careful when investigating the scaling properties of the TF model.

When we were looking for bounds on B, we found it advantageous to exploit certain scaling properties of the respective functionals. The scaling transformations that we considered then, were, Eq. (119):

$$f(x) \rightarrow f(\mu x) \tag{2-208}$$

and, Eq. (146) with $g(x)=f'(x)+q/x_o=f'(x)$ for $q=o$:

$$f(x) \rightarrow \frac{\mu_1}{\mu_2} f(\mu_2 x) \quad , \tag{2-209}$$

where μ, μ_1, and μ_2 were arbitrary (positive) numbers. Let us now examine the implications of transformations as general as (209) applied to the TF potential functional.

We return to Eq. (45),

$$E_{TF} = \int (d\vec{r}) \left(-\frac{1}{15\pi^2}\right) [-2(V+\zeta)]^{5/2} - \frac{1}{8\pi}\int (d\vec{r}) [\vec{\nabla}(V+\frac{Z}{r})]^2 - \zeta N$$
$$= E_1 + E_2 - \zeta N \quad , \tag{2-210}$$

and consider

$$V(r) \rightarrow \mu^{\nu} V(\mu r) \quad , \quad (\mu>o) \quad . \tag{2-211}$$

Since, for the existence of E_2, we need [Eq.(52)]

$$rV(r) \rightarrow -Z \qquad \text{for} \qquad r \rightarrow o \quad , \tag{2-212}$$

such a scaling transformation of V has to be accompanied by a scaling of Z,

$$Z \rightarrow \mu^{\nu-1} Z \quad . \tag{2-213}$$

For convenience, we also scale ζ by

$$\zeta \rightarrow \mu^{\nu} \zeta \quad , \tag{2-214}$$

so that the structure $V+\zeta$ is conserved.

In terms of f(x), (211) and (214), <u>without</u> (213), read

$$f(x) \rightarrow \mu^{\nu-1} f(\mu x) \quad , \tag{2-215}$$

which identifies (207) and (208) as the special situations ν=4 and ν=1, whereas (209) is realized by $\mu_2=\mu$ and $\mu_1=\mu^{\nu}$. However, <u>with</u> (213), we just have (208), as we should have, since (212) is equivalent to requiring f(o)=1; and only (208) is consistent with this constraint.

Under (211), (213), and (214) the various contributions to E_{TF} scale according to

$$E_1 \rightarrow E_1(\mu) = \int (d\vec{r}) \left(-\frac{1}{15\pi^2}\right) [-2\mu^{\nu}(V(\mu r)+\zeta)]^{5/2}$$
$$= \mu^{5\nu/2-3} E_1 \quad , \tag{2-216}$$

and

$$E_2 \rightarrow E_2(\mu) = -\frac{1}{8\pi}\int (d\vec{r}) [\vec{\nabla}(\mu^{\nu}V(\mu r) + \frac{\mu^{\nu}Z}{\mu r})]^2 =$$

$$= \mu^{2\nu-1} E_2 \quad , \tag{2-217}$$

as well as

$$\zeta N \to \mu^{\nu} \zeta N \quad . \tag{2-218}$$

Consequently

$$E_{TF} \to E_{TF}(\mu) = \mu^{5\nu/2-3} E_1 + \mu^{2\nu-1} E_2 - \mu^{\nu}\zeta N \quad . \tag{2-219}$$

For $\mu=1$, this is just E_{TF}; for $\mu=1+\delta\mu$, we have $E_{TF}+\delta_\mu E_{TF}$. Now, since the energy is stationary under infinitesimal variations of V and ζ, all first order changes must originate in the scaling of Z. That is

$$\delta_\mu E_{TF} = \frac{\partial E_{TF}}{\partial Z} \delta_\mu (\mu^{\nu-1} Z) = (\nu-1) Z \frac{\partial E_{TF}}{\partial Z} \delta\mu \quad . \tag{2-220}$$

On the other hand, from (219) we get

$$\delta_\mu E_{TF} = [(\tfrac{5}{2}\nu-3)E_1 + (2\nu-1)E_2 - \nu\zeta N]\delta\mu \quad , \tag{2-221}$$

so that we conclude

$$(\tfrac{5}{2}\nu-3)E_1 + (2\nu-1)E_2 - \nu\zeta N = (\nu-1)Z\frac{\partial}{\partial Z}E_{TF} \quad . \tag{2-222}$$

This is a linear equation in ν. It has to hold for any ν. So we obtain two independent relations among the different energy quantities - two "virial theorems." Besides $\nu=1$, when

$$-\frac{1}{2} E_1 + E_2 - \zeta N = 0 \quad , \tag{2-223}$$

the other natural choice is the TF scaling $\nu=4$ [see the comment to Eq. (215)], for which

$$7E_1 + 7E_2 - 4\zeta N = 3Z\frac{\partial}{\partial Z}E_{TF} \quad . \tag{2-224}$$

The latter combines with $E_{TF}=E_1+E_2-\zeta N$ and

$$\frac{\partial}{\partial N} E_{TF} = -\zeta \tag{2-225}$$

to give

$$7E_{TF}(Z,N) = 3(Z\frac{\partial}{\partial Z}+N\frac{\partial}{\partial N})E_{TF}(Z,N) \quad . \tag{2-226}$$

We have made explicit here, that the energy of an atom is a function of Z and N.

For N=Z, Eq.(226) has the simple implication

$$7E_{TF}(Z,Z) = 3Z\frac{d}{dZ}E_{TF}(Z,Z) \quad , \tag{2-227}$$

or

$$E_{TF}(Z,Z) = -C\,Z^{7/3} \quad , \tag{2-228}$$

where the constant C is yet undetermined. It is found by combining our knowledge of $\zeta=-\partial E_{TF}/\partial N=o$ for $N=Z$ with

$$\begin{aligned} Z\frac{\partial}{\partial Z}E_{TF} &= Z\frac{\partial}{\partial Z}E_2 = -\frac{1}{4\pi}\int(d\vec{r})\,\vec{\nabla}(V+\frac{Z}{r})\cdot\vec{\nabla}\frac{Z}{r} \\ &= \int(d\vec{r})\,(V+\frac{Z}{r})\frac{1}{4\pi}\nabla^2\frac{Z}{r} \\ &= -Z(V+\frac{Z}{r})\Big|_{r=o} ;= -\frac{B}{a}Z^{7/3} \quad \text{for} \quad N=Z \quad . \end{aligned} \tag{2-229}$$

The third step here is a partial integration; the fourth one recognizes $-Z\delta(\vec{r})$ as the source of the Coulomb potential Z/r; the last one, valid for N=Z only, uses Eq.(65). [The comment to that equation says that (229) identifies the interaction energy between the nuclear charge and the electrons:

$$E_{Ne} = Z\frac{\partial}{\partial Z}E_{TF} \quad , \tag{2-230}$$

which, according to (1-96), is a general statement, not limited to the TF model in its validity.] Now,

$$\begin{aligned} E_{TF}(Z,Z) &= \frac{3}{7}(Z\frac{\partial}{\partial Z}+N\frac{\partial}{\partial N})E_{TF}(Z,N)\Big|_{N=Z} = \frac{3}{7}Z\frac{\partial}{\partial Z}E_{TF}(Z,N)\Big|_{N=Z} \\ &= -\frac{3}{7}\frac{B}{a}Z^{7/3} \quad , \end{aligned} \tag{2-231}$$

which identifies the constant C. This is, of course, the result that we had found earlier in Eq.(67).

The first of our "virial" theorems, Eq.(223), has the consequence

$$E_1 = 2(E_2-\zeta N) = 2(E_{TF} - E_1) \quad , \tag{2-232}$$

or,

$$E_1 = \frac{2}{3} E_{TF} \quad , \tag{2-233}$$

and

$$E_2 - \zeta N = \frac{1}{3} E_{TF} \; . \tag{2-234}$$

For a Coulomb system, like the one we are considering, one expects the usual theorems about the kinetic and the potential energy:

$$E_{kin} = - E_{TF} \quad , \quad E_{pot} = 2E_{TF} \; . \tag{2-235}$$

Indeed, they emerge from the relations that we have found so far. It is essential to remember how E_1 and E_2 are composed of E_{kin} and $E_{pot} = E_{Ne} + E_{ee}$:

$$\begin{aligned} E_1 &= E_{kin} + E_{Ne} + 2E_{ee} + \zeta N \quad , \\ E_2 &= - E_{ee} \quad . \end{aligned} \tag{2-236}$$

Note in particular the double counting of the electron-electron energy in E_1. With (230) we have

$$\begin{aligned} E_{pot} &= E_{Ne} + E_{ee} = Z\frac{\partial}{\partial Z}E_{TF} - E_2 \\ &= (Z\frac{\partial}{\partial Z} + N\frac{\partial}{\partial N})E_{TF} - (E_2-\zeta N) \quad , \end{aligned} \tag{2-237}$$

which makes use of (225). Now Eqs.(226) and (234) can be employed to produce the second statement of (235), which then implies the first one immediately.

The relative sizes are

$$\begin{aligned} E_{ee} : E_{kin} : (-E_{Ne}) &= (-\tfrac{1}{3}E_{TF}-\zeta N):(-E_{TF}):(-\tfrac{7}{3}E_{TF}+\zeta N) \quad ; \\ &= 1 : 3 : 7 \quad \text{for } N = Z, \text{ when } \zeta = o \; . \end{aligned} \tag{2-238}$$

In words: the electron-electron energy of a neutral TF atom is one third of the kinetic energy and one seventh of the (negative of the) nucleus-electron energy.

For ions, there is less specific information in Eq.(226). It merely implies that $E_{TF}(Z,N)$ can be written in the form

$$E_{TF}(Z,N) = Z^{7/3} \times [\text{function of } \frac{N}{Z}] \quad . \tag{2-239}$$

This invites introducing a reduced binding energy, e(q), that is a function of q=1-N/Z, the degree of ionization:

$$E_{TF}(Z,N) = -\frac{Z^{7/3}}{a} e(q) \quad . \tag{2-240}$$

We know e(q) for q=o, i.e., N=Z :

$$e(o) = \frac{3}{7} B \quad , \tag{2-241}$$

which is simply Eq.(231). The factor multiplying e(q) in (240) is the same as the one in Eqs.(108) and (138). The maximum (minimum) property of the TF potential (density) functional is, therefore, here expressed as

$$\begin{aligned} \{\frac{2}{5}\int_o^\infty dx \frac{[f(x)]^{5/2}}{x^{1/2}} + \frac{1}{2}\int_o^\infty dx[f'(x)+\frac{q}{x_o}]^2 + \frac{q(1-q)}{x_o}\} \\ \geq e(q) \geq \\ -\{\frac{3}{5}\int_o^\infty dx\, x^{1/3}[g'(x)]^{5/3} + g(o) + \frac{1}{2}\int_o^\infty dx[g(x)]^2\} \ . \end{aligned} \tag{2-242}$$

The competing g(x)'s are hereby restricted by [Eq.(98)]

$$\int_o^\infty dx\, xg'(x) = 1-q \quad , \tag{2-243}$$

whereas

$$f(x=o) = 1 \quad . \tag{2-244}$$

The equal signs in (242) hold only for $g(x)=f'(x)+q/x_o$, when f(x) obeys Eqs.(59) and (60), which also determine x_o.

We can relate e(q) to $x_o(q)$ by recognizing that Eq.(225) says

$$\frac{\partial}{\partial N}\left(-\frac{Z^{7/3}}{a}e(q)\right) = \frac{Z^{4/3}}{a}\frac{d}{dq}e(q)$$
$$= -\zeta = -\frac{Z^{4/3}}{a}\frac{q}{x_o(q)} \quad , \tag{2-245}$$

thus,

$$\frac{d}{dq}e(q) = -\frac{q}{x_o(q)} \quad . \tag{2-246}$$

Consequently,

$$e(q) = \frac{3}{7}B - \int_o^q dq' \frac{q'}{x_o(q')} \quad , \tag{2-247}$$

and

$$e(q) = \int_q^1 dq' \frac{q'}{x_o(q')} \quad , \tag{2-248}$$

of which the first one should be applied to weakly ionized systems ($N \lesssim Z$, $q \gtrsim o$), whereas the second one is designed for highly ionized atoms ($N \ll Z$, $q \lesssim 1$). In Eq. (248) the obvious statement

$$e(q=1) = o \tag{2-249}$$

has been used; it says: no electrons - no binding energy.

For ions, Eq. (229) gives

$$Z\frac{\partial}{\partial Z}E_{TF} = -Z\left(V+\zeta+\frac{Z}{r}\right)\Big|_{r=o} + \zeta Z$$
$$= \frac{Z^{7/3}}{a}\left[f_q'(o) + \frac{q}{x_o}\right] \quad , \tag{2-250}$$

so that Eq. (226) translates into

$$7e(q) = 3\left[-f'_q(o) - \frac{q^2}{x_o(q)}\right] \quad . \tag{2-251}$$

By writing a subscript q we emphasize the q dependence of $f_q(x)$ and its initial slope $f_q'(o)$. The comparison of (251) with (247) results in

$$
\begin{aligned}
-f'_q(o) &= B + \frac{q^2}{x_o(q)} - \frac{7}{3}\int_o^q dq' \, \frac{q'}{x_o(q')} \\
&= B + \int_o^q dq' \, q'^{7/3} \frac{d}{dq'}\Big[\frac{1}{q'^{1/3} x_o(q')}\Big] \quad .
\end{aligned}
\tag{2-252}
$$

The latter equality is verified by performing a partial integration. Since $-f'_q(o)$ increases with increasing q, we learn here that

$$
\frac{d}{dq}[q^{1/3}x_o(q)] < o \quad . \tag{2-253}
$$

We observe in these relations, that in studying ions the central quantity is $x_o(q)$. It is basically available from solving numerically the differential equation for $f_q(x)$, Eq.(59):

$$
f''_q(x) = \frac{[f_q(x)]^{3/2}}{x^{1/2}} \tag{2-254}
$$

with the boundary conditions (60):

$$
f_q(o) = 1 \ , \quad f_q(x_o(q)) = o \ , \quad -x_o(q)f'_q(x_o(q)) = q \quad . \tag{2-255}
$$

Nevertheless, in the two limiting situations $q \lesssim 1$ and $q \gtrsim o$ it is possible to make precise statements about the analytic dependence of $x_o(q)$ on $1-q(=N/Z)$, or, q, respectively. Let us first concentrate on highly ionized atoms, $q \lesssim 1$.

<u>Highly ionized TF atoms.</u> In the limit of extremely high ionization, $N/Z \to o$, the interelectronic interaction becomes insignificant as compared to the nucleus-electron interaction. In this situation V is simply the Coulomb potential $-Z/r$, and we are dealing with Bohr atoms, which have been studied in the first Chapter. We concluded already, in Eq.(46), that then

$$
E_{TF}(Z,N) = -Z^2(\tfrac{3}{2}N)^{1/3} \quad \text{for} \quad N/Z \to o \quad . \tag{2-256}
$$

As a statement about e(q) this reads

$$
e(q) = a[\tfrac{3}{2}(1-q)]^{1/3} = \tfrac{3}{4}(\tfrac{\pi}{2})^{2/3}(1-q)^{1/3}
$$

$$\text{for } q \to 1 \quad . \tag{2-257}$$

After employing Eq.(246) we find

$$x_o(q) = [\frac{16}{\pi}(1-q)]^{2/3} \quad \text{for} \quad q \to 1 \quad , \tag{2-258}$$

which is recognized to be Eq.(1-35) when $\zeta=(Z^{4/3}/a)(q/x_o)$, etc., is inserted there.

If N/Z is not that small, Eq.(257) and (258) acquire corrections that account for the repulsion among the electrons. A systematic treatment proceeds from noting that f(x) is not the best suited parametrization of the potential for the present purpose. It is advantageous to introduce another function $\phi(t)$ by means of[15]

$$V(r) + \zeta = -\frac{Z-N}{r}\phi(r/r_o) \quad , \tag{2-259}$$

or, recalling $\zeta = (Z-N)/r_o$ [Eq.(55)] ,

$$V + \zeta = -\zeta\frac{\phi(t)}{t} \quad , \quad t=r/r_o = x/x_o \quad . \tag{2-260}$$

Because of the great similarity between the definition of f(x) in (57) and the one of $\phi(t)$ in (259), the two functions are simply related to each other:

$$f(x) = q\,\phi(x/x_o) \quad , \quad \phi(t) = \frac{1}{q}f(tx_o) \quad . \tag{2-261}$$

Consequently, $\phi(t)$ obeys the differential equation

$$\phi''(t) = \lambda\frac{[\phi(t)]^{3/2}}{t^{1/2}} \quad , \tag{2-262}$$

and is subject to

$$\phi(o) = \frac{1}{q} \quad , \quad \phi(1) = o \quad , \quad \phi'(1) = -1 \quad , \tag{2-263}$$

where $\lambda=\lambda(q)$ is given by

$$\lambda(q) = q^{1/2}[x_o(q)]^{3/2} \quad . \tag{2-264}$$

As a consequence of (258), λ is small for $q \lesssim 1$:

$$\lambda \cong \frac{16}{\pi}(1-q) = \frac{16}{\pi}\frac{N}{Z} \quad \text{for} \quad \frac{N}{Z} = 1-q \gtrsim 0 \quad . \tag{2-265}$$

Why is it fitting to switch from $f(x)$ to $\phi(t)$? The reason is that the appearance of λ in (262) offers the possibility of expanding $\phi(t)$ in powers of λ, whereby the smallness of λ promises a good convergence of the expansion.

The differential equation (262) and the boundary conditions at $t>1$ in (263) can be combined into the integral equation

$$\phi_\lambda(t) = 1 - t + \lambda \int_t^1 dt'(t'-t)\, \frac{[\phi_\lambda(t')]^{3/2}}{t'^{1/2}} \quad . \tag{2-266}$$

where we wrote $\phi_\lambda(t)$ in order to emphasize the λ dependence of ϕ. After solving this equation for a chosen λ, the corresponding value of q emerges from

$$\frac{1}{q} = \phi_\lambda(0) = 1 + \lambda \int_0^1 dt\, t^{1/2}[\phi_\lambda(t)]^{3/2} \quad . \tag{2-267}$$

In the first place, one obtains $1/q$ as a function of λ, from which $\lambda(q)$ is to be found in an additional step. Then Eq.(248), here in the form

$$e(q) = \int_q^1 dq' \frac{q'^{7/3}}{[\lambda(q')]^{2/3}} \quad , \tag{2-268}$$

supplies the desired $e(q)$. The evaluation of this integral is eased by writing it as an integration over λ instead of q, since then $q(\lambda)$ enters, not $\lambda(q)$. The rewriting begins with

$$\begin{aligned} e(q) &= \int_{\lambda(q)}^{0} d\lambda' \,\frac{dq(\lambda')}{d\lambda'}\, \frac{[q(\lambda')]^{4/3}}{\lambda'^{2/3}} \\ &= \frac{3}{7}\int_0^{\lambda(q)} d\lambda'\, \lambda'^{-2/3} \frac{d}{d\lambda'}\left(1-[q(\lambda')]^{7/3}\right) \quad ; \end{aligned} \tag{2-269}$$

in view of this implication of Eq.(265) :

$$\lambda^{-2/3}\left(1-[q(\lambda)]^{7/3}\right) = \lambda^{-2/3}\left(1-[1-\frac{\pi}{16}\lambda+\ldots]^{7/3}\right) =$$

$$= \frac{7\pi}{48}\lambda^{1/3} + O(\lambda^{4/3}) \quad ; \; = o \quad \text{for} \quad \lambda = o \quad , \tag{2-270}$$

a partial integration can be performed with the outcome

$$e(q) = \frac{3}{7}\frac{1-q^{7/3}}{[\lambda(q)]^{2/3}} + \frac{2}{7}\int_o^{\lambda(q)} d\lambda' \, \frac{1-[q(\lambda')]^{7/3}}{\lambda'^{5/3}} \quad . \tag{2-271}$$

An alternative expression makes use of (251); here it reads

$$e(q) = \frac{3}{7}\frac{q^{4/3}}{[\lambda(q)]^{2/3}}(-\phi'_\lambda(o)-q) \quad , \tag{2-272}$$

where

$$-\phi'_\lambda(o) = 1 + \lambda\int_o^1 dt \, \frac{[\phi_\lambda(t)]^{3/2}}{t^{1/2}} \tag{2-273}$$

is the initial slope of $\phi_\lambda(t)$. Note that the equivalence of (271) and (272) allows to relate $\phi'_\lambda(o)$ to $\phi_\lambda(o) = 1/q(\lambda)$ [this is Eq.(267)]:

$$-\phi'_\lambda(o) = [\phi_\lambda(o)]^{4/3}\{1 + \frac{2}{3}\lambda^{2/3}\int_o^\lambda d\lambda' \, \frac{1-[\phi_{\lambda'}(o)]^{-7/3}}{\lambda'^{5/3}}\} \; , \tag{2-274}$$

which is a useful equation for checking against algebraic mistakes.

Let us now, indeed, expand $\phi_\lambda(t)$ in powers of λ,

$$\phi_\lambda(t) = 1 - t + \sum_{k=1}^{\infty}\lambda^k \, \phi_k(t) \quad . \tag{2-275}$$

This, inserted into (266), implies

$$\sum_{\ell=1}^{\infty}\lambda^\ell\phi_\ell(t) = \lambda\int_t^1 dt'(t'-t) \; t'^{-1/2}[1-t'+\sum_{k=1}^{\infty}\lambda^k \; \phi_k(t')]^{3/2} . \tag{2-276}$$

The technique that produced the recurrence relation for the s_ℓ in Eq. (174) can be applied here, too. We find

$$\phi_1(t) = \int_t^1 dt'(t'-t)t'^{-1/2}(1-t')^{3/2} \quad , \tag{2-277}$$

and

$$\phi_\ell(t) = \int_t^1 dt'(t'-t)t'^{-1/2} \sum_{j=1}^{\ell-1} \binom{3/2}{j} (1-t')^{3/2-j}$$

$$\times \sum_{k_1=1}^{\ell-j} \cdots \sum_{k_j=1}^{\ell-j} \phi_{k_1}(t') \cdots \phi_{k_j}(t') \delta_{\ell-1, k_1+\cdots+k_j} \tag{2-278}$$

for $\ell=2, 3, \ldots$.

The first few $\phi_\ell(t)$ can be expressed in terms of elementary functions; unfortunately, the degree of algebraic complexity increases rapidly. Let us, therefore, confine ourselves to explicitly stating only $\phi_1(t)$ as it emerges from (277) :

$$\phi_1(t) = (\tfrac{1}{8} - \tfrac{3}{4}t)\arccos(t^{1/2}) + (\tfrac{1}{8} + \tfrac{2}{3}t - \tfrac{1}{6}t^2)t^{1/2}(1-t)^{1/2}. \tag{2-279}$$

In particular, we have

$$\phi_1(o) = \int_o^1 dt\ t^{1/2}(1-t)^{3/2} = \frac{\pi}{16} , \tag{2-280}$$

$$-\phi_1'(o) = \int_o^1 dt\ t^{-1/2}(1-t)^{3/2} = \frac{3\pi}{8} .$$

Equation (279) is utilized in

$$\phi_2(o) = \frac{3}{2} \int_o^1 dt\ t^{1/2}(1-t)^{1/2}\ \phi_1(t) = \frac{2}{15} - \frac{3\pi^2}{256} , \tag{2-281}$$

and

$$-\phi_2'(o) = \frac{3}{2} \int_o^1 dt\ t^{-1/2}(1-t)^{1/2}\ \phi_1(t) = \frac{1}{3} - \frac{3\pi^2}{256} . \tag{2-282}$$

We are now prepared to employ Eq.(267) in order to find the leading corrections to (265), (257), and (258). From

$$\frac{1}{q} = \frac{1}{1-(1-q)} = 1 + (1-q) + (1-q)^2 + \ldots$$

$$= \phi_\lambda(o) = 1 + \phi_1(o)\lambda + \phi_2(o)\lambda^2 + \ldots , \tag{2-283}$$

we get

$$\lambda(q) = \frac{1}{\phi_1(o)}(1-q)[1+(1-\frac{\phi_2(o)}{[\phi_1(o)]^2})(1-q) + \ldots] \tag{2-284}$$

$$= \frac{16}{\pi}(1-q)[1+(4-\frac{512}{15\pi^2})(1-q) + \ldots] \quad ,$$

or

$$\lambda(q) = \frac{16}{\pi}\frac{N}{Z}[1+(4-\frac{512}{15\pi^2})\frac{N}{Z} + O((\frac{N}{Z})^2)] \quad \text{for } N/Z<<1 \quad . \tag{2-285}$$

Then, using either one of the Eqs.(268),(269), or (271), we find

$$e(q) = \frac{3}{4}(\frac{\pi}{2})^{2/3}(1-q)^{1/3}[1-(1-\frac{256}{45\pi^2})(1-q) + \ldots]$$

$$= \frac{3}{4}(\frac{\pi}{2})^{2/3}(\frac{N}{Z})^{1/3}[1-(1-\frac{256}{45\pi^2})\frac{N}{Z} + O((\frac{N}{Z})^2)] \quad . \tag{2-286}$$

Also, from (264) or (246),

$$x_o(q) = [\frac{16}{\pi}(1-q)]^{2/3}[1+(3-\frac{1024}{45\pi^2})(1-q) + \ldots]$$

$$= (\frac{16}{\pi}\frac{N}{Z})^{2/3}[1+(3-\frac{1024}{45\pi^2})\frac{N}{Z} + O((\frac{N}{Z})^2)] \quad . \tag{2-287}$$

The numerical versions thereof are

$$\lambda = 5.093\,\frac{N}{Z}(1+0.5416\,\frac{N}{Z} + \ldots) \quad ,$$

$$e = 1.0135(\frac{N}{Z})^{1/3}(1-0.4236\,\frac{N}{Z} + ..) \quad , \tag{2-288}$$

$$x_o = 2.960(\frac{N}{Z})^{2/3}(1+0.6944\,\frac{N}{Z} + \ldots) \quad .$$

Here then is the modification of Eq.(46) that we promised at that time:

$$E_{TF}(Z,N) = -Z^2(\frac{3}{2}N)^{1/3}[1-(1-\frac{256}{45\pi^2})\frac{N}{Z} + \ldots] \tag{2-289}$$

for $N<<Z$;

it is obtained by inserting (287) into (240).

A simple check of consistency is provided by (253). This states

$$\frac{d}{dq}\,\lambda(q) < 0 \quad , \tag{2-290}$$

or,

$$\frac{d}{d(N/Z)}\,\lambda(q{=}1-N/Z) > 0 \quad . \tag{2-291}$$

A quick look at (285) shows that this is true, indeed.

We close this section on highly ionized TF atoms with a discussion of the relative sizes of E_{kin}, E_{ee}, and E_{Ne}. In order to be able to apply Eq. (238), we need ζN. It is given by

$$\begin{aligned}\zeta N &= -N\frac{\partial}{\partial N}\,E_{TF}(Z,N)\\ &= \frac{1}{3}Z^2\left(\frac{3}{2}N\right)^{1/3}\left[1-\left(4-\frac{1024}{45\pi^2}\right)\frac{N}{Z}+\ldots\right] \quad ,\end{aligned} \tag{2-292}$$

so that Eqs. (236) and (234) produce

$$\begin{aligned}E_{ee} &= -\frac{1}{3}E_{TF} - \zeta N\\ &= Z^2\left(\frac{3}{2}N\right)^{1/3}\left[\left(1-\frac{256}{45\pi^2}\right)\frac{N}{Z}+\ldots\right] \quad .\end{aligned} \tag{2-293}$$

The interaction energy of the electrons with the nucleus is given by [see Eq. (230)]

$$\begin{aligned}E_{Ne} &= Z\frac{\partial}{\partial Z}\,E_{TF}(Z,N)\\ &= -2Z^2\left(\frac{3}{2}N\right)^{1/3}\left[1-\left(\frac{1}{2}-\frac{128}{45\pi^2}\right)\frac{N}{Z}+\ldots\right] \quad ,\end{aligned} \tag{2-294}$$

whereas the kinetic energy is simply the negative of E_{TF}, as is expressed in Eq. (235). Consequently,

$$\frac{E_{ee}}{-E_{Ne}} = \left(\frac{1}{2}-\frac{128}{45\pi^2}\right)\frac{N}{Z} + O\left(\left(\frac{N}{Z}\right)^2\right) \quad , \tag{2-295}$$

which states that E_{ee} is negligible in the limit of extremely high ionization. [We have already made use of this (physically obvious) fact repeatedly; see, for example, Eq. (256).] Together with the neutral-atom statement of (238), we have, therefore,

$$\frac{E_{ee}}{-E_{Ne}} = \begin{cases} 1/7 & \text{for} \quad N = Z \\ 0.2118\frac{N}{Z} & \text{for} \quad N \ll Z \end{cases} . \tag{2-296}$$

Likewise, one obtains

$$\frac{E_{kin}}{-E_{Ne}} = \begin{cases} 3/7 & \text{for} \quad N = Z \\ \frac{1}{2}(1-0.2118\frac{N}{Z}) & \text{for} \quad N \ll Z \end{cases} , \tag{2-297}$$

and

$$\frac{E_{ee}}{E_{kin}} = \begin{cases} 1/3 & \text{for} \quad N = Z \\ 0.4236\frac{N}{Z} & \text{for} \quad N \ll Z \end{cases} . \tag{2-298}$$

Weakly ionized TF atoms. As λ increases, $\phi_\lambda(t)$ grows larger for all $t<1$, as is evident from Eq. (262), or Eq. (266). Thus $q=1/\phi_\lambda(o)$ decreases, finally reaching $q=o$ for the critical value

$$\Lambda = \lambda(q)\Big|_{q\to o} . \tag{2-299}$$

Consequently,

$$\begin{aligned} x_o(q) &= q^{-1/3}[\lambda(q)]^{2/3} \\ &\cong \Lambda^{2/3}\, q^{-1/3} \quad \text{for} \quad q\to o , \end{aligned} \tag{2-300}$$

so that Eq. (247) implies

$$e(q) \cong \frac{3}{7}B - \frac{3}{7}\Lambda^{-2/3}\, q^{7/3} \quad \text{for} \quad q\to o . \tag{2-301}$$

Accordingly, we have in the limit of very weak ionization

$$E_{TF}(Z,N) \cong -\frac{3}{7}\frac{Z^{7/3}}{a}[B - \Lambda^{-2/3}\, q^{7/3}] =$$

$$= -\frac{3}{7}\frac{1}{a}[B\,Z^{7/3} - \Lambda^{-2/3}(Z-N)^{7/3}] \tag{2-302}$$

$$\text{for} \quad N \lesssim Z \quad .$$

If we insert Eq.(301) into the inequalities of (242), suitably chosen trial functions f(x) and g(x) supply bounds on $\Lambda^{-2/3}$. Details are given in Problem 10, from where we cite

$$0.0946 < \Lambda^{-2/3} < 0.1008 \quad , \tag{2-303}$$

or,

$$\Lambda^{-2/3} = 0.0977 \pm 0.0031 \quad , \tag{2-304}$$

which tells us that $\Lambda^{-2/3}$ is about six percent of B. A weakly ionized TF atom has, therefore, practically the same binding energy that has the neutral atom. In other words: the outermost electrons contribute very little to the total binding energy of the atom.

In the limit $q \to o$, the relation between f(x) and $\phi(t)$ becomes singular. We cannot give sense to the right hand side of

$$f_q(x)\Big|_{q\to o} = F(x) = q\,\phi_{\lambda(q)}(x/x_o(q))\Big|_{q\to o} \tag{2-305}$$

[Eq.(261)], because $x_o(q\to\infty) = \infty$ squeezes $t=x/x_o(q)$ into an infinitesimal vicinity of t=o. There is, nevertheless, a sensible limit to $\phi_\lambda(t)$ as λ approaches Λ. We write $\Phi(t)$ for this $\phi_\Lambda(t)$. It obeys the differential equation

$$\Phi''(t) = \Lambda\,\frac{[\Phi(t)]^{3/2}}{t^{1/2}} \quad , \tag{2-306}$$

and is subject to

$$\Phi(1) = o \quad , \quad \Phi'(1) = -1 \quad , \tag{2-307}$$

and

$$\Phi(t) \to \infty \quad \text{as} \quad t \to o \quad . \tag{2-308}$$

Although $\Phi(t)$ is somehow corresponding to the situation of neutral TF atoms, the trouble of Eq.(305) signifies that it cannot be used as a parametrization of the potential $V(r)$. Fortunately, there is still a use for Eqs.(306) through (308), in as much as they offer a simple and highly precise method for calculating Λ . Here is how it goes: $\Phi(t)$ possesses an expansion in powers of αt^{σ}, with a yet undetermined constant α and

$$\sigma = 7 + \gamma = \frac{1}{2}(7+\sqrt{73}) \quad , \tag{2-309}$$

of the form

$$\Phi(t) = \frac{144/\Lambda^2}{t^3}[1-\alpha t^{\sigma} + \frac{9}{12+4\sigma^2}(\alpha t^{\sigma})^2 + \ldots] \quad , \tag{2-310}$$

which is, of course, an immediate analog to Eqs.(180), (193), (192), and (190). The coefficients of the powers of αt^{σ} obey the recurrence relation (196) after replacing γ by σ. The (numerical) range of convergence of this series is $\alpha t^{\sigma} \lesssim 0.6$, or, anticipating that α is close to unity, $t \lesssim 0.94$. On the other hand, $\Phi(t)$ can also be expanded in powers of $(1-t)^{1/2}$,

$$\begin{aligned}\Phi(t) = (1-t) &+ \frac{4\Lambda}{35}(1-t)^{7/2} + \frac{2\Lambda}{63}(1-t)^{9/2} \\ &+ \frac{\Lambda}{66}(1-t)^{11/2} + \frac{\Lambda^2}{175}(1-t)^{6} + \ldots \quad ,\end{aligned} \tag{2-311}$$

this series being convergent for $(1-t)^{1/2} \lesssim 0.35$, or, $t \gtrsim 0.88$, when Λ is within the bounds of (303). There is a range of t around $t=0.9$ where both expansions are converging. This allows to determine Λ and α numerically by forcing the two expansions to agree within the accuracy to which they represent solutions to Eq.(306). Such a calculation[16] resulted in

$$\Lambda = 32.729416116173 \tag{2-312}$$

and

$$\alpha = 1.0401806573862 \quad . \tag{2-313}$$

Naturally, physics does not need this many decimals; they are reported only in order to demonstrate the marvelous precision of this simple method. Please note that one cannot compute B and β in a similar way, because the expansions (164) and (193) converge for $x \lesssim 0.15$ and $x \gtrsim 30$, respectively. There is no overlap.

The Λ of (312) yields

$$\Lambda^{-2/3} = 0.0977330 \quad , \tag{2-314}$$

so that we obtain, from Eq. (302),

$$-E_{TF}(Z,N) \cong 0.768745\, Z^{7/3} - 0.047310(Z-N)^{7/3}$$

$$\text{for} \quad N \lesssim Z \quad . \tag{2-315}$$

The correction to the neutral atom binding energy is rather small; even for N=Z/2 it is only about one percent.

Since Λ is large, the series of Eq. (275) does not converge rapidly (if at all) for weakly ionized atoms, and the switching from $f(x)$ to $\phi(t)$ is pointless in this situation. Here we make use of the fact that the difference between $F(x)$ and $f_q(x)$ is small, when $q \gtrsim o$ and $x < x_o(q)$. In particular, $f_q'(o)$ does not differ significantly from $-B$, so that $f_q'(o)+B$ is a possibly useful expansion parameter. We use it in making the ansatz

$$f_q(x) = F(x) + \sum_{k=1}^{\infty} [f_q'(o) + B]^k \, f_k(x) \quad , \tag{2-316}$$

where the $f_k(x)$ are subject to

$$f_k(o) = o \quad \text{for} \quad k=1,2,\ldots \tag{2-317}$$

and

$$f_1'(o) = 1 \quad , \quad f_k'(o) = o \quad \text{for} \quad k=2,3,\ldots \quad . \tag{2-318}$$

To first order in $f_q'(o)+B$, the differential equation obeyed by $f_q(x)$ requires[17]

$$\left[\frac{d^2}{dx^2} - \frac{3}{2}\left[\frac{F(x)}{x}\right]^{1/2}\right] f_1(x) = o \quad . \tag{2-319}$$

One solution is

$$f_o(x) = F(x) + \frac{1}{3}x\,F'(x) = \frac{1}{3x^2}\frac{d}{dx}(x^3F(x)) \quad , \tag{2-320}$$

because

$$\left[\frac{d^2}{dx^2} - \frac{3}{2}[\frac{F(x)}{x}]^{1/2}\right]\left[\frac{1}{3x^2}\frac{d}{dx}x^3\right]F(x)$$

$$= (\frac{1}{3x^2}\frac{d}{dx}x^3)\left[\frac{d^2}{dx^2} - [\frac{F(x)}{x}]^{1/2}\right]F(x) = o \quad . \tag{2-321}$$

However, inasmuch as

$$f_o(1) = 1 \quad , \quad f_o'(1) = -\frac{4}{3}B \quad , \tag{2-322}$$

$f_o(x)$ is not proportional to $f_1(x)$. The Wronskian of the differential equation (319) relates the two functions to each other:

$$f_o(x)\ f_1'(x) - f_o'(x)\ f_1(x) = 1 \quad . \tag{2-323}$$

This is equivalent to

$$\frac{d}{dx}\frac{f_1(x)}{f_o(x)} = [f_o(x)]^{-2} \quad , \tag{2-324}$$

which has the consequence

$$f_1(x) = f_o(x)\int_o^x \frac{dx'}{[f_o(x')]^2} \quad . \tag{2-325}$$

This does, indeed, satisfy the requirements $f_1(o)=o$ and $f_1'(o)=1$, so that we need not add a multiple of $f_o(x)$ on the right hand side.

For large x, we have

$$f_o(x) = \frac{1}{3x^2}\frac{d}{dx}[144\,G\,(y(x))] = \frac{(12)^2}{3x^2}\frac{d}{dx}[1-y(x) + \ldots]$$

$$\cong \frac{(12)^2}{3}\frac{\gamma y(x)}{x^3} = \frac{(12)^2}{3}\gamma\beta\, x^{-(\gamma+3)} \quad , \tag{2-326}$$

which uses Eqs.(180), (187), (184), and (192). This inserted into (323) or, equivalently, (326) produces

$$f_1(x) \cong \frac{1}{48\beta}\frac{1}{\gamma(\gamma+\sigma)}x^{\sigma-3} \quad \text{for large } x \tag{2-327}$$

[$\sigma = 7+\gamma$, Eq.(309)].

In deriving Eq.(319) the first order approximation $(1+\varepsilon)^{3/2} \cong 1+\frac{3}{2}\varepsilon$ has been used for $\varepsilon=[f_q'(o)+B]f_1(x)/F(x)$. Consequently,

$$f_q(x) \cong F(x) + [f_q'(o)+B]f_1(x) \tag{2-328}$$

must not be applied to $x \cong x_o$, where $\varepsilon \cong -1$. We can, however, supplement (328) with

$$f_q(x) \cong q(1-\frac{x}{x_o}) \tag{2-329}$$

which is valid for $x \cong x_o$: [This is obviously no more than the first term of Eq.(311) as it analogously appears in $\phi_\lambda(t)$]. Let us now join the two approximations for $f_q(x)$, Eqs.(328) and (329), at a certain $x=x_1$. The three unknown quantities $x_o(q)$, $x_1(q)$, and $f_q'(o)+B$ are determined from the requirement that $f_q(x)$ and its two first derivatives are continuous at $x=x_1$. This can be done explicitly in the limiting situation of very small q, since then both $x_o(q)$ and $x_1(q)$ are large, which allows to employ the large-x forms of $F(x)$ and $f_1(x)$. Thus, we have the three algebraic equations

$$\begin{aligned} f_q&: \quad \frac{144}{x_1^3} + [f_q'(o)+B]\frac{x_1^{\sigma-3}}{48\beta\gamma(\gamma+\sigma)} = q(1-\frac{x_1}{x_o}) \quad , \\ f_q'&: \quad -\frac{432}{x_1^4} + [f_q'(o)+B]\frac{(\sigma-3)x_1^{\sigma-4}}{48\beta\gamma(\gamma+\sigma)} = -q/x_o \quad , \\ f_q''&: \quad \frac{(12)^3}{x_1^5} + [f_q'(o)+B]\frac{18\,x_1^{\sigma-5}}{48\beta\gamma(\gamma+\sigma)} = o \quad ; \end{aligned} \tag{2-330}$$

the last one uses $(\sigma-3)(\sigma-4)=18$. These three equations imply

$$\frac{x_1}{x_o} = \frac{\sigma+3/2}{\sigma+2} = \frac{\gamma+17/2}{\gamma+9} = \frac{37+\sqrt{73}}{48} = 0.9488 \quad , \tag{2-331}$$

and

$$x_o = \frac{\sigma+2}{\sigma+3/2}\,[96(\sigma+2)]^{1/3}\,q^{-1/3} = 10.32\,q^{-1/3} \quad , \tag{2-332}$$

as well as

$$-f_q'(o) = B + \frac{1}{2}(96)^{-(\gamma+1)/3}\,\beta\gamma(\gamma+\sigma)(\sigma+2)^{-\sigma/3}\,q^{\sigma/3} =$$

$$= B + 8.05 \times 10^{-3}\ q^{2.59} \quad , \tag{2-333}$$

whereby identities like $\sigma=\gamma+7$ and $\gamma\sigma=6$ have been used. Of course, since these results are based upon the simple approximations (328) and (329) we should not take them too seriously. Nevertheless, their structure is certainly right. For instance, Eq. (332) says that the combination $q^{1/3}x_o(q)$ approaches a constant as $q\to o$. This much we know already - the constant is $\Lambda^{2/3}=10.2320$. The estimate for $\Lambda^{2/3}$ obtained in (332) differs from the actual value by less than one percent.

Something new is to be learned from Eq. (333). As a preparation, we differentiate Eq. (252) with respect to q:

$$\frac{d}{dq}[-f_q'(o)] = q^{7/3}\ \frac{d}{dq}\left[\frac{1}{q^{1/3}x_o(q)}\right] = q^{7/3}\ \frac{d}{dq}[\lambda(q)]^{-2/3} \ ; \tag{2-334}$$

the latter equality is a consequence of the definition of $\lambda(q)$ in Eq. (264). Now Eq. (333) implies that

$$q\ \frac{d}{dq}[\lambda(q)]^{-2/3} \sim q^{(\sigma-7)/3} = q^{\gamma/3} \quad . \tag{2-335}$$

We infer that

$$\lambda(q) = \Lambda\,[1+(\text{powers of } q^{\gamma/3})] \tag{2-336}$$

for values of q not too large. Then, of course,

$$x_o(q) = \Lambda^{2/3}\ q^{-1/3}[1+(\text{powers of } q^{\gamma/3})] \quad , \tag{2-337}$$

and

$$e(q) = \frac{3}{7}B - \frac{3}{7}\ \Lambda^{-2/3}\ q^{7/3}[1+e_1q^{\gamma/3} + e_2q^{2\gamma/3} + \ldots] \ , \tag{2-338}$$

which is an implication of (337) when it is inserted into Eq. (247). The challenge consists in calculating the coefficients e_1, e_2, ..., which determine the corresponding coefficients in (336) and (337). In particular, Eq. (246) supplies

$$x_o(q) = \Lambda^{2/3}\ q^{-1/3}[1+ \frac{\sigma}{7}e_1q^{\gamma/3}+ \frac{\sigma+\gamma}{7}e_2q^{2\gamma/3}+ \ldots]^{-1} = \tag{2-339}$$

$$= \Lambda^{2/3}\, q^{-1/3}[1- \tfrac{\sigma}{7}e_1 q^{\gamma/3} + ((\tfrac{\sigma}{7}e_1)^2 - \tfrac{\sigma+\gamma}{7}e_2)q^{2\gamma/3} + \ldots] \;.$$

Then

$$\lambda(q) = [q^{1/3}x_o(q)]^{3/2} \tag{2-340}$$

$$= \Lambda[1- \tfrac{3\sigma}{14}e_1\, q^{\gamma/3} + \tfrac{3}{2}(\tfrac{5}{4}(\tfrac{\sigma}{7}e_1)^2 - \tfrac{\sigma+\gamma}{7}e_2)q^{2\gamma/3} + \ldots] \;,$$

and from combining (251) and (246) with (338)

$$-f_q'(o) = \tfrac{7}{3}e(q) + \frac{q^2}{x_o(q)} = (\tfrac{7}{3} - q\tfrac{d}{dq})e(q) \tag{2-341}$$

$$= -q^{10/3}\,\frac{d}{dq}[q^{-7/3}\; e(q)]$$

$$= q^{10/3}\,\frac{d}{dq}(-\tfrac{3}{7}B\,q^{-7/3} + \tfrac{3}{7}\,\Lambda^{-2/3}[1+e_1q^{\gamma/3}+e_2q^{2\gamma/3}+\ldots]),$$

or,

$$-f_q'(o) = B + \tfrac{\gamma}{7}\,\Lambda^{-2/3}\, q^{\sigma/3}[e_1 + 2e_2\, q^{\gamma/3} + \ldots] \;. \tag{2-342}$$

The comparison with (333) yields a first estimate for e_1:

$$e_1 \cong 0.75 \;, \tag{2-343}$$

which, in view of the crudeness of the approximation used in arriving at (333), cannot be expected to have more significance than stating the order of magnitude. (We shall see below that the actual value is about ten percent larger.)

A systematic computation of e_1, e_2, ... starts from the expansion (316). Comparing powers of $f_q'(o)+B$ in the differential equation obeyed by $f_q(x)$ produces

$$f_\ell''(x) = \sum_{j=1}^{\ell}\binom{3/2}{j}\frac{[F(x)]^{3/2-j}}{x^{1/2}}\sum_{k_1=1}^{\ell-j+1}\cdots\sum_{k_j=1}^{\ell-j+1} f_{k_1}(x)\ldots$$

$$\times\ldots f_{k_j}(x)\,\delta_{\ell,k_1+\ldots+k_j} \;. \tag{2-344}$$

The j=1 term on the right hand side is brought over to the left, so that

$$[\frac{d^2}{dx^2} - \frac{3}{2}\sqrt{F(x)/x}]\, f_\ell(x)$$

$$= \sum_{j=2}^{\ell} \binom{3/2}{j} \frac{[F(x)]^{3/2-j}}{x^{1/2}} \sum_{k_1=1}^{\ell-j+1} \cdots \sum_{k_j=1}^{\ell-j+1} f_{k_1}(x) \ldots$$

$$\times \ldots f_{k_j}(x)\, \delta_{\ell, k_1+\ldots+k_j} \quad . \tag{2-345}$$

This right hand side contains $f_1(x)$, ..., $f_{\ell-1}(x)$ but not $f_\ell(x)$. The solutions to the corresponding homogeneous differential equation are $f_o(x)$ and $f_1(x)$, given in Eqs.(320) and (325). With their aid we can construct Green's function G(x,x') which satisfies

$$[\frac{d^2}{dx^2} - \frac{3}{2}\sqrt{F(x)/x}]\; G(x,x') = \delta(x-x') \quad ,$$

$$G(x,x') = o \quad \text{and} \quad \frac{\partial}{\partial x} G(x,x') = o \quad \text{for} \quad x=o \quad . \tag{2-346}$$

It is given by

$$G(x,x') = [f_o(x')f_1(x) - f_o(x)f_1(x')]\,\eta(x-x') \quad . \tag{2-347}$$

Thus

$$f_\ell(x) = \int_o^x dx'\, [f_o(x')f_1(x) - f_o(x)f_1(x')]$$

$$\times [(\frac{d^2}{dx'^2} - \frac{3}{2}\sqrt{F(x')/x'})\; f_\ell(x')] \quad , \tag{2-348}$$

where we refrained from explicitly inserting the right hand side of (345).

The use of Eq.(348) does not lie primarily in explicitly calculating $f_2(x)$, $f_3(x)$, etc. but in studying their structure. Recall that F(x) can be written as

$$F(x) = \frac{144}{x^3}\, G(y(x)) = \frac{144}{x^3}\left[1 + \sum_{k=1}^{\infty} c_k [y(x)^k]\right] \tag{2-349}$$

which is repeating Eqs. (180) and (193). As a consequence, $f_o(x)$ has the form

$$f_o(x) = \frac{144}{x^3}\frac{1}{3}\gamma\, y(x)[1+(\text{powers of } y(x))] \quad . \tag{2-350}$$

Inserted into (325) this implies

$$f_1(x) = \frac{144}{x^3}\frac{3}{(12)^4}\frac{1}{\beta\gamma(\gamma+\sigma)}x^\sigma[1+(\text{powers of } y(x))] \quad , \tag{2-351}$$

of which we have seen the leading term in (327). Now we employ the recurrence formula (348) to conclude that

$$f_\ell(x) = \frac{144}{x^3}[-\frac{3}{(12)^4}\frac{1}{\beta\gamma(\gamma+\sigma)}x^\sigma]^\ell d_\ell[1+(\text{powers of } y(x))] \quad , \tag{2-352}$$

where the constants d_ℓ obey

$$d_1 = -1 \tag{2-353}$$

and

$$d_\ell = \frac{2\gamma}{(\ell\sigma+\gamma)(\ell-1)}\sum_{j=2}^{\ell}\binom{3/2}{j}\sum_{k_1=1}^{\ell-j+1}\cdots\sum_{k_j=1}^{\ell-j+1}d_{k_1}\cdots d_{k_j}\delta_{\ell,k_1+\ldots+k_j} \quad . \tag{2-354}$$

This we recognize to be the recursion for the c_k of Eq. (196), after γ and σ are interchanged. Consequently, the d_ℓ's are the coefficients that appear in the expansion of Eq. (310). That is

$$\Phi(t) = \frac{144/\lambda_o^2}{t^3}[1+\sum_{k=1}^{\infty}d_k(\alpha t^\sigma)^k] \quad . \tag{2-355}$$

This connection between the $f_\ell(x)$'s and $\Phi(t)$, which is, of course, not accidental, is the clue to computing e_1, e_2, ... of Eq. (338). We reveal its significance by inserting Eqs. (349) and (352) into the ansatz (316),

$$f_q(x) = \frac{144}{x^3}[1+\sum_{k=1}^{\infty}(-[f_q'(0)+B]\frac{3}{(12)^4}\frac{[x_o(q)]^\sigma}{\beta\gamma(\gamma+\sigma)})^k(\frac{x}{x_o(q)})^{k\sigma}d_k] +$$

$$+\ldots \quad , \tag{2-356}$$

where the ellipsis indicates the terms containing "powers of $y(x)$." After introducing h_q by

$$-f_q'(o) = B + \frac{1}{3}(12)^4\alpha\beta\gamma(\gamma+\sigma)[h_q/x_o(q)]^\sigma \quad , \tag{2-357}$$

Eq.(355) is employed:

$$f_q(x) = \frac{144}{x^3}[1+\sum_{k=1}^{\infty} d_k(\alpha[h_q \frac{x}{x_o(q)}]^\sigma)^k]$$

$$+\ldots \tag{2-358}$$

$$= q[\frac{\Lambda}{\lambda(q)}]^2 h_q^3 \; \Phi(h_q \frac{x}{x_o(q)}) + \ldots \quad ,$$

where $[\lambda(q)]^2=q[x_o(q)]^3$ is used. What is exhibited in (358) is the part of $f_q(x)$ that goes with the zeroth power of $y(x)$. Likewise, an arbitrary power of $y(x)$, say $[y(x)]^m$ contributes

$$[y(x)]^m \frac{144}{x^3}[c_m + (\text{powers of } \alpha(h_q x/x_o(q))^\sigma)]$$

$$\equiv q[\frac{\Lambda}{\lambda(q)}]^2 \; h_q \; c_m[y(x)]^m \; \psi_m(h_q \frac{x}{x_o(q)}) \tag{2-359}$$

to $f_q(x)$. The functions $\psi_m(t)$ thus defined are such that

$$\psi_m(t) = \frac{144/\Lambda^2}{t^3}[1 + (\text{powers of } \alpha t^\sigma)] \quad . \tag{2-360}$$

We can now make explicit what supplements Eq.(358):

$$f_q(x) = q[\frac{\Lambda}{\lambda(q)}]^2 \; h_q^3[\Phi(h_q \frac{x}{x_o(q)}) + \sum_{m=1}^{\infty} c_m[y(x)]^m \psi_m(h_q \frac{x}{x_o(q)})] \tag{2-361}$$

Whereas (316) is an expansion that is expected to converge rapidly for x not too close to $x_o(q)$, the series of (361) is the faster convergent the smaller $y(x)$ is. This identifies large values of x (i.e., $x \lesssim x_o$) as the domain of application.

It is instructive to make contact with the original definition of $\Phi(t)$,

$$\Phi(t) = \phi_{\lambda(q)}(t)\Big|_{q\to o} = \frac{1}{q} f_q(tx_o(q))\Big|_{q\to o} \qquad (2\text{-}362)$$

[see the comment to Eq.(305)]. Since

$$y(tx_o(q))\Big|_{q\to o} = \beta(tx_o(q))^{-\gamma}\Big|_{q\to o} = o \quad , \qquad (2\text{-}363)$$

the combination of Eqs.(361) and (362) reads

$$\Phi(t) = h_q^3\,\Phi(h_q t)\Big|_{q\to o} \quad , \qquad (2\text{-}364)$$

from which we learn that

$$h_q\Big|_{q\to o} = 1 \;. \qquad (2\text{-}365)$$

This tells us what e_1 is. Equations (342) and (357) together say

$$\begin{aligned} e_1 + 2e_2\, q^{\gamma/3} + \ldots &= -[f_q'(o)+B]\,\frac{7}{\gamma}\,\Lambda^{2/3}\, q^{-\sigma/3} \\ &= \frac{7}{3}(12)^4\alpha\beta(\gamma+\sigma)\,\Lambda^{2/3}\Big[\frac{h_q}{q^{1/3}x_o(q)}\Big]^{\sigma} \quad , \end{aligned} \qquad (2\text{-}366)$$

or, after making use of $q^{1/3}x_o(q)=[\lambda(q)]^{2/3}$,

$$e_1 + 2e_2\, q^{\gamma/3} + \ldots = \frac{7}{3}\Big(\frac{12}{\Lambda}\Big)^4\alpha\beta(\gamma+\sigma)\,\Lambda^{-2\gamma/3}\Big(h_q\Big[\frac{\Lambda}{\lambda(q)}\Big]^{2/3}\Big)^{\sigma} \;. \qquad (2\text{-}367)$$

Now the limit $q\to o$ identifies

$$e_1 = \frac{7}{3}\Big(\frac{12}{\Lambda}\Big)^4\,\alpha\beta(\gamma+\sigma)\,\Lambda^{-2\gamma/3} \quad . \qquad (2\text{-}368)$$

With α, β, and Λ from (313), (204), and (312), respectively, the numerical value of e_1 is roughly 10% larger than the estimate of (343):

$$e_1 = 0.825908 \quad . \tag{2-369}$$

Note that Eq.(368) reveals the physical significance of α and β; that of B and Λ has been clear since Eqs.(67) and (301).

The requirement $f_q(x=x_o(q))=o$ relates h_q to $y_o(q)$, given by

$$\begin{aligned} y_o(q) &\equiv y(x_o(q)) = \beta[x_o(q)]^{-\gamma} \\ &= \beta\Lambda^{-2\gamma/3} q^{\gamma/3}[\frac{\Lambda}{\lambda(q)}]^{2\gamma/3} \quad , \end{aligned} \tag{2-370}$$

inasmuch as $x = x_o(q)$ in Eq.(361) yields

$$\Phi(h_q) + \sum_{m=1}^{\infty} c_m[y_o(q)]^m \psi_m(h_q) = o. \tag{2-371}$$

With the aid of $\Phi'(1)= -1$ [which is Eq.(307)], we find to first order in $y_o(q)$, or, $q^{\gamma/3}$, respectively:

$$\begin{aligned} h_q &= 1 + c_1\ y_o(q)\ \psi_1(1) + O(y_o(q))^2) \\ &= 1 - \beta\ \Lambda^{-2\gamma/3}\ \psi_1(1)q^{\gamma/3} + O(q^{2\gamma/3}) \quad , \end{aligned} \tag{2-372}$$

where $c_1= -1$ has been used. In conjunction with Eq.(340) this has the consequence

$$\begin{aligned} (h_q[\frac{\Lambda}{\lambda(q)}]^{2/3})^\sigma &= (1+(\frac{\sigma}{7}e_1 -\beta\Lambda^{-2\gamma/3}\ \psi_1(1))q^{\gamma/3} +\ldots)^\sigma \\ &= 1 + \sigma(\frac{\sigma}{7}e_1 -\beta\Lambda^{-2\gamma/3}\ \psi_1(1))q^{\gamma/3} +\ldots \ , \end{aligned} \tag{2-373}$$

so that the order $q^{\gamma/3}$ in Eq.(367) is

$$2e_2 = \sigma e_1(\frac{\sigma}{7}e_1 -\beta\Lambda^{-2\gamma/3}\ \psi_1(1)) \quad . \tag{2-374}$$

To proceed further, we need to know $\psi_1(1)$.

The insertion of Eq.(361) into the differential equation obeyed by $f_q(x)$ produces

$$[\frac{d^2}{dt^2} - \frac{3}{2}\ \Lambda\sqrt{\Phi(t)/t}]t^{-\gamma}\ \psi_1(t) = o \quad , \tag{2-375}$$

when terms linear in $y(x)$ are identified. This is quite analogous to Eq.(319), so that

$$\psi_o(t) \equiv t^\gamma[\Phi(t) + \frac{1}{3}t\,\Phi'(t)] \qquad (2\text{-}376)$$

is one solution of (375), the one corresponding to $f_o(x)$ of (320). The Wronskian of $\psi_1(t)$ with $\psi_o(t)$ is

$$\psi_o(t)\psi_1'(t) - \psi_o'(t)\psi_1(t) = \frac{1}{3}(\frac{12}{\Lambda})^4 \alpha\sigma(\sigma+\gamma)t^{2\gamma} \quad , \qquad (2\text{-}377)$$

which makes use of Eq.(360) and the small t form of $\psi_o(t)$,

$$\psi_o(t) = t^\gamma \frac{1}{3t^2}\frac{d}{dt}(t^3\Phi(t))$$

$$= \frac{1}{3}t^{-(2-\gamma)}\ (\frac{12}{\Lambda})^2\frac{d}{dt}[1-\alpha t^\sigma+d_2(\alpha t^\sigma)^2+\ldots] \qquad (2\text{-}378)$$

$$= -\frac{1}{3}(\frac{12}{\Lambda})^2\alpha\sigma\, t^{\sigma+\gamma-3}[1-2d_2\,\alpha\, t^\sigma+..] \quad .$$

Equation (377) now implies

$$\psi_1(t) = \psi_o(t)[\psi_1(1)/\psi_o(1) - \frac{1}{3}(\frac{12}{\Lambda})^4\,\alpha\sigma(\sigma+\gamma)\int_t^1 dt'\,(\frac{t'^\gamma}{\psi_o(t')})^2] \quad , \qquad (2\text{-}379)$$

where $\psi_1(1)$ is determined by the $t\to o$ form of $\psi_1(t)$, statet in Eq.(360). In connection with (378) this requires that the square brackets in (379) possess the form

$$-\frac{3}{\alpha\sigma}t^{-(\sigma+\gamma)}[1+\ (\text{powers of } \alpha\, t^\sigma)]$$

$$= \psi_1(1)/\psi_o(1) - \frac{3}{\alpha\sigma}(\sigma+\gamma)\int_t^1 dt'[-\frac{1}{3}(\frac{12}{\Lambda})^2\,\alpha\sigma\,\frac{t'^\gamma}{\psi_o(t')}]^2$$

$$= \psi_1(1)/\psi_o(1) - \frac{3}{\alpha\sigma}(\sigma+\gamma)\int_t^1 dt'\ t^{-2(\sigma-3)}\,(1+4d_2\alpha\, t^\sigma) \qquad (2\text{-}380)$$

$$- \frac{3}{\alpha\sigma}(\sigma+\gamma)\int_t^1 dt'\{[-\frac{1}{3}(\frac{12}{\Lambda})^2\,\alpha\sigma\,\frac{t'^\gamma}{\psi_o(t')}]^2 - t^{-2(\sigma-3)}\,(1+4d_2\alpha\, t^\sigma)\}.$$

In the latter version, the second integral is no longer singular at t=o, since the integrand has the structure

$$\{\ \ldots\} = \alpha^2 t'^6 [\text{const.} + (\text{powers of } \alpha t^\sigma)] \quad , \tag{2-381}$$

which integrates to

$$\int_t^1 dt' \{\ \ldots\} = \int_o^1 dt' \{\ \ldots\} + \alpha^2 t^7 [\text{const.} + (\text{powers of } \alpha t^\sigma)]$$

$$= \int_o^1 dt' \{\ldots\} + t^{-(\sigma+\gamma)} (\alpha t^\sigma)^2 [\text{const.} + (\text{powers of } \alpha t^\sigma)] \ . \tag{2-382}$$

The first integral in (380) is

$$-\frac{3}{\alpha\sigma}(\sigma+\gamma)\left[\frac{1-t^{7-2\sigma}}{7-2\sigma} + 4d_2\alpha\,\frac{1-t^{7-\sigma}}{7-\sigma}\right]$$

$$= \frac{3}{\alpha\sigma}(1+4d_2\,\frac{\sigma+\gamma}{\gamma}\,\alpha) - \frac{3}{\alpha\sigma}\,t^{-(\sigma+\gamma)}(1+4d_2\,\frac{\sigma+\gamma}{\gamma}\,\alpha t^\sigma) \quad . \tag{2-383}$$

The consequence of (380) is therefore

$$\psi_1(1)/\psi_o(1) + \frac{3}{\alpha\sigma}(1+4d_2\,\frac{\sigma+\gamma}{\gamma}\,\alpha)$$

$$-\frac{3}{\alpha\sigma}(\sigma+\gamma)\int_o^1 dt\{[-\frac{1}{3}(\frac{12}{\Lambda})^2\alpha\sigma\,\frac{t^\gamma}{\psi_o(t)}]^2 - \frac{1+4d_2\alpha t^\sigma}{t^{\sigma+\gamma+1}}\} = o \quad . \tag{2-384}$$

With the aid of $\psi_o(1) = -1/3$ and, from (310) or (354),

$$d_2 = \frac{9}{12+4\sigma^2} = \frac{9/2}{\sigma(2\sigma+\gamma)} = \frac{3}{4}\,\frac{\gamma}{2\sigma+\gamma} \quad , \tag{2-385}$$

this says

$$\psi_1(1) = \frac{1}{\alpha\sigma}(1+3\,\frac{\sigma+\gamma}{2\sigma+\gamma}\,\alpha)$$

$$+\frac{\sigma+\gamma}{\alpha\sigma}\int_o^1 dt\,\{\frac{1+3\frac{\sigma+\gamma}{2\sigma+\gamma}\alpha t^\sigma}{t^{\sigma+\gamma-1}} - [\frac{1}{3}(\frac{12}{x_o})^2\alpha\sigma\,\frac{t^\gamma}{\psi_o(t)}]^2\} \ . \tag{2-386}$$

This expression does not lend itself to further algebraic simplifications.

The numerical value of $\psi_1(1)$, obtained by a method analogous to the one that produced Λ and α in Eqs.(312) and (313), is

$$\psi_1(1) = 0.3216868353717 \quad , \tag{2-387}$$

which illustrates once more the high precision of the algorithm. The Wronskian (377), at t=1, is employed in finding

$$\psi_1'(1) = (4+\gamma)\psi_1(1). - (\frac{12}{\Lambda})^4 \alpha\sigma(\sigma+\gamma)$$
$$= 0.2869164052321 \quad , \tag{2-388}$$

whereas the differential equation (375) supplies

$$\psi_1''(1) = 2\gamma\psi_1'(1) - \gamma(\gamma+1)\psi_1(1)$$
$$= 0.002936027410 \quad . \tag{2-389}$$

The algebraic statements of (388) and (389) can be combined into

$$\psi_1(1) = 2(\frac{12}{\Lambda})^4 \alpha(\sigma+\gamma) + \frac{1}{6}\psi_1''(1) \quad , \tag{2-390}$$

where, because of the smallness of $\psi_1''(1)$, the latter part is only about 0.15% of the sum. In conjunction with Eq.(368), this implies

$$\beta\Lambda^{-2\gamma/3}\psi_1(1) = \frac{6}{7} e_1 + \frac{1}{6}\beta\Lambda^{-2\gamma/3}\psi_1''(1) \quad , \tag{2-391}$$

which we insert into (374) to find

$$e_2 = \frac{\sigma+6}{14} e_1^2[1 - \frac{1}{\sigma-6}\frac{\psi_1''(1)}{2(\frac{12}{\Lambda})^4\alpha(\sigma+\gamma)}] \quad ; \tag{2-392}$$

here, the $\psi_1''(1)$ term represents a 0.5% correction to

$$e_2 \cong \frac{\sigma+6}{14} e_1^2 = 0.671015 \quad , \tag{2-393}$$

resulting in

$$e_2 = 0.667554 \quad . \tag{2-394}$$

Naturally, the subsequent coefficients in (338) can be computed the same way.

The results of this section are summarized in

$$-E_{TF}(Z,N) = 0.768745\ Z^{7/3}$$
$$-0.047310(Z-N)^{7/3}[1+0.825908(1-N/Z)^{0.257334}$$
$$+0.667554(1-N/Z)^{0.514668}$$
$$+\ldots] \quad , \qquad (2\text{-}395)$$

which is the weak-ionization analog to the high-ionization result of Eq.(289)[and its supplement of Problem 7].

One last remark is in order. How could we get around without making explicit use of the requirement $-x_o(q)f_q'(x_o(q)) = q$? As applied to (361) it reads

$$h_q\Phi'(h_q) + \sum_{m=1}^{\infty} c_m[y_o(q)]^m[h_q\psi_m'(h_q) - \gamma m\,\psi_m(h_q)]$$
$$= -\left[\frac{\lambda(q)}{\Lambda}\right]^2 h_q^{-3} \quad . \qquad (2\text{-}396)$$

Indeed, this together with (371) gives $\lambda(q)$ as a function of $y_o(q)$, which can be converted into $\lambda(q)$ as a function of q, whereafter Eq.(340) identifies e_1, e_2, etc. Fortunately, we came to know the relation between $f_q'(o)$ and h_q in Eq.(356), so that we could avoid the more tedious (though, of course, equivalent) procedure based upon Eq.(396).

Arbitrarily ionized TF atoms. We have spent some time on studying the analytic form of such quantities like e(q), $x_o(q)$, and $-f_q'(o)$ as functions of q - both for $q \lesssim 1$ and for $q \gtrsim o$, which are the situations of highly and weakly ionized atoms, respectively. These considerations, however, did not tell us how good are few-terms approximations as in Eqs. (289) and (395). Let us, therefore, make the comparison with the results of numerical integrations of the differential equation obeyed by $f_q(x)$ for various values of q.

We present in Table 2 the outcome of such calculations for the nineteen q values 0.95,0.90,...,0.05 , supplemented by what we know for q=1 and q=o. The fractional binding energy

$$e(q)/e(o) = E_{TF}(Z,N)/E_{TF}(Z,Z) \tag{2-397}$$

is additionally plotted, as a function of q, in Fig.4. We observe that

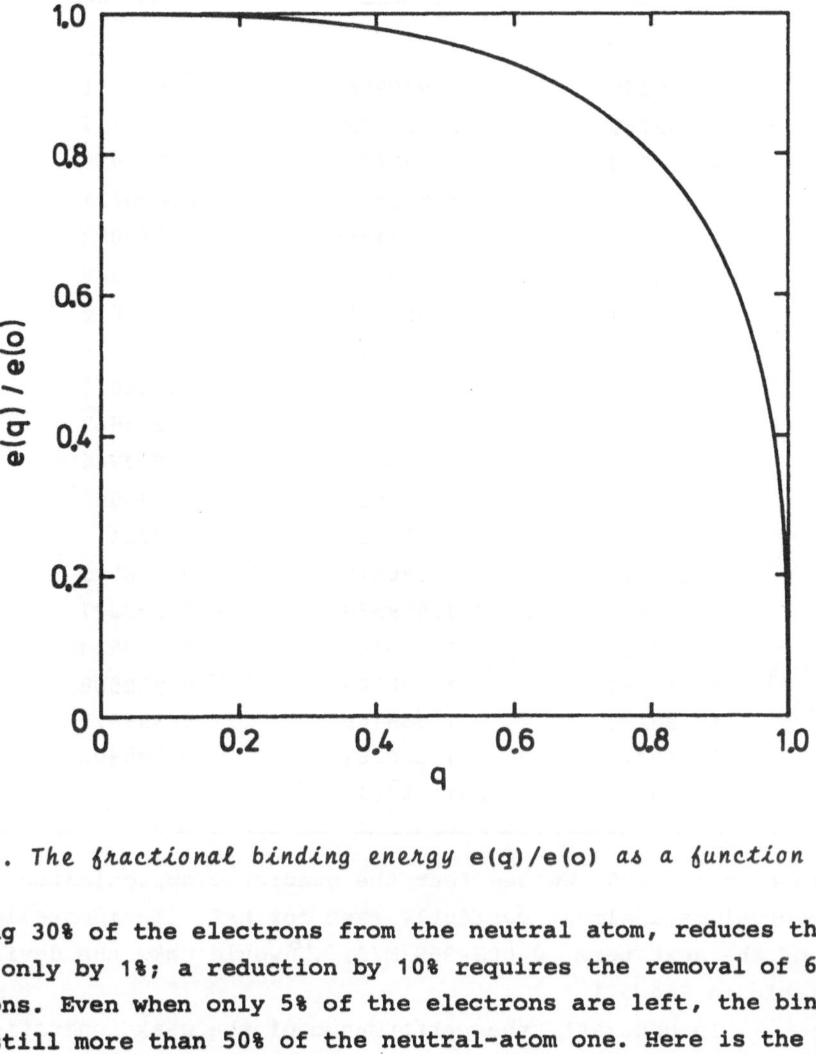

Fig.2-4. The fractional binding energy e(q)/e(o) *as a function of* q.

removing 30% of the electrons from the neutral atom, reduces the binding energy only by 1%; a reduction by 10% requires the removal of 65% of the electrons. Even when only 5% of the electrons are left, the binding energy is still more than 50% of the neutral-atom one. Here is the quantitative version of the qualitative remark that the innermost electrons contribute most to the binding energy, the outermost least.

From Eq.(286), (241) and Problem 7 we find that, for $N<<Z$,

$$\frac{e(q)/e(o)}{1.489(N/Z)^{1/3}} = 1 - 0.4236\,\frac{N}{Z} + 0.0909\left(\frac{N}{Z}\right)^2 + \ldots \quad . \tag{2-398}$$

The successive approximations that this represents are compared to the

Table 2-2. TF quantities $x_o(q)$, $-f_q'(o)$, and $e(q)/e(o)$ for N/Z= 1-q = 0, 0.05, ..., 1.

N/Z	$x_o(q)$	$-f_q'(o)$	$e(q)/e(o)$
0	0	∞	0
0.05	0.416269	3.020996	0.537084
0.10	0.685790	2.233243	0.662517
0.15	0.934348	1.952470	0.742539
0.20	1.179253	1.813524	0.800221
0.25	1.428919	1.734116	0.844082
0.30	1.689292	1.684993	0.878380
0.35	1.965691	1.653119	0.905616
0.40	2.263681	1.631819	0.927406
0.45	2.589715	1.617337	0.944875
0.50	2.951825	1.607410	0.958847
0.55	3.360561	1.600602	0.969946
0.60	3.830452	1.595965	0.978668
0.65	4.382486	1.592853	0.985410
0.70	5.048683	1.590815	0.990503
0.75	5.881272	1.589530	0.994227
0.80	6.973385	1.588763	0.996824
0.85	8.513784	1.588345	0.998508
0.90	10.92728	1.588149	0.999475
0.95	16.10273	1.588081	0.999908
1	∞	B=1.588071	1

actual values in Fig.2-5. We see that the quadratic approximation reproduces the actual data almost perfectly even for $N \lesssim Z$. [Incidentally, the inclusion of the next term, $0.0024369(N/Z)^3$,[18] would make the deviation unrecognizable in Fig.5.]

In contrast, the performance of the weak ionization expansion [Eqs.(338), (369), and (394)],

$$\frac{1-e(q)/e(o)}{0.06154q^{7/3}} = 1 + 0.8259\, q^{\gamma/3} + 0.6676\, q^{2\gamma/3} + \ldots \quad ,(2\text{-}399)$$

is significantly worse; see Fig.6. Obviously, the coefficients in this expansions do not get small as rapidly as the ones in (398). One needs a few more terms in (399) for a high quality approximation over a large

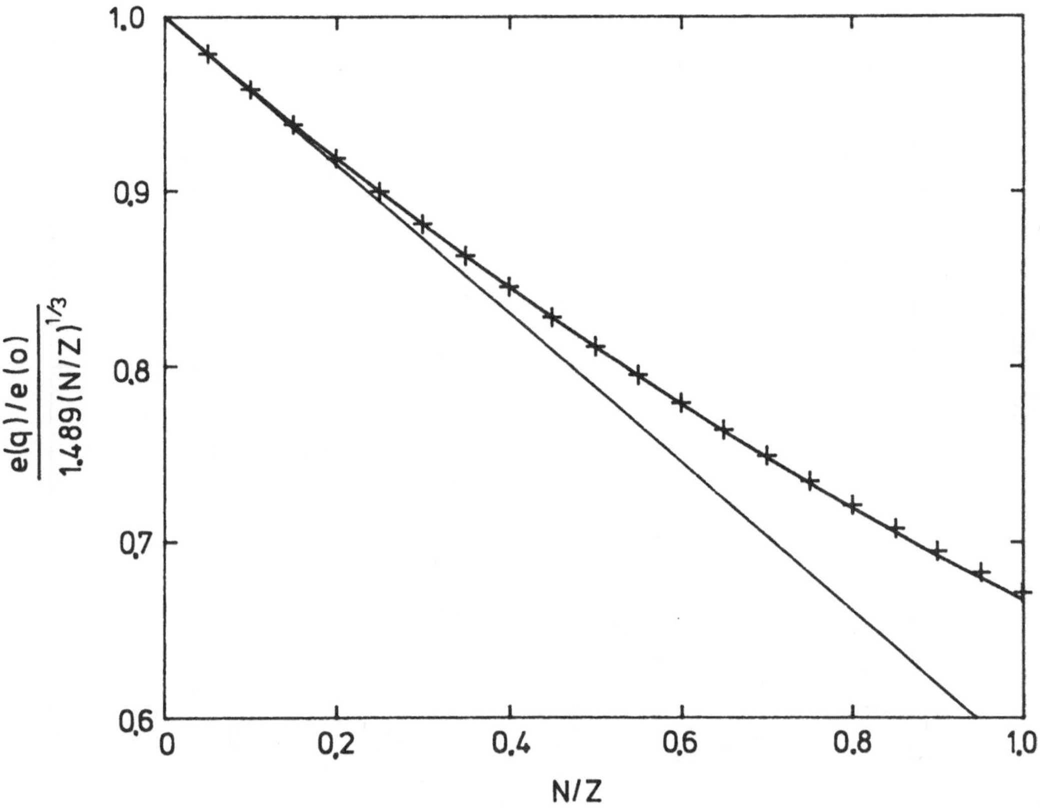

Fig.2-5. Comparison of the linear and quadratic approximations of Eq.(398) with the actual numbers (crosses).

range of $q^{\gamma/3}$. At this time, only the numerical value of $e_3 = 0.550086$ has been calculated[19], whereas e_4, e_5, ... are not known as yet. This value for e_3 leads to the dashed curve in Fig.6.

Validity of the TF model. The detailed discussion of the TF model, which touched upon all its important aspects, has made us familiar with the properties of TF atoms. In order to improve the description we must now find out what the deficiencies of the model are.

The approximations that define the model are those which brought us from Eq.(40) to Eq.(41). They are: (i) the (highly) semi-classical evaluation of the trace in E_{IP} according to the recipy of Eq. (1-43); and (ii) the disregard of electron-electron interactions except for the (direct) electrostatic one (in particular, we did not care for the exchange energy). Of the two, the first one is the more serious one, because it leads to an incorrect treatment of the most strongly

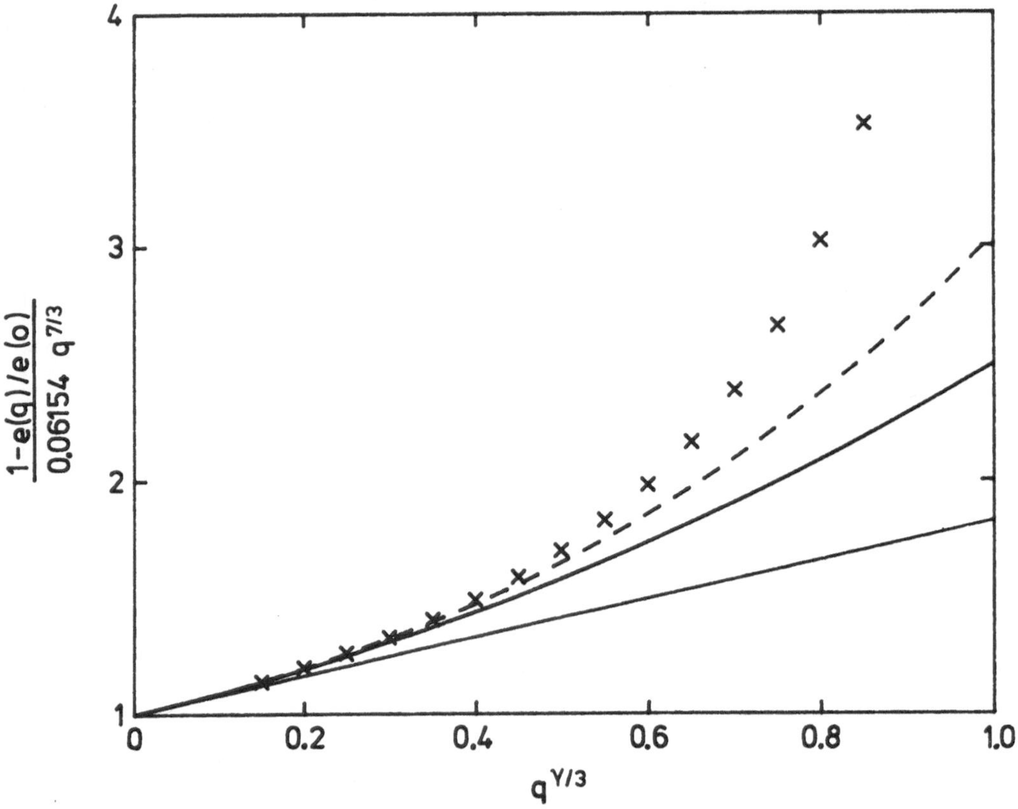

Fig.2-6. Comparison of the linear and quadratic approximations of Eq.(399) with the actual numbers (crosses). The dashed curve is the corresponding cubic approximation.

bound electrons, the ones close to the nucleus that contribute most to the energy. To make this point, let us recall that the application of Eq.(1-43)(i.e., the evaluation of traces of unordered operators by phase space integrals) is justified when commutator terms, as they appear in the ordering process, are negligible. In the present context this requires that the commutator of the momentum and the potential, which equals i times the gradient of the potential, be "small." Small compared to what? Physically, this gradient is small if the potential does not change significantly over the range important for an electron. Since the quantum standard of length, associated with an individual electron, is its deBroglie wave length, λ, a small gradient means

$$|\lambda \vec{\nabla} V| \ll |V| \quad . \tag{2-400}$$

Substantial changes in V occur on a scale set by the distance r, so that criterion (400) requires that

$$\lambda << r \quad . \tag{2-401}$$

On the other hand, λ is the inverse momentum (we ignore factors of two and pi for this kind of reasoning), which in turn is given by the square root of the potential, see Eq.(42). In short, we have, as criterion for the validity of the TF model, the relation

$$r \sqrt{|V+\zeta|} >> 1 \quad . \tag{2-402}$$

Upon introducing TF variables, this reads

$$Z^{1/3} \; |x \; f_q(x)|^{1/2} >> 1 \quad . \tag{2-403}$$

First, we learn here that, for a given x, the TF model is reliable only if Z is large enough. Second, there is information about the regions where the approximation cannot be trusted.

At short distances, $f_q(x)$ practically equals unity, and the left-hand side of (403) is of the order of one, when $x \sim Z^{-2/3}$, or $r \sim 1/Z$. Consequently, there is an inner region of strong binding where the TF approximation fails. Indeed, the innermost electrons are described incorrectly in the TF model.

Then, near the edge of the atom at $x=x_o$, $f_q(x)$ has the linear form of Eq.(329). Now the left-hand side of (403) is of the order of one, when $|x-x_o| \sim Z^{-2/3}/q^2$, or $|r-r_o| \sim 1/(Zq^2)$. Thus we find the outer region of weak binding to be also treated inadequately in the TF model. The situation is, of course not basically different for neutral atoms, although the argument has to be modified. For q=o, the TF function F(x) appears in (403). Its large-x form $F(x) \sim 1/x^3$ implies that the criterion is not satisfied, once x is of the order $Z^{1/3}$, or $r \sim 1$.

In Figs.7 and 8 plots of the radial densities

$$D(r) = 4\pi r^2 n(r) \tag{2-404}$$

are used to illustrate these observations concerning the validity of the TF model. Please note that the regions of failure shrink with increasing Z. We conlcude that (in some sense) the TF approximation becomes exact for $Z \to \infty$.[20]

Nice, but in the real world Z isn't that large, the more so $Z^{1/3}$, which obviously is the relevant parameter. It ranges merely from

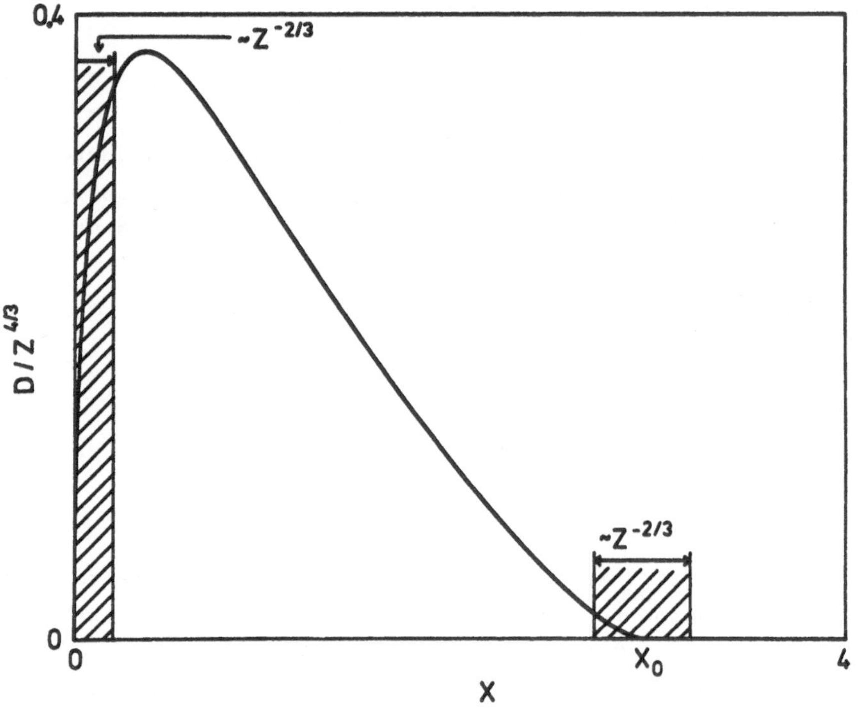

Fig.2-7. Regions of relability and failure of the TF model in an ionized atom (q=1/2), *illustrated by the radial density as a function of the TF variable* x.

one to roughly five over the whole Periodic Table. Clearly, modifications aimed at improving the TF model are called for. All following Chapters are devoted to their discussion. The TF atom is thereby the leading approximation, and the supplements to the TF model will all be regarded as small corrections. For this reason it was necessary to spend so much time with a detailed study of the TF model.

It is important to appreciate that the density, which was used in Figs.7 and 8, is the right quantity to plot for this purpose. The TF prediction (51)

$$\begin{aligned} n_{TF}(r) &= \frac{1}{3\pi^2}[-2(V+\zeta)]^{3/2} \\ &\cong \frac{1}{3\pi^2}(2Z/r)^{3/2} \quad \text{for} \quad r\to o \end{aligned} \tag{2-405}$$

is clearly very much in error at small distances. Also, for an ion of

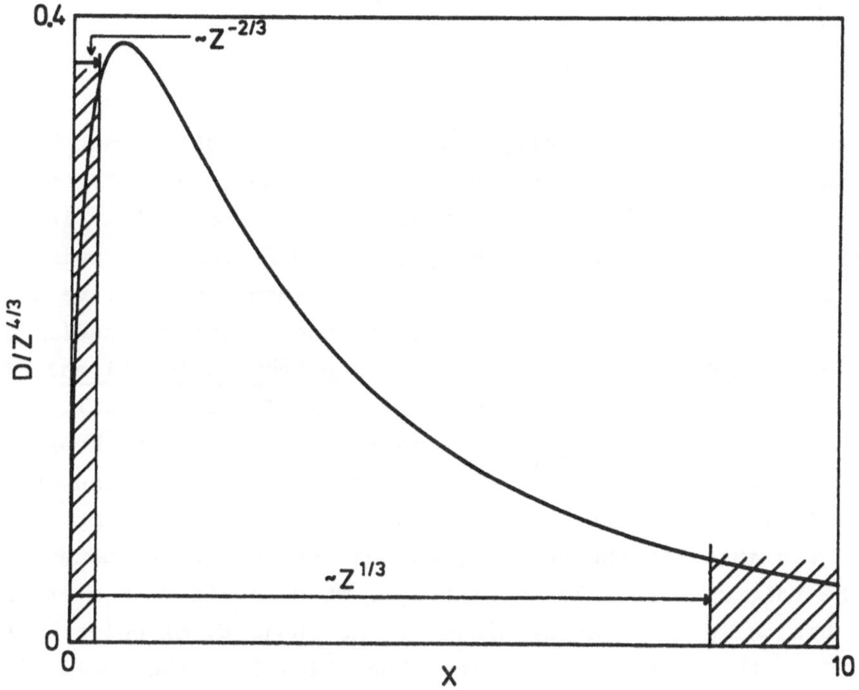

Fig. 2-8. Like Fig. 7, for a neutral atom.

degree of ionization q, we obtain

$$n_{TF}(r) \cong \frac{1}{3\pi^2}[2Zq(1-r/r_o)]^{3/2} \quad \text{for} \quad r \lesssim r_o \ ,$$

$$n_{TF}(r) = o \quad \text{for} \quad r > r_o \tag{2-406}$$

for the density around the edge of the atom. This is a sharp boundary instead of the quantum-mechanically correct smooth transition into the classically forbidden domain, where the real density decreases exponentially. In the situation of neutral atoms, the large-r behavior of the density is

$$n_{TF}(r) \cong \frac{1}{3\pi^2}[\, 2\frac{Z}{r}\,\frac{144}{x^3}\,]^{3/2} = \frac{243}{8}\pi\frac{1}{r^6} \ , \tag{2-407}$$

where the large-x form of F(x) is employed. Again, this is not the correct exponential dependence on the distance.

The principal lesson consists in stating that the real density is not of the form

$$n = n_{TF} + (\text{ a small correction}) \ . \tag{2-408}$$

As a consequence, the TF density functional of Eq.(95) cannot be used as the starting point when looking for corrections. In contrast, the TF potential is very much like the real effective potential, inasmuch as it behaves like $-Z/r$ for $r \to o$ and like $-(Z-N)/r$ for $r \to \infty$, being structureless in between. The structure is in the second derivative of the potential (related to the density), not in the potential itself. Therefore, we must find modifications of the TF potential functional of Eq. (45) in order to overcome the insufficiencies of the TF model. If this is so, why does the vast majority of people working on TF theory use the language of density functionals? As far as I can see, the reasons are historical ones. In the original work by Thomas and Fermi[21] the principal variable was the density, whereas the effective potential played the role of an auxiliary quantity. This remained so over the years in basically all presentations of the subject, of which Gombás' textbook[22] is the most prominent one. Then, in 1964, the socalled Hohenberg-Kohn theorem[23] (of which we shall sketch a proof in the next section) triggered the development of a density functional formalism. Because of this theorem, density functionals appear to be well founded theoretically, in contrast to formulations based upon the concept of the effective potential, which is widely regarded as an intuitive approach (our introduction certainly is in this spirit) lacking a "rigorous" theoretical foundation. In the following section, which continues the "general formalism" that we left after Eq.(40), we shall see that this preconception is wrong. The potential functional is as well defined as the density functional, and for the reasons given above it is the preferable formulation in atomic physics.

Density and potential functionals. For a proof of the aforementioned Hohenberg-Kohn theorem (we shall state it below), we return to the many particle Hamilton operator of Eq.(1-7), where we replace the nucleus-electron potential $-Z/r$ by an arbitrary external potential $V_{ext}(\vec{r})$, and split H_{mp} into the kinetic energy operator H_{kin}, and the interaction energy operators H_{ext} and H_{ee}:

$$H_{mp} = \sum_{j=1}^{N} \frac{1}{2} p_j^2 + \sum_{j=1}^{N} V_{ext}(r_j) + \frac{1}{2}\sum_{j,k=1}^{N}{}' \frac{1}{r_{jk}} =$$

$$= H_{kin} + H_{ext} + H_{ee} \quad . \tag{2-409}$$

Different ground states $|\psi_o\rangle$ will correspond to differing choices of V_{ext}. In order to simplify the argument, we shall assume that, except for the irrelevant possibility of a reorientation of all spins, the ground states are unique (a slight, and otherwise innocuous, change of the external potential would destroy any degeneracy anyhow). Thus, the density in the ground state,

$$\begin{aligned} n(\vec{r}') = &\int (d\vec{r}_2')(d\vec{r}_3')\ldots(d\vec{r}_N') \ |\langle\vec{r}',\vec{r}_2',\vec{r}_3',..,\vec{r}_N'|\psi_o\rangle|^2 \\ &+ \int (d\vec{r}_1')(d\vec{r}_3')\ldots(d\vec{r}_N') \ |\langle\vec{r}_1',\vec{r}',\vec{r}_3',..,\vec{r}_N'|\psi_o\rangle|^2 \\ &+ \ldots \\ &+ \int (d\vec{r}_1')\ldots(d\vec{r}_{N-1}') \ |\langle\vec{r}_1',\vec{r}_2',\ldots,\vec{r}_{N-1}',\vec{r}'|\psi_o\rangle|^2 \\ = &\ N\int (d\vec{r}_2')\ldots(d\vec{r}_N') \ |\langle\vec{r}',\vec{r}_2',..,\vec{r}_N'|\psi_o\rangle|^2 \end{aligned} \tag{2-410}$$

(the latter equality makes use of the antisymmetry of the wave function, and a trace affecting the spin indices only is understood implicitly), is a functional of the external potential V_{ext}.

Two different external potentials, V_{ext} and $\tilde{V}_{ext}$, will lead to two different ground states $|\psi_o\rangle$ and $|\tilde{\psi}_o\rangle$, since the respective Schrödinger equations are different. (The situation $\tilde{V}_{ext} = V_{ext}$ + const. is not interesting, since we consider only potentials that are <u>physically</u> different). The expectation values of H_{mp} and $\tilde{H}_{mp}$ are minimized by $|\psi_o\rangle$ and $|\tilde{\psi}_o\rangle$, respectively, so that

$$\langle\psi_o|H_{mp}|\psi_o\rangle - \langle\tilde{\psi}_o|\tilde{H}_{mp}|\tilde{\psi}_o\rangle < \langle\tilde{\psi}_o|H_{mp}-\tilde{H}_{mp}|\tilde{\psi}_o\rangle \tag{2-411}$$

and

$$\langle\psi_o|H_{mp}|\psi_o\rangle - \langle\tilde{\psi}_o|\tilde{H}_{mp}|\tilde{\psi}_o\rangle > \langle\psi_o|H_{mp}-\tilde{H}_{mp}|\psi_o\rangle \quad , \tag{2-412}$$

which are combined into

$$\langle\psi_o|H_{mp}-\tilde{H}_{mp}|\psi_o\rangle < \langle\tilde{\psi}_o|H_{mp}-\tilde{H}_{mp}|\tilde{\psi}_o\rangle \quad . \tag{2-413}$$

Now we insert

$$H_{mp} - \tilde{H}_{mp} = H_{ext} - \tilde{H}_{ext} = \sum_{j=1}^{N} (V_{ext}(\vec{r}_j) - \tilde{V}_{ext}(\vec{r}_j)) \qquad (2\text{-}414)$$

and obtain

$$\langle\psi_o|H_{mp} - \tilde{H}_{mp}|\psi_o\rangle = \sum_{j=1}^{N} \int (d\vec{r}_1{}') \ldots (d\vec{r}_N{}') (V_{ext}(\vec{r}_j{}') - \tilde{V}_{ext}(\vec{r}_j{}'))$$

$$\times \; |\langle\vec{r}_1{}', \ldots, \vec{r}_N{}'|\psi_o\rangle|^2$$

$$= \int (d\vec{r}\,') [V_{ext}(\vec{r}\,') - \tilde{V}_{ext}(\vec{r}\,')] n(\vec{r}\,') \quad , \qquad (2\text{-}415)$$

and likewise for $|\tilde{\psi}_o\rangle$. The implication of (413) is therefore

$$\int (d\vec{r}\,') [V_{ext}(\vec{r}\,') - \tilde{V}_{ext}(\vec{r}\,')] [n(\vec{r}\,') - \tilde{n}(\vec{r}\,')] < o \quad , \qquad (2\text{-}416)$$

from which we conclude that $n \neq \tilde{n}$. Different external potentials not only produce different ground states but also different ground-state densities. Consequently, a given n corresponds to a certain V_{ext} which is uniquely determined by n. In other words: V_{ext} is a functional of n. And since the ground state $|\psi_o\rangle$ is a functional of V_{ext}, it can be regarded as a functional of n as well. Then the expectation values of H_{kin} and H_{ee} in the ground states are also functionals of the density. Here then is the Hohenberg-Kohn theorem: there exist universal (i.e., independent of V_{ext}) functionals of the density $E_{kin}(n)$ and $E_{ee}(n)$, so that the ground-state energy equals

$$E(n) = E_{kin}(n) + \int (d\vec{r}\,') V_{ext}(\vec{r}\,') n(\vec{r}\,') + E_{ee}(n) \quad , \qquad (2\text{-}417)$$

where n is the ground-state density. The minimum property of $\langle\psi|H_{mp}|\psi\rangle$ implies that the energy E(n) is minimized by the correct ground-state density; trial densities $\tilde{n}$, which must be subject to the normalization

$$\int (d\vec{r}\,') \tilde{n}(\vec{r}\,') = N \quad , \qquad (2\text{-}418)$$

yield larger energies $E(\tilde{n})$ than the ground-state energy E(n). It is useful to include the constraint (418) into the energy functional by means of

$$E(n,\zeta) = E_{kin}(n) + \int (d\vec{r}\,') V_{ext}(\vec{r}\,') n(\vec{r}\,') + E_{ee}(n) -$$

$$-\zeta(\,N - \int(d\vec{r})n\,) \quad , \tag{2-419}$$

since this $E(n,\zeta)$ is stationary under arbitrary variations of both the density n and the Lagrangian multiplier ζ.

Before proceeding to construct the related potential functional, a few remarks are in order. The Hohenberg-Kohn theorem is a very general one; in particular, the specific forms of H_{kin} and H_{ee} never enter. The price for the generality is paid in form of a total lack of knowledge concerning the structure of the density functionals $E_{kin}(n)$ and $E_{ee}(n)$. The theorem states no more than their existence. Obviously, the detailed form of these functionals must depend upon the specific H_{kin} and H_{ee} that are investigated [one could, for instance, consider relativistic corrections to the kinetic energy, or, in applications to nuclear physics, reflect upon fermion-fermion interactions different from the Coulomb form of (409)]. Also, no technical procedure is known that would enable us to perform the step from H_{mp} to $E(n)$. One must rely upon some physical insight, when constructing functionals that approximate the actual $E(n)$.

The kinetic energy in the ground state of H_{mp} of (409) is the expectation value

$$E_{kin} = \langle\psi_o|H_{kin}|\psi_o\rangle = \langle\psi_o|\sum_{j=1}^{N}\frac{1}{2}p_j^2|\psi_o\rangle \quad , \tag{2-420}$$

which, in configuration space, appears as

$$\begin{aligned} E_{kin} &= \frac{1}{2}N\int(d\vec{r}\,')\int(d\vec{r}_2')\ldots(d\vec{r}_N')\vec{\nabla}'\psi_o^*(\vec{r}\,',\vec{r}_2',\ldots,\vec{r}_N') \\ &\qquad\cdot\vec{\nabla}'\psi_o(\vec{r}\,',\vec{r}_2',\ldots,\vec{r}_N') \\ &= \int(d\vec{r}\,')(d\vec{r}\,'')\frac{1}{2}\delta(\vec{r}\,'-\vec{r}\,'')\;\vec{\nabla}'\cdot\vec{\nabla}''\;n^{(1)}(\vec{r}\,';r'') \;. \end{aligned} \tag{2-421}$$

Here, once more, the antisymmetry of the wave function has been used, and we have introduced the one-particle density matrix

$$n^{(1)}(\vec{r}\,';\vec{r}\,'') = N\int(d\vec{r}_2')\ldots(d\vec{r}_N')\psi_o^*(\vec{r}\,'',\vec{r}_2',\ldots,\vec{r}_N')\psi_o(\vec{r}\,',\vec{r}_2',\ldots,\vec{r}_N') \quad , \tag{2-422}$$

which is an immediate generalization of (410), so that the density itself is the diagonal part of $n^{(1)}(\vec{r}\,';\vec{r}\,'')$,

$$n(\vec{r}') = n^{(1)}(\vec{r}';\vec{r}') \quad . \tag{2-423}$$

Let us now attempt to interpret $n^{(1)}(\vec{r}';\vec{r}")$ as the matrix element of an effective density operator,

$$n^{(1)}(\vec{r}';\vec{r}") = 2\langle\vec{r}'|\eta(-\tfrac{1}{2}p^2 - V(\vec{r}) - \zeta)|\vec{r}"\rangle \quad . \tag{2-424}$$

(A more careful discussion hereof will be presented in Chapter Four.) The effective potential $V(\vec{r})$ that appears here is unspecified at this stage, except for remarking that it is a functional of the density, $n(\vec{r})$, because the density matrix on the left-hand side is such a functional. The factor of two is the spin multiplicity which we now choose to make explicit instead of further assuming that a trace on spin indices is left implicit. Note that V is determined without the option of adding a constant, since Eq. (423) has to hold for the given density.

The diagonal version of Eq. (424) showed up earlier, in Eq. (20). We are clearly back to the picture of particles moving independently in an effective potential V. The notation established then is useful here, too. In particular, we introduce the independent particle Hamilton operator

$$H(\vec{r},\vec{p}) = \frac{1}{2}p^2 + V(\vec{r}) \tag{2-425}$$

just as in Eq. (3). The kinetic energy of (421) is then rewritten as

$$\begin{aligned} E_{kin} &= \int(d\vec{r}')(d\vec{r}")[\tfrac{1}{2}\vec{\nabla}'\cdot\vec{\nabla}"\ \delta(\vec{r}'-\vec{r}")]n^{(1)}(\vec{r}';\vec{r}") \\ &= \int(d\vec{r}')(d\vec{r}")\ \langle\vec{r}"|\tfrac{1}{2}p^2|\vec{r}'\rangle\ 2\ \langle\vec{r}'|\eta(-H-\zeta)|\vec{r}"\rangle \\ &= \mathrm{tr}\ \tfrac{1}{2}p^2\ \eta(-H-\zeta) \quad , \end{aligned} \tag{2-426}$$

where we remember that the trace operation includes multiplying by the spin factor. The quantity E_1 of Eq. (7),

$$E_1 = \mathrm{tr}(H+\zeta)\eta(-H-\zeta) \tag{2-427}$$

is a functional of $V+\zeta$, thus a functional of n, as $V=V(n)$. The kinetic energy (426) is contained in (427),

$$\begin{aligned} E_{kin} &= E_1 - \mathrm{tr}(V+\zeta)\eta(-H-\zeta) \\ &= E_1 - \int(d\vec{r}')(V(\vec{r}')+\zeta)n(\vec{r}') \quad . \end{aligned} \tag{2-428}$$

This we insert into (419) and arrive at

$$E(n,\zeta) = E_1(V+\zeta) - \int(d\vec{r}\,')(V(\vec{r}\,') - V_{ext}(\vec{r}\,'))n(\vec{r}\,') + E_{ee}(n) - \zeta N \quad . \tag{2-429}$$

In the present context, V is still regarded as a functional of n. Therefore, (429) is the same functional as in (419), we have done no more than reorganize the right-hand side. Consequently, the functional (429) is stationary under variations of n and ζ around their correct values, just as (419) is stationary. An infinitesimal variation of ζ induces a change in $E(n,\zeta)$ given by

$$\delta_\zeta E(n,\zeta) = (\frac{\partial}{\partial\zeta}E_1(V+\zeta) - N)\delta\zeta = o \quad ; \tag{2-430}$$

it is, indeed, zero for the same reasons that implied Eq.(13). Now consider a variation of the density:

$$\begin{aligned}\delta_n E(n,\zeta) &= \int(d\vec{r}\,')\delta_n V(\vec{r}\,')n(\vec{r}\,') - \int(d\vec{r}\,')\delta_n V(\vec{r}\,')n(\vec{r}\,') \\ &\quad - \int(d\vec{r}\,')(V(\vec{r}\,') - V_{ext}(\vec{r}\,'))\delta n(\vec{r}\,') \\ &\quad + \int(d\vec{r}\,')\delta n(\vec{r}\,')\; V_{ee}(\vec{r}\,') \\ &= \int(d\vec{r}\,')[-V(\vec{r}\,') + V_{ext}(\vec{r}\,') + V_{ee}(\vec{r}\,')]\delta n(\vec{r}\,') \quad ,\end{aligned} \tag{2-431}$$

where Eqs.(14) and (25) have been employed. The stationary property of $E(n,\zeta)$ thus implies

$$V = V_{ext} + V_{ee} \quad . \tag{2-432}$$

In words: the effective potential equals the sum of the external potential and the effective interaction potential, V_{cc}, defined by Eq.(25),

$$\delta_n E_{ee}(n) = \int(d\vec{r}\,')\delta n(\vec{r}\,')V_{ee}(\vec{r}\,') \quad . \tag{2-433}$$

Note in particular, that V is always a local (i.e., momentum independent) potential.

So far, V has been regarded as a functional of the density. Because of the circumstance that the contributions in (431), that originate in variations of V, take care of themselves, we can equally well treat V as an independent variable. The energy functional

$$E(V,n,\zeta) = E_1(V+\zeta) - \int (d\vec{r}')(V-V_{ext})n + E_{ee}(n) - \zeta N \qquad (2\text{-}434)$$

is obviously stationary under independent infinitesimal variations of V,n, and ζ. If we do not want to have both V and n as independent quantities, we have the option of eliminating one of the two. The step from (434) back to (419) is done by first solving Eq.(20) for V, thereby expressing the potential in terms of the density, and then using this V(n) in (434). Likewise, to obtain a functional of the potential alone, one has to use Eq.(432), in which V_{ee} is a functional of the density, to express n as a functional of V. This n(V) then eliminates the density from (434) leaving us with a potential functional $E(V,\zeta)$.

Let us illustrate these ideas with the respective TF functionals. Starting from

$$E_{TF}(V,n,\zeta) = \int (d\vec{r})\left(-\frac{1}{15\pi^2}\right)[-2(V+\zeta)]^{5/2} - \int (d\vec{r})\left(V+\frac{Z}{r}\right)n$$

$$+ \frac{1}{2}\int (d\vec{r})(d\vec{r}')\,\frac{n(\vec{r})n(\vec{r}')}{|\vec{r}-\vec{r}'|} - \zeta N \quad , \qquad (2\text{-}435)$$

where V_{ext} is now the potential energy of an electron with the nucleus, $-Z/r$, we get the density funcitonal of Eq.(95), after first inverting [Eq.(51)]

$$n = \frac{1}{3\pi^2}[-2(V+\zeta)]^{3/2} \qquad (2\text{-}436)$$

to

$$V = -\frac{1}{2}(3\pi^2 n)^{2/3} - \zeta \quad , \qquad (2\text{-}437)$$

which then allows to rewrite the first and second term in (435) accordingly. On the other hand, if

$$V = -\frac{Z}{r} + \int (d\vec{r}')\,\frac{n(\vec{r}')}{|\vec{r}-\vec{r}'|} \qquad (2\text{-}438)$$

is solved for n,

$$n = -\frac{1}{4\pi}\nabla^2\left(V+\frac{Z}{r}\right) \qquad (2\text{-}439)$$

(this is, of course, Poisson's equation), we can eliminate n from the second and third term in (435) and are led to the TF potential functional of Eq.(45). Of course, within the framework of the TF model, the three functionals $E(V,n,\zeta)$, $E(n,\zeta)$, and $E(V,\zeta)$ are perfectly equivalent, but I repeat: as a basis for improvements over the TF approximation, the potential functional is the preferable one.

We have investigated earlier the scaling properties of the TF model. Let us now see, what one can state about the behavior of the exact density functionals $E_{kin}(n)$ and $E_{ee}(n)$ under scale transformations of the density,

$$n(\vec{r}')\rightarrow n_\mu(\vec{r}') = \mu^3\, n(\mu\vec{r}') \quad . \tag{2-440}$$

The TF approximations to E_{kin} and E_{ee} scale in the manner that one would intuitively expect:

$$\begin{aligned}\left(E_{kin}(n)\right)_{TF} &= \int(d\vec{r}')\,\frac{1}{10\pi^2}(3\pi^2 n(\vec{r}'))^{5/3}\\ &\rightarrow \mu^{-3}\int(d\mu\vec{r}')\,\frac{1}{10\pi^2}(3\pi^2\mu^3 n(\mu\vec{r}'))^{5/3}\\ &= \mu^2\left(E_{kin}(n)\right)_{TF} \quad ,\end{aligned} \tag{2-441}$$

and likewise

$$\begin{aligned}\left(E_{ee}(n)\right)_{TF} &= \frac{1}{2}\int(d\vec{r}')(d\vec{r}'')\,\frac{n(\vec{r}')n(\vec{r}'')}{|\vec{r}'-\vec{r}''|}\\ &\rightarrow \mu\left(E_{ee}(n)\right)_{TF} \quad .\end{aligned} \tag{2-442}$$

For the exact functionals, the equations

$$E_{kin}(n_\mu) = \mu^2\, E_{kin}(n) \tag{2-443}$$

and

$$E_{ee}(n_\mu) = \mu\, E_{ee}(n) \tag{2-444}$$

do <u>not</u> hold, however; even their combination

$$E_{kin}(n_\mu) + E_{ee}(n_\mu) = \mu^2\, E_{kin}(n) + \mu\, E_{ee}(n) \tag{2-445}$$

is only true if $\mu=1+\varepsilon$ with an infinitesimal ε. This surprising observation has been made only recently, by Levy and Perdew.[24] Please note that the statement (445) is, indeed, only needed for such $\mu\cong 1$, in order to derive the virial theorem

$$2\,E_{kin}(n) = -\,E_{ee}(n) + \int(d\vec{r}\,')n(\vec{r}\,')\vec{r}\,'\cdot\vec{\nabla}'\,V_{ext}(\vec{r}\,') \qquad (2\text{-}446)$$

from the minimum property of the density functional (417).

As a first step towards proving these remarks about Eqs.(443) through (445), we recall that to any given density there correspond uniquely a certain external potential and a certain ground state. Let us keep the notation V_{ext} and $|\psi_o\rangle$ for the ones related to the actual ground-state density n, and write V^{μ}_{ext} and $|\psi_o^{\mu}\rangle$ for the ones that go with the scaled density n_μ. Thus $|\psi_o^{\mu}\rangle$ obeys

$$(H_{kin} + H^{\mu}_{ext} + H_{ee})|\psi_o^{\mu}\rangle = E_o^{\mu}|\psi_o^{\mu}\rangle \quad , \qquad (2\text{-}447)$$

where E_o^{μ} is the ground-state energy for n_μ. Clearly, if we transform $|\psi_o\rangle$ according to

$$\langle\vec{r}_1',\ldots,\vec{r}_N'|\psi_o\rangle \rightarrow \mu^{3N/2}\,\langle\mu\vec{r}_1',\ldots,\mu\vec{r}_N'|\psi_o\rangle \quad , \qquad (2\text{-}448)$$

then the density is scaled as in Eq.(440). Inasmuch as

$$\mu^{3N/2}\,\langle\mu\vec{r}_1',\ \ldots,\ \mu\vec{r}_N'| = \langle\vec{r}_1',\ \ldots,\vec{r}_N'|U(\mu) \quad , \qquad (2\text{-}449)$$

where the unitary operator $U(\mu)$ is given by

$$U(\mu) = \exp\{\,i\,\tfrac{1}{2}\sum_{j=1}^{N}(\vec{r}_j\cdot\vec{p}_j+\vec{p}_j\cdot\vec{r}_j)\ \log\mu\,\} \quad , \qquad (2\text{-}450)$$

we can read (448) as

$$|\psi_o\rangle \rightarrow U(\mu)|\psi_o\rangle \quad . \qquad (2\text{-}451)$$

The point is that this scaled $|\psi_o\rangle$ is not equal to $|\psi_o^{\mu}\rangle$. This emerges from considering the Schrödinger equation obeyed by $U(\mu)|\psi_o\rangle$, which is immediately obtained from the one satisfied by $|\psi_o\rangle$. We have

$$U(\mu)(H_{kin} + H_{ext} + H_{ee})\,U^{-1}(\mu)\,U(\mu)|\psi_o\rangle = E_oU(\mu)|\psi_o\rangle. \qquad (2\text{-}452)$$

The action of U upon $\vec{r}_j$ and $\vec{p}_j$ is simply

$$U(\mu)\ \vec{r}_j\ U^{-1}(\mu) = \mu\ \vec{r}_j \quad ,$$
$$U(\mu)\ \vec{p}_j\ U^{-1}(\mu) = \frac{1}{\mu}\ \vec{p}_j \quad , \tag{2-453}$$

so that

$$U(\mu)\ H_{kin}\ U^{-1}(\mu) = \frac{1}{\mu^2}\ H_{kin} \quad ,$$
$$U(\mu)\ H_{ee}\ U^{-1}(\mu) = \frac{1}{\mu}\ H_{ee} \quad , \tag{2-454}$$

and

$$U(\mu)\ H_{ext}\ U^{-1}(\mu) = \sum_{j=1}^{N} V_{ext}(\mu\vec{r}_j) \quad . \tag{2-455}$$

Consequently,

$$(H_{kin} + \mu^2 U(\mu) H_{ext} U^{-1}(\mu) + \mu H_{ee}) U(\mu)|\psi_o\rangle = \mu^2 E_o U(\mu)|\psi_o\rangle \ , \tag{2-456}$$

which, in view of the factor μ multiplying H_{ee}, is not of the form required for $|\psi_o^\mu\rangle$ in Eq.(447). Thus, indeed

$$U(\mu)|\psi_o\rangle \neq |\psi_o^\mu\rangle \quad , \tag{2-457}$$

for $\mu\neq 1$. Nevertheless, $|\psi_o^\mu\rangle$ and $U(\mu)|\psi_o\rangle$ are not unrelated. In particular, they give rise to the same density, $n_\mu(\vec{r})$, when inserted into (410). This implies the equality

$$\langle\psi_o^\mu\ |\ H_{ext}^\mu\ |\ \psi_o^\mu\rangle = \langle\psi_o|\ U^{-1}(\mu)\ H_{ext}^\mu\ U(\mu)|\psi_o\rangle \quad , \tag{2-458}$$

and for the same reason

$$\langle\psi_o\ |\ H_{ext}\ |\psi_o\rangle = \langle\psi_o^\mu\ |\ U(\mu)\ H_{ext}\ U^{-1}(\mu)|\psi_o^\mu\rangle \quad . \tag{2-459}$$

We are now prepared to employ the minimum property of the expectation value of the Hamilton operator of Eq.(447) in the form

$$\langle\psi_o^\mu\ |(H_{kin} + H_{ext}^\mu + H_{ee})|\psi_o^\mu\rangle \ \leq$$

$$\leq \langle\psi_o| \; U^{-1}(\mu)(H_{kin}+H_{ext}^{\mu}+H_{ee})U(\mu)|\psi_o\rangle \quad , \qquad (2\text{-}460)$$

which, as a consequence of Eqs.(454) and (458), says

$$E_{kin}(n_\mu) + E_{ee}(n_\mu) \leq \mu^2 E_{kin}(n) + \mu E_{ee}(n) \quad . \qquad (2\text{-}461)$$

The equal sign holds only for $\mu=1$, in the first place. Since the right hand side always exceeds the left hand one for $\mu\neq1$, however, the two sides must agree up to first order in $\varepsilon=\mu-1$, at least, so that the equal sign actually applies to $\mu=1+\varepsilon$ with an infinitesimal ε. This is the statement we made at Eq.(445). Another way of expressing the same fact is

$$\frac{d}{d\mu}[E_{kin}(n_\mu) + E_{ee}(n_\mu)]\Big|_{\mu=1} = 2\,E_{kin}(n) + E_{ee}(n) \quad . \qquad (2\text{-}462)$$

We can also exploit the minimum property of the expectation value of the Hamilton operator of Eq.(456). Here we have

$$\langle\psi_o \,|\, U^{-1}(\mu)(H_{kin}+ \mu^2\, U(\mu)H_{ext}\, U^{-1}(\mu) + \mu H_{ee})U(\mu)\,|\psi_o\rangle$$

$$\leq \langle\psi_o^{\mu}\,|(H_{kin}+ \mu^2 U(\mu)H_{ext}\, U^{-1}(\mu)+\mu H_{ee})\,|\psi_o^{\mu}\rangle \quad , \qquad (2\text{-}463)$$

or with (454) and (459),

$$\mu^2\, E_{kin}(n) + \mu^2\, E_{ee}(n) \leq E_{kin}(n_\mu) + \mu\, E_{ee}(n_\mu) \quad , \qquad (2\text{-}464)$$

where, again, the equal sign is true for all μ's that differ from unity at most infinitesimally.

Equations (461) and (464) can be combined into two statements about E_{kin} and E_{ee} individually, namely

$$(\mu-1)[E_{ee}(n_\mu) - \mu\, E_{ee}(n)] \geq o \qquad (2\text{-}465)$$

and

$$(\mu-1)[E_{kin}(n_\mu) - \mu^2\, E_{kin}(n)] \leq o \quad . \qquad (2\text{-}466)$$

It seems natural to assume that the left-hand sides in (465) and (466) are of second order in $\varepsilon=\mu-1$ for small ε. If this were true, these equations would mean that

$$\frac{d}{d\mu} E_{ee}(n_\mu)\Big|_{\mu=1} - E_{ee}(n) > o \tag{2-467}$$

and

$$\frac{d}{d\mu} E_{kin}(n_\mu)\Big|_{\mu=1} - 2\, E_{kin}(n) < o \quad . \tag{2-468}$$

As a matter of fact, we shall see below that equal signs have to be written in (467) and (468) instead of ">" and "<". Consequently, the left-hand sides in (465) and (466) are, at least, of order ε^4, the respective square brackets of order ε^3. Therefore, also in Eq.(461) the equality sign holds up to order ε^3, at least. These remarks go beyond the results of Ref.24, where Levy and Perdew stopped at stating (465) and (466).

For a proof of what has just been said, we have to turn to the potential functional $E_1(V+\zeta)$. In Eq.(216) we found that the TF approximation to E_1 responds like

$$\left(E_1(V+\zeta)\right)_{TF} \rightarrow \mu^{5\nu/2-3} \left(E_1(V+\zeta)\right)_{TF} \quad , \tag{2-469}$$

when V and ζ are scaled according to Eqs.(211) and (214),

$$V(\vec{r}) \rightarrow \mu^\nu\, V(\mu\vec{r}) \quad , \quad \zeta \rightarrow \mu^\nu\, \zeta \quad . \tag{2-470}$$

Although the exact $E_1(V+\zeta)$ does not behave like (469) for arbitrary ν, it does so for $\nu=2$:

$$E_1(V+\zeta) \rightarrow \mu^2\, E_1(V+\zeta) \tag{2-471}$$

for

$$\begin{aligned} V(\vec{r}) &\rightarrow V_\mu(\vec{r}) = \mu^2\, V(\mu\vec{r}) \quad , \\ \zeta &\rightarrow \mu^2\, \zeta \quad . \end{aligned} \tag{2-472}$$

We demonstrate this by first observing that

$$\begin{aligned} &\eta(-\tfrac{1}{2}\, p^2 - \mu^2\, V(\mu\vec{r}) - \mu^2\, \zeta) \\ &= \eta(-\tfrac{1}{2}(\vec{p}/\mu)^2 - V(\mu\vec{r}) - \zeta) \\ &= U(\mu)\, \eta(-\tfrac{1}{2}\, p^2 - V(\vec{r}) - \zeta)\, U^{-1}(\mu) \quad , \end{aligned} \tag{2-473}$$

where $U(\mu)$ now denotes the one-particle version of (450),

$$U(\mu) = \exp\{i\,\tfrac{1}{2}(\vec{r}\cdot\vec{p} + \vec{p}\cdot\vec{r})\log\mu\} \quad . \tag{2-474}$$

This is used in

$$\begin{aligned} E_1(V+\zeta) &= \mathrm{tr}(\tfrac{1}{2}p^2 + V(\vec{r})+\zeta)\eta(-\tfrac{1}{2}p^2 - V(\vec{r})-\zeta) \\ &\to \mathrm{tr}(\tfrac{1}{2}p^2 + \mu^2 V(\mu\vec{r})+\mu^2\zeta)\eta(-\tfrac{1}{2}p^2 - \mu^2 V(\mu\vec{r})-\mu^2\zeta) \\ &= \mu^2 \mathrm{tr}\, U(\mu)(\tfrac{1}{2}p^2 + V(\vec{r})+\zeta)\eta(-\tfrac{1}{2}p^2 - V(\vec{r})-\zeta)U^{-1}(\mu) \\ &= \mu^2\, E_1(V+\zeta) \quad , \end{aligned} \tag{2-475}$$

or

$$E_1(V_\mu + \mu^2\zeta) = \mu^2\, E_1(V+\zeta) \quad , \tag{2-476}$$

indeed. [The invariance of the trace under cyclic permutations has been employed in the last step of (475).]

Before proceeding, it is instructive to show where the attempt of repeating the argument for $\nu\neq 2$ fails. The analog of (473) would require an operator (not necessarily a unitary one), $U_\nu(\mu)$, such that

$$U_\nu(\mu)\,\vec{r}\,U_\nu^{-1}(\mu) = \mu\vec{r} \quad , \tag{2-477}$$

$$U_\nu(\mu)\,\vec{p}\,U_\nu^{-1}(\mu) = \mu^{-\nu/2}\,\vec{p} \quad .$$

Unfortunately, there is no such operator, except for $\nu=2$, as emerges from considering the commutator of the transformed quantities:

$$\mu^{1-\nu/2}\, i\,\overleftrightarrow{1} = [\mu\vec{r}, \mu^{-\nu/2}\vec{p}] = U_\nu[\vec{r},\vec{p}]U_\nu^{-1} = i\,\overleftrightarrow{1} \quad . \tag{2-478}$$

This is a contradiction, unless $\nu=2$.

In the section about the scaling properties of the TF model we remarked that a scaling transformation of the effective potential $V(\vec{r})$ must be accompanied by a corresponding transformation of the external potential $V_{ext}(\vec{r})$. In that earlier context, this was achieved by changing Z appropriately [Eq. (213)], because the only V_{ext} considered then was the Coulomb potential $-Z/r$. In the more general present discussion, we preserve the structure $V-V_{ext}$ by scaling V_{ext} like V in Eq. (473),

$$V_{ext}(\vec{r}) \to V_{ext,\mu}(\vec{r}) = \mu^2\, V_{ext}(\mu\vec{r}) \quad . \tag{2-479}$$

The density is, of course, scaled as in (440). Under these simultaneous transformations of V,n,ζ, and V_{ext}, the potential-density functional of (434) behaves as described by

$$E(V,n,\zeta) \to E_\mu(V,n,\zeta) = E(V_\mu,n_\mu,\mu^2\zeta)$$
$$= \mu^2\{E_1(V+\zeta) - \int(d\vec{r}\,')(V(\vec{r}\,') - V_{ext}(\vec{r}\,'))n(\vec{r}\,')-\zeta N\}$$
$$+ E_{ee}(n_\mu) \quad . \tag{2-480}$$

Since E(V,n,ζ) is stationary under infinitesimal variations of V,n, and ζ, all first order changes must originate in the scaling of V_{ext}. [The same argument was also applied to $E_{TF}(\mu)$ of Eq.(219).] Thus,

$$\frac{d}{d\mu} E_\mu(V,n,\zeta)\Big|_{\mu=1} = \int(d\vec{r}\,')n(\vec{r}\,') \frac{\partial}{\partial\mu} V_{ext,\mu}(\vec{r}\,')\Big|_{\mu=1} \quad , \tag{2-481}$$

or with (479),

$$\frac{d}{d\mu} E_\mu(V,n,\zeta)\Big|_{\mu=1} = \int(d\vec{r}\,')n(\vec{r}\,')[2\,V_{ext}(\vec{r}\,')+\vec{r}\,'\cdot\vec{\nabla}'V_{ext}(\vec{r}\,')]. \tag{2-482}$$

On the other hand, Eq.(480) implies

$$\frac{d}{d\mu} E_\mu(V,n,\zeta)\Big|_{\mu=1} = 2\,\{E_1(V+\zeta)- \int(d\vec{r}\,')(V(\vec{r}\,')-V_{ext}(\vec{r}\,'))$$
$$\times\, n(\vec{r}\,')-\zeta N\}+ \frac{d}{d\mu} E_{ee}(n_\mu)\Big|_{\mu=1} \quad . \tag{2-483}$$

The equivalence of these two right-hand sides, combined with the virial theorem (446), yields

$$\frac{d}{d\mu} E_{ee}(n_\mu)\Big|_{\mu=1} - E_{ee}(n) = 2\,E_{kin}(n)-2\{E_1(V+\zeta)-\int(d\vec{r}\,')Vn-\zeta N\} \quad . \tag{2-484}$$

The last step consists of recognizing that for the actual V,n, and ζ, the contents of the curly brackets equals the kinetic energy. This emerges from Eq.(428). Consequently, the right-hand side is zero. We arrive at

$$\frac{d}{d\mu} E_{ee}(n_\mu)\Big|_{\mu=1} = E_{ee}(n) \quad , \tag{2-485}$$

and as a consequence of (462),

$$\frac{d}{d\mu} E_{kin}(n_\mu)\Big|_{\mu=1} = 2\, E_{kin}(n) \quad . \tag{2-486}$$

Indeed, the statements following Eq.(468) are justified.

Please be aware of the following mental trap. If the density is eliminated from $E(V,n,\zeta)$, so that we are left with the potential functional $E(V,\zeta)$, one could think that the resulting kinetic energy,

$$E_{kin}(V,\zeta) = E_1(V+\zeta) - \int(d\vec{r}\,')V(\vec{r}\,')n(\vec{r}\,') - \zeta N \tag{2-487}$$

scales according to

$$E_{kin}(V_\mu,\mu^2\zeta) = \mu^2\, E_{kin}(V,\zeta) \quad , \tag{2-488}$$

inasmuch as [Eq.(14)]

$$\delta_V\, E_1(V+\zeta) = \int(d\vec{r}\,')\delta V(\vec{r}\,')n(\vec{r}\,') \quad , \tag{2-489}$$

together with (471), implies

$$n(\vec{r}\,') \rightarrow \mu^3\, n(\mu\vec{r}\,') \quad , \tag{2-490}$$

if V and ζ are scaled as in (472). This is <u>not</u> so, however, because the potential functional that is to be inserted into Eq.(487) for $n(\vec{r}\,')$ is not the one obtained from (489), but the one that emerges from

$$\delta_n\, E_{ee}(n) = \int(d\vec{r}\,')\delta n(\vec{r}\,')[V(\vec{r}\,') - V_{ext}(\vec{r}\,')] \tag{2-491}$$

[Eqs.(432) and (433)]. In the TF approximation, for instance, this is the Poisson equation

$$n(\vec{r}\,') = -\frac{1}{4\pi}\nabla'^2(V(\vec{r}\,') - V_{ext}(\vec{r}\,')) \tag{2-492}$$

in which the scaling of V and V_{ext} [Eqs.(472) and (479)] produces

$$n(\vec{r}\,') \rightarrow \mu^4\, n(\mu\vec{r}\,') \quad , \tag{2-493}$$

different from the desired form of (490). Therefore, Eq.(488) is not true, not even for μ's that differ from unity by an infinitesimal amount.

The kinetic energy by itself is not a central quantity in the potential functional formalism. What we have just seen is an illustration of this remark.

Relation between the TF approximation and Hartree's method. Somewhere at the beginning of Chapter One there is the promise to discuss the connection between TF theory and HF theory "to some extent in Chapter Two." This time has finally come.

Hartree's[25] basic idea consists in approximating the ground-state wave-function by a product in which each factor refers to just one of the electrons:

$$\langle \vec{r}_1{}', \vec{r}_2{}', \ldots, \vec{r}_N{}' | \psi_o \rangle \cong \psi_1(\vec{r}_1)\psi_2(\vec{r}_2)\ldots\psi_N(\vec{r}_N) \quad . \tag{2-494}$$

The ψ_j's are supposed to be orthonormal,

$$\int (d\vec{r}\,')\ \psi_j{}^*(\vec{r}\,')\ \psi_k(\vec{r}\,') = \delta_{jk} \quad , \tag{2-495}$$

so that the wave function (494) is properly normalized to unity. The requirement of antisymmetry is not satisfied by (494). Consequently, exchange effects are not treated correctly. In the present context, where we want to make contact with the original TF model, neglecting exchange is consistent. We are actually talking about Hartree's approximation, not about the Hartree-Fock model, which does include exchange. This restriction is not essential for the discussion. The argument can be repeated for a comparison of HF theory with the proper extension of the TF model that includes the exchange interaction, which will be derived in Chapter Four. At this moment we are content with the simple TF model and Hartree's ansatz (494).

With Eq. (494) we obtain approximations to the expectation values of the three parts of the many-particle Hamilton operator (409). These are given by

$$\begin{aligned} E_{kin} &= \langle \psi_o | H_{kin} | \psi_o \rangle \\ &\cong \sum_{j=1}^{N} \int (d\vec{r}\,')\ \frac{1}{2}\ \vec{\nabla}'\psi_j{}^*(\vec{r}\,')\cdot\vec{\nabla}'\psi_j(\vec{r}\,') \quad , \end{aligned} \tag{2-496}$$

and

$$E_{ext} = \langle \psi_o | H_{ext} | \psi_o \rangle \cong$$

$$\cong \sum_{j=1}^{N} \int (d\vec{r}\,') \ \psi_j{}^*(\vec{r}\,') \ V_{ext}(\vec{r}\,') \ \psi_j(\vec{r}\,') \tag{2-497}$$

as well as

$$E_{ee} = \langle\psi_o \mid H_{ee} \mid \psi_o\rangle$$

$$\cong \sum_{j=1}^{N} \tfrac{1}{2}\int (d\vec{r}\,')\,\psi_j{}^*(\vec{r}\,')\,[\sum_{\substack{k=1\\k\neq j}}^{N} \int (d\vec{r}\,'')\,\psi_k{}^*(\vec{r}\,'')\frac{1}{|\vec{r}\,'-\vec{r}\,''|}\psi_k(\vec{r}\,'')]$$

$$\times\ \psi_j(\vec{r}\,') \quad . \tag{2-498}$$

Since the description does not pay attention to the exchange energy, we do not have to be pedantic either when it comes to excluding the self-energy. In other words: it is perfectly consistent to include the k=j term in Eq.(498). The approximation to the ground-state energy is then

$$E = \langle\psi_o|H_{mp}|\psi_o\rangle \cong E_{Hartree}$$

$$= \sum_{j=1}^{N} \int (d\vec{r}\,')\,\{\tfrac{1}{2}\vec{\nabla}'\psi_j{}^*(\vec{r}\,')\cdot\vec{\nabla}'\psi_j(\vec{r}\,')+\psi_j{}^*(\vec{r}\,')V_{ext}(\vec{r}\,')\psi_j(\vec{r}\,')$$

$$+\ \tfrac{1}{2}\psi_j{}^*(\vec{r}\,')\,[\sum_{k=1}^{N}\int (d\vec{r}\,'')\,\psi_k{}^*(\vec{r}\,'')\frac{1}{|\vec{r}\,'-\vec{r}\,''|}\ \psi_k(\vec{r}\,'')]\,\psi_j(\vec{r}\,')\} \quad . \tag{2-499}$$

The as yet undetermined ψ_j's are now chosen such that $E_{Hartree}$ is stationary under infinitesimal variations of them. Thus

$$\sum_{j=1}^{N}\int (d\vec{r}\,')\,\delta\psi_j{}^*(\vec{r}\,')\,\{-\ \tfrac{1}{2}\nabla'^{\,2} + V_{ext}(\vec{r}\,') + \sum_{k=1}^{N}\int (d\vec{r}\,'')\,\psi_k{}^*(\vec{r}\,'')\frac{1}{|\vec{r}\,'-\vec{r}\,''|}$$

$$\times\ \psi_k(\vec{r}\,'')\}\,\psi_j(\vec{r}\,') = o \quad . \tag{2-500}$$

The variations $\delta\psi_j{}^*$ are not arbitrary but subject to

$$\int (d\vec{r}\,')\,\delta\psi_j{}^*(\vec{r}\,') \ \psi_k(\vec{r}\,') = o \quad , \tag{2-501}$$

which is a consequence of the orthonormalization (495). Therefore, Eq. (500) implies

$$\{-\tfrac{1}{2}\nabla'^2 + V_{ext}(\vec{r}') + \sum_{k=1}^{N} \int (d\vec{r}'') \psi_k{}^*(\vec{r}'') \frac{1}{|\vec{r}'-\vec{r}''|} \psi_k(\vec{r}'')\} \psi_j(\vec{r}')$$

$$= \sum_{\ell=1}^{N} \varepsilon_{j\ell}\, \psi_\ell(\vec{r}') \quad , \qquad (2\text{-}502)$$

where the constants $\varepsilon_{j\ell}$ are the Lagrange mulitpliers of the constraints (495). The single-particle wave-functions ψ_j and the $\varepsilon_{j\ell}$ are to be determined simultaneously from Eqs. (502) and (495).

The hermitian property of the differential operator $\{\ldots\}$ in (502) is employed in demonstrating that the matrix $(\varepsilon_{j\ell})$ is hermitian:

$$\varepsilon_{j\ell} = \int (d\vec{r}')\ \psi_\ell{}^*(\vec{r}')\{\ldots\}\ \psi_j(\vec{r}')$$

$$= \int (d\vec{r}')\ \psi_j(\vec{r}')\{\ldots\}\ \psi_\ell{}^*(\vec{r}') = \varepsilon^*_{\ell j} \quad . \qquad (2\text{-}503)$$

Another observation is the nonuniqueness of the solution to (502) and (495). If ψ_j and $\varepsilon_{j\ell}$ are one solution, then

$$\tilde{\psi}_j = \sum_{\ell=1}^{N} u_{j\ell}\ \psi_\ell \qquad (2\text{-}504)$$

and

$$\tilde{\varepsilon}_{j\ell} = \sum_{k,m=1}^{N} u_{jk}\ \varepsilon_{km}\ u^*_{\ell m} \qquad (2\text{-}505)$$

is another one, whereby $(u_{j\ell})$ is any unitary matrix,

$$\sum_{m=1}^{N} u^*_{mj}\ u_{mk} = \delta_{jk} \quad . \qquad (2\text{-}506)$$

It is essential here that the density, that appears in (502), is invariant under such a unitary transformation:

$$\tilde{n}(\vec{r}') = \sum_{k=1}^{N} \tilde{\psi}_k{}^*(\vec{r}')\ \tilde{\psi}_k(\vec{r}') \quad =$$

$$= \sum_{k=1}^{N} \psi_k^*(\vec{r}\,') \; \psi_k(\vec{r}\,') = n(\vec{r}\,') \quad . \tag{2-507}$$

[The approximate wave-function (494) is obviously not invariant under (504). This is nothing to worry about, because as soon as (494) is antisymmetrized, the effect of (504) reduces to the mulitplication by a phase-factor.]

Since $(\varepsilon_{j\ell})$ is hermitian, we can choose $(u_{j\ell})$ such that $(\tilde{\varepsilon}_{j\ell})$ is diagonal,

$$\tilde{\varepsilon}_{j\ell} = \tilde{\varepsilon}_j \; \delta_{j\ell} \quad . \tag{2-508}$$

Then Eq.(502) is Schrödinger's equation in appearance,

$$\{-\tfrac{1}{2}\nabla'^2 + V(\vec{r}\,')\}\, \tilde{\psi}_j(\vec{r}\,') = \tilde{\varepsilon}_j \; \tilde{\psi}_j(\vec{r}\,') \quad , \tag{2-509}$$

where the effective single-particle potential V is

$$V(\vec{r}\,') = V_{ext}(\vec{r}\,') + \sum_{k=1}^{N} \int (d\vec{r}\,'')\, \psi_k^*(\vec{r}\,'') \frac{1}{|\vec{r}\,'-\vec{r}\,''|} \psi_k(r'') \quad , \tag{2-510}$$

which is equivalent to

$$\sum_{k=1}^{N} \psi_k^*(\vec{r}\,')\psi_k(\vec{r}\,') = -\frac{1}{4\pi} \nabla'^2 (V(\vec{r}\,') - V_{ext}(\vec{r}\,')) \quad . \tag{2-511}$$

Let us now look at the Hartree energy. It is

$$\begin{aligned} E_{Hartree} &= \sum_{j=1}^{N} \int (d\vec{r}\,')\psi_j^*(\vec{r}\,')\{-\tfrac{1}{2}\nabla'^2 + V_{ext}(\vec{r}\,') \\ &\qquad + \tfrac{1}{2}(V(\vec{r}\,') - V_{ext}(\vec{r}\,'))\}\psi_j(\vec{r}\,') \\ &= \sum_{j=1}^{N} \int (d\vec{r}\,')\psi_j^*(\vec{r}\,') \sum_{\ell=1}^{N} \varepsilon_{j\ell}\, \psi_\ell(\vec{r}\,') \\ &\qquad - \tfrac{1}{2}\int (d\vec{r}\,')(V(\vec{r}\,') - V_{ext}(\vec{r}\,')) \sum_{j=1}^{N} \psi_j^*(\vec{r}\,')\psi_j(\vec{r}\,') \quad , \end{aligned} \tag{2-512}$$

where both (502) and (510) have been used. With the aid of the orthonormality of the ψ_j's and with Eq. (511) we obtain

$$E_{\text{Hartree}} = \sum_{j=1}^{N} \varepsilon_{jj} - \frac{1}{8\pi}\int (d\vec{r}\,')\,[\vec{\nabla}'(V(\vec{r}\,') - V_{\text{ext}}(\vec{r}\,'))]^2 \ . \quad (2\text{-}513)$$

This will look even more like the TF potential functional after we use

$$\sum_{j=1}^{N} \varepsilon_{jj} = \sum_{j=1}^{N} \tilde{\varepsilon}_{jj} = \sum_{j=1}^{N} \tilde{\varepsilon}_{j} \quad , \quad (2\text{-}514)$$

in conjunction with the fact that the $\tilde{\varepsilon}_j$ are the N smallest eigenvalues of the single-particle Hamilton operator

$$H = \frac{1}{2}p^2 + V(\vec{r}) \quad , \quad (2\text{-}515)$$

to write

$$\sum_{j=1}^{N} \varepsilon_{jj} = \text{tr}\, H\, \eta(-H-\zeta) \quad , \quad (2\text{-}516)$$

where, of course, ζ is such that the count of occupied states equals the number of electrons:

$$N = \text{tr}\ \eta(-H-\zeta) \ . \quad (2\text{-}517)$$

If we combine (516) and (517) in the now familiar way,

$$\sum_{j=1}^{N} \varepsilon_{jj} = \text{tr}(H+\zeta)\eta(-H-\zeta) - \zeta N$$
$$= E_1(V+\zeta) - \zeta N \quad , \quad (2\text{-}518)$$

then

$$E_{\text{Hartree}} = E_1(V+\zeta) - \frac{1}{8\pi}\int (d\vec{r}\,')\,[\vec{\nabla}'(V(\vec{r}\,') - V_{\text{ext}}(\vec{r}\,'))]^2 - \zeta N. \quad (2\text{-}519)$$

It becomes clear now what the fundamental difference is between the TF approach and Hartree's method. The latter asks: what are the optimal single-particle wave functions to be used in (494)?

The answer is given by the Hartree equations (502).[26] But suppose we do not care that much for the ψ_j's. Then we can equally well put the question: what is the best effective potential in (519)? We reply immediately: the TF potential, if $E_1(V+\zeta)$ is evaluated in the semiclassical limit. Does this mean that the TF model is an approximation to Hartree's description? No, it is rather the other way round: the Hartree picture contains more detail than it should. In view of all the approximations made before arriving at (519), there is absolutely no point in being extremely precise when evaluating $E_1(V+\zeta)$.

Summing up: the TF model and Hartree's method are really two independent, though related, approaches. None is a priori the better or worse one. Whereas I do not want to go as far as Lieb does ["... TF theory is well defined.(...) - a state of affairs in marked contrast to that of HF theory."[27]], I do have the impression that in applying TF methods one is more conscious about the physical approximations that enter the development.

In one respect the Hartree detour over the single-particle wave functions is superior to the TF phase-space integral: the Schrödinger equation (509) treats the strongly bound electrons correctly without any further ado. We shall see in the next Chapter how the TF model can be modified, in a simple way, in order to handle these innermost electrons properly. With this improvement the TF description is in no way inferior to Hartree's.

Please do not miss how naturally we have been led to a potential functional, Eq.(519), not to a density functional. Here is, once more, support for our view that TF theory is best thought of as formulated in terms of the effective potential. Then the density is not a fundamental but a derived quantity.

Problems

2-1. For the generalization of the independent-particle Hamilton operator of Eq.(3) to

$$H = \frac{1}{2}(\vec{p} - \alpha\vec{A}(\vec{r}))^2 + V(\vec{r}) \quad ,$$

where $\alpha = \frac{e^2}{\hbar c} = 1/137.036...$ is Sommerfeld's fine structure constant and $\vec{A}$ is an effective vector potential (in atomic units), show that the analogs of (14) and (20) are

$$\delta_{\vec{A}} E_{IP} = \delta_{\vec{A}} E_1 = -\alpha\int (d\vec{r}\,')\delta\vec{A}(\vec{r}\,')\cdot\vec{j}(\vec{r}\,') \quad ,$$

and

$$\vec{j}(\vec{r}\,') = 2\langle\vec{r}\,'|\tfrac{1}{2}[(\vec{p}-\alpha\vec{A})\eta(-H-\zeta)+\eta(-H-\zeta)(\vec{p}-\alpha\vec{A})]|\vec{r}\,'\rangle \ .$$

Then generalize Eq.(25) to read

$$\delta E_{ee} = \int (d\vec{r}\,')[\delta n(\vec{r}\,')V_{ee}(\vec{r}\,') - \alpha\delta\vec{j}\cdot\vec{A}_{ee}] \ .$$

Next conclude that, instead of (30), the stationary energy expression is now

$$E = E_{IP} - \int (d\vec{r}\,')V_{ee}(\vec{r}\,')n(\vec{r}\,') + \alpha\int (d\vec{r}\,')\vec{A}_{ee}(\vec{r}\,')\cdot\vec{j}(\vec{r}\,') + E_{ee} \ ,$$

since

$$\vec{A} = \vec{A}_{ext} + \vec{A}_{ee} \ ,$$

with a given external vector potential $\vec{A}_{ext}$. How does the TF version of E_1 depend on $\vec{A}$?

2-2. Another application of the stationary property of the electrostatic potential functional of Eq.(78). Instead of $\Phi(\vec{r})$, insert $\Phi(\vec{r}\,')$, where $\vec{r}\,'$ is related to $\vec{r}$ through an infinitesimal translation by $\delta\vec{\varepsilon}$ and an infinitesimal rotation around $\delta\vec{\omega}$,

$$\vec{r} = \vec{r}\,' + \delta\vec{\varepsilon} + \delta\vec{\omega}\times\vec{r}\,' \ .$$

Use $(d\vec{r}) = (d\vec{r}\,')$ and $(\vec{\nabla}\Phi(\vec{r}\,'))^2 = (\vec{\nabla}'\Phi(\vec{r}\,'))^2$ to write the primed energy as

$$E' = \int (d\vec{r}\,')[\rho(\vec{r}\,'+\delta\vec{\varepsilon}+\delta\vec{\omega}\times\vec{r}\,')\Phi(\vec{r}\,') - \frac{1}{8\pi}(\vec{\nabla}'\Phi(\vec{r}\,'))^2]$$

$$\leq E = \int (d\vec{r})[\rho(\vec{r})\Phi(\vec{r}) - \frac{1}{8\pi}(\vec{\nabla}\Phi(\vec{r}))^2] \ ,$$

where the equal sign holds to first order in $\delta\vec{\varepsilon}$ and $\delta\vec{\omega}$. Conclude, that the self force vanishes [Eq.(86)], and also the self torque,

$$\int (d\vec{r})\rho(\vec{r})\,\vec{r}\times(-\vec{\nabla}\Phi(\vec{r})) = 0 \ .$$

2-3. Write a computer program for the TF function F(x) as outlined around Eq.(200). Use it to confirm

$$\int_0^\infty dx\, F(x) = 1.80006394 \quad ,$$

$$\int_0^\infty dx [F(x)]^2 = 0.61543464 \quad ,$$

$$\int_0^\infty dx [-F'(x)]^3 = 0.35333456 \quad ,$$

$$\int_0^\infty dx \sqrt{F(x)/x} = 3.915933 \quad .$$

2-4. With the computer program of Problem 3 check that the maximum of $xF(x)$ occurs at $x=2.104025280$, where $F(x)=0.2311514708$.

2-5. This maximum of $xF(x)$ is relatively broad, so that

$$F''(x)/[F(x)]^2 = [x F(x)]^{-1/2} \cong \text{constant} \quad .$$

An approximation to $F(x)$ is therefore represented by the solution of

$$\tilde{F}''(x) = \frac{6}{\tilde{x}^2}[\tilde{F}(x)]^2 \quad , \quad \tilde{x} = \text{const.} \; ,$$

subject to $\tilde{F}(o)=1$, $\tilde{F}(\infty)=o$, and (to fix the value of $\tilde{x}$)

$$\int_0^\infty dx\; x^{1/2} [\tilde{F}(x)]^{3/2} = 1 \quad .$$

Find this $\tilde{F}(x)$.[28] How good is this approximation when it is employed in calculating the numbers of Problems 3 and 4?

2-6. Insert $\phi_\lambda(o)$ of Eq.(283) into Eq.(274) to find $\phi_1'(o)$ and $\phi_2'(o)$ as the coefficients in

$$\phi_\lambda'(o) = -1 + \phi_1'(o)\lambda + \phi_2'(o)\lambda^2 + \ldots \quad .$$

Compare with Eqs.(280) and (281).

2-7. Find $\phi_2(t)$ from Eq.(278); then evaluate $\phi_3(o)$. Use it to show that

(i) in Eq. (285):

$$O((N/Z)^2) = \left(12 - \frac{21067}{60\pi^2} + \frac{524288}{255\pi^4}\right)\left(\frac{N}{Z}\right)^2 + O((N/Z)^3) \quad ;$$

(ii) in Eqs. (286) and (289):

$$O((N/Z)^2) = \left(\frac{2}{3} + \frac{223}{30\pi^2} - \frac{262144}{2025\pi^4}\right)\left(\frac{N}{Z}\right)^2 + O((N/Z)^3) \quad ;$$

(iii) in Eq. (287):

$$O((N/Z)^2) = \left(\frac{22}{3} - \frac{19019}{90\pi^2} + \frac{2883584}{2025\pi^4}\right)\left(\frac{N}{Z}\right)^2 + O((N/Z)^3) \quad .$$

2-8. Use the recurrence relation (278) to show that

$$\phi_\ell(t) \sim (1-t)^{(5\ell+2)/2} \quad \text{for} \quad t \lesssim 1 \quad .$$

Look back at Eq. (311) and notice that, indeed, the first occurence of Λ is in the $(1-t)^{7/2}$-term, and of Λ^2 in the $(1-t)^6$-term.

2-9. Show that, for $\lambda > \Lambda$, $\phi_\lambda(t)$ has a pole, at $t = t_\lambda > o$, of the form

$$\phi(t) \cong \frac{400}{\lambda^2} \frac{t_\lambda}{(t-t_\lambda)^2} \quad \text{for} \quad t \cong t_\lambda \quad .$$

As $\lambda \to \infty$, $t_\lambda \to 1$, so that $\phi_\lambda''(t) \cong \lambda[\phi(t)]^{3/2}$ for $t_\lambda < t \leq 1$. Use this to demonstrate that

$$1 - t_\lambda \cong \int_o^\infty \frac{d\phi}{\sqrt{1 + \frac{4}{5}\lambda\phi^{5/2}}} = \left(\frac{61.9}{\lambda}\right)^{2/5} \quad ,$$

for $\lambda \gg \Lambda$.

2-10. Upper and lower bounds to $\Lambda^{-2/3}$ can be obtained from Eq. (301) when it is combined with the inequalities of (242). A suitable trial function $f(x)$ is given by

$$f(x) = \begin{cases} F(x) & \text{for} \quad o \leq x \leq x_1 \\ \frac{q}{x_o}(x_2 - x) & \text{for} \quad x_1 \leq x \end{cases} \quad ,$$

where q is fixed, x_o is arbitrary, and x_1 and x_2 are such that $f(x)$ and its derivative are continuous. For $q \gtrsim o$, x_1 is sufficiently large to justify the use of the asymptotic form (179) for $F(x \geq x_1)$. Show that this implies

$$x_2 = \frac{4}{3} x_1 \qquad \text{and} \qquad x_1^4 = \frac{432}{q/x_o} \quad .$$

Then derive

$$\frac{2}{5} \int_o^\infty dx \, \frac{[f(x)]^{5/2}}{x^{1/2}} = \frac{2}{7} B - \frac{2}{35} \frac{(12)^5}{x_1^7} + \frac{2}{5} (\frac{q}{x_o})^{5/2} (\frac{4}{3})^3 x_1^3 (\frac{5\pi - 9\sqrt{3}}{48}) \; ,$$

and

$$\frac{1}{2} \int_o^\infty dx [f'(x) + \frac{q}{x_o}]^2 = \frac{1}{7} B - \frac{3}{56} \frac{(12)^5}{x_1^7} - \frac{q}{x_o} + \frac{5}{6} (\frac{q}{x_o})^2 x_1 \quad .$$

Putting everything together you should have

$$\frac{3}{7} B - \frac{(12)^5}{x_1^7} [\frac{30}{7} - \frac{8\pi}{3\sqrt{3}} + \frac{3}{4} \frac{x_o}{x_1}]$$

$$> \frac{3}{7} B - \frac{3}{7} \Lambda^{-2/3} q^{7/3} \quad , \qquad \text{for } q \to o \; .$$

It is then useful to switch from x_o to a new independent parameter, λ, by setting $x_o \equiv \lambda^{2/3} q^{-1/3}$. Check that then $x_1 = (432)^{1/4} \lambda^{1/6} q^{-1/3}$, so that, for all $\lambda > o$,

$$\Lambda^{-2/3} > \frac{7}{3} \lambda^{-2/3} - \frac{8}{3^{5/4}} (\frac{56\pi}{3\sqrt{3}} - 30) \lambda^{-7/6} \quad .$$

Optimize λ and find the lower bound on $\Lambda^{-2/3}$ of (303). Show that, for this optimal λ, the ratio x_2/x_o does not equal unity. Consequently, the trial $f(x)$ does not change its sign at $x = x_o$, as the actual $f(x)$ does. Impose $x_2 = x_o$ and demonstrate that a lower bound on $\Lambda^{-2/3}$ emerges, which is worse than the previous one.

For an upper bound on $\Lambda^{-2/3}$ use the trial function

$$g(x) = \begin{cases} F'(x) + q/x_o & \text{for } o \leq x \leq x_1 \\ -q/x_o (1 - x/x_2) & \text{for } x_1 \leq x \leq x_2 \\ o & \text{for } x_2 \leq x \end{cases} \quad .$$

Make sure that g is continuous and obeys Eq. (243). Then evaluate the g-functional of (242). You should get

$$\Lambda^{-2/3} < \frac{1}{60}\left[\frac{6t}{(5t-1)(3-t)}\right]^{7/3}\left(1-\frac{1}{2}t\right)^{1/3}\left[\frac{1}{9}\left(\frac{35}{t}-16+191t-74t^2\right)\right.$$

$$\left.+\frac{7}{4}t^{1/3}\left(1-\frac{1}{2}t\right)^{1/3}\left(1-t^{4/3}\right)\right] ,$$

where the range of $t=x_1/x_2$ is $\frac{1}{5}<t\leq 1$. Find (numerically) the optimal value for t and thus the upper bound on $\Lambda^{-2/3}$ of (303).

2-11. Insert Eq. (316) into

$$\Phi(t) = \frac{1}{q} f_q(t\, x_o(q))\Big|_{q\to o}$$

and derive (352).

2-12. Derive Eq. (462) directly from Eqs. (433), (432), and (428).

2-13. Because of the homogeneity and isotropy of the physical three-dimensional space, the density functionals $E_{kin}(n)$ and $E_{ee}(n)$, which appear in Eq. (417), have the same numerical value for $n(\vec{r}\,')$ and the infinitesimally translated and rotated $\tilde{n}(\vec{r}\,') = n(\vec{r}\,' + \delta\vec{\varepsilon} + \delta\vec{\omega}\times\vec{r}\,')$. Combine this with the stationary property of (417) to show that there is no net force,

$$\int(d\vec{r}\,')n(\vec{r}\,')(-\vec{\nabla}'V_{ext}(\vec{r}\,')) = o \quad ,$$

and no net torque,

$$\int(d\vec{r}\,')n(\vec{r}\,')\vec{r}\,'\times(-\vec{\nabla}'V_{ext}(\vec{r}\,')) = o \quad ,$$

exerted on the system by the external potential. Are you reminded of Problem 2?

2-14. Show that the density functional of the kinetic energy is given by

$$E_{kin}(n) = \int(d\vec{r})\,\frac{1}{8}\,(\vec{\nabla}n)^2/n \quad ,$$

if there is only one electron. This does scale like (443). Why is there no contradiction to the general statement that E_{kin} does not obey (443)?

Chapter Three

STRONGLY BOUND ELECTRONS

In the preceding Chapter there was a section entitled "Validity of the TF model," in which we found two regions of failure of the TF model: (i) the inner region of strong binding, where r does not exceed $\sim 1/Z$; and (ii) the outer region of weak binding around the edge of the atom (r larger than ~ 1, for a neutral atom). Of these the first one is more important because of the enormous binding energies of electrons that are close to the nucleus. Consequently, the leading correction to the TF model consists of an improved handling of the strongly bound electrons. This is the topic of the present Chapter.

<u>Qualitative argument.</u> If we simply exclude the critical vicinity of the nucleus when evaluating the r-integral of Eq.(2-44), then the TF version of E_1 is replaced by

$$\begin{aligned}(E_1)_{TFS} &\cong \int_{r\gtrsim 1/Z} (d\vec{r})\left(-\frac{1}{15\pi^2}\right)[-2(V+\zeta)]^{5/2} \\ &\cong (E_1)_{TF} - \int_{r\lesssim 1/Z} (d\vec{r})\left(-\frac{1}{15\pi^2}\right)\left[-2\left(-\frac{Z}{r}\right)\right]^{5/2} \qquad (3\text{-}1) \\ &= (E_1)_{TF} + C\, Z^2 \quad ,\end{aligned}$$

with C a constant of order unity. Here we have made use of $V \cong -Z/r$ for small r, which states, once more, that the dynamics are dominated by the nucleus-electron interaction if r is sufficiently small. The third initial of the subscript TFS stands for Scott, who (in 1952) was the first to present a discussion of this leading correction to the TF energy.[1]

Our simple qualitative argument (which is a variant of the one given by Schwinger[2]) says that the TF energy is supplemented by an additive term proportional to Z^2, i.e., of relative order $Z^2/Z^{7/3} = Z^{-1/3}$ as compared to the TF contribution. This is consistent with the observations of the Introduction, where we have seen such terms in Eqs.(1-22) and (1-84), the numerical value of C being 1/2 in both equations. More evidence in favor of a Z^2 term is supplied by Fig.2-2, where the smooth TF curve would be shifted down by the amount of 2 C , in which event the agreement with the HF crosses would be significantly improved.

Scott's result,

$$E_{TFS}(Z,N) = E_{TF}(Z,N) + \frac{1}{2} Z^2 \quad , \tag{3-2}$$

is identical with the one of the Introduction. In view of the primitive models of Bohr atoms, with or without shielding, that are used there, it may be puzzling that the numerical values of the coefficients agree. This mystery is easily resolved: all that matters is the Coulombic shape of the effective potential for $r \to o$. The models of the Introduction are, certainly, realistic at these small distances. But there is even more to it: since one can easily imagine that the slight deviation of the effective potential from its limiting form $-Z/r$ + constant is irrelevant, Scott's result is anticipated to remain valid, when his reasoning is abandoned in favor of a more convincing one. We postpone the presentation of Scott's original argument until later.

One remarkable feature of Scott's correction is its independence of the number of electrons, N. This is, of course, a consequence of the circumstance that the small-r shape of the effective potential does not depend on N, or, again, the most strongly bound electrons are hardly aware of the more weakly bound ones because the Coulomb forces of the nucleus are so strong.

First quantitative derivation of Scott's correction. For a quantitative treatment of the strongly bound electrons we split $E_1(\zeta)$ [cf. Eq.(2-7)] into two parts,

$$\begin{aligned} E_1(\zeta) &= \mathrm{tr}(H+\zeta)\eta(-H-\zeta) \\ &= \mathrm{tr}(H+\zeta)\eta(-H-\zeta_s) \\ &\quad +\mathrm{tr}(H+\zeta)[\eta(-H-\zeta)-\eta(-H-\zeta_s)] \\ &\equiv E_s + E_{\zeta\zeta_s} \quad . \end{aligned} \tag{3-3}$$

The separating binding-energy ζ_s distinguishes the strongly bound electrons from the rest of the atom, the respective contributions to E_1 being E_s and $E_{\zeta\zeta_s}$. This ζ_s is not a uniquely defined quantity, but it is not arbitrary either. It has to be small compared to the typical single-electron Coulomb energy ($\sim Z^2$) but large on the TF scale:

$$Z^{4/3} \ll \zeta_s \ll Z^2 \quad . \tag{3-4}$$

In this first quantitative discussion we simplify matters by assuming that for the evaluation of E_s the Coulombic approximation $V(r) \cong -Z/r$ suffices. The effects of the deviation of $V(r)$ from this limiting form will be dealt with later. At the present stage we are content with the remark that said approximation can always be justified if ζ_s is chosen large enough.

Thus we have[3]

$$E_s \cong \mathrm{tr}(\tfrac{1}{2}p^2 - \tfrac{Z}{r} + \zeta)\eta(-\tfrac{1}{2}p^2 + \tfrac{Z}{r} - \zeta_s) \quad . \tag{3-5}$$

The step function in Eq.(5) selects the states with binding energy larger than ζ_s. Since we are back to the Bohr atom (without shielding), this means that a certain number of Bohr shells is summed over. If the last one included in the sum has principal quantum number n_s, then its single-electron binding-energy $\frac{1}{2}Z^2/n_s^2$ exceeds ζ_s whereas that of the next shell does not:

$$\tfrac{1}{2}Z^2/n_s^2 \quad > \quad \zeta_s \quad > \quad \tfrac{1}{2}Z^2/(n_s+1)^2 \quad . \tag{3-6}$$

This situation is illustrated by the sketch presented as Fig.1. Another way of writing (6) is

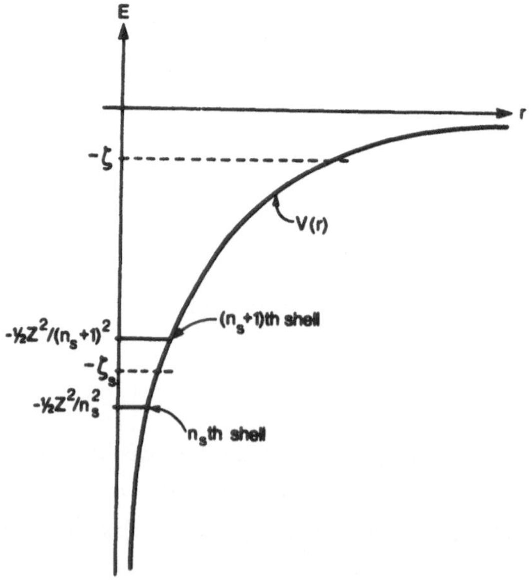

Fig.3-1. Concerning corrections for strongly bound electrons; see text.

$$n_s < \left(\frac{Z^2}{2\zeta_s}\right)^{1/2} < n_s + 1 \quad , \tag{3-7}$$

or, when we introduce a kind of continuous version of n_s,

$$\nu_s \equiv Z/\sqrt{2\zeta_s} \quad , \tag{3-8}$$

then

$$n_s = [\nu_s] = [Z/\sqrt{2\zeta_s}] \quad , \tag{3-9}$$

which uses the Gaussian notation for the largest integer contained in ν_s. Now we look back at Eqs. (1-9) and (1-10) to find the contribution to E_1 from the strongly bound electrons. It is [with $m=n_s$ and $\mu=o$ in Eqs. (1-9) and (1-10)]

$$E_s = -Z^2[\nu_s] + \zeta N_s \quad , \tag{3-10}$$

where

$$N_s = \frac{2}{3}([\nu_s]+\frac{1}{2})^3 - \frac{1}{6}([\nu_s]+\frac{1}{2}) \tag{3-11}$$

is the total number of specially treated strongly bound electrons. Its approximate connection with ζ_s,

$$N_s \cong \frac{2}{3}\nu_s^3 = \frac{2}{3}\left(\frac{Z^2}{2\zeta_s}\right)^{3/2} \quad , \tag{3-12}$$

or

$$\zeta_s \cong \frac{1}{2} Z^2 \left(\frac{3}{2} N_s\right)^{-2/3} \quad , \tag{3-13}$$

inserted into the relations (4), says

$$1 \ll N_s \ll Z \quad ; \tag{3-14}$$

N_s is a small fraction of Z. In particular, if $N \ll Z$, then all electrons should be regarded as strongly bound, and the interelectronic interactions should be treated as a small perturbation.

Now we turn to $E_{\zeta\zeta_s}$ of Eq. (3), the contribution to E_1 from the more weakly bound electrons. For these we expect the TF evaluation of the trace by means of the highly semiclassical phase-space integral to be justified. Let us therefore check in detail that there is no significant contribution to $E_{\zeta\zeta_s}$ from the vicinity of the nucleus. First observe that $E_{\zeta\zeta_s}$ can be written as

$$E_{\zeta\zeta_s} = \mathrm{tr}(H+\zeta)\eta(-H-\zeta) - \mathrm{tr}(H+\zeta_s)\eta(-H-\zeta_s)$$
$$+ (\zeta_s-\zeta)\mathrm{tr}(-H-\zeta_s) \quad , \tag{3-15}$$

or with Eqs.(2-7) and (2-9),

$$\begin{aligned} E_{\zeta\zeta_s} &= E_1(\zeta) - E_1(\zeta_s) + (\zeta_s-\zeta)N(\zeta_s) \\ &= -\int_\zeta^{\zeta_s} d\zeta'\, N(\zeta') + (\zeta_s-\zeta)N(\zeta_s) \quad . \end{aligned} \tag{3-16}$$

Then employ the identity

$$(\zeta_s-\zeta)N(\zeta_s) = \int_\zeta^{\zeta_s} d\zeta' \frac{d}{d\zeta'}\{(\zeta'-\zeta)N(\zeta')\} \tag{3-17}$$

to arrive at

$$E_{\zeta\zeta_s} = \int_\zeta^{\zeta_s} d\zeta'(\zeta'-\zeta)\frac{d}{d\zeta'}N(\zeta') \quad . \tag{3-18}$$

The TF version of $N(\zeta')$, given in Eq.(2-50), is easily differentiated, producing

$$E_{\zeta\zeta_s} = -\int(d\vec{r}) \int_\zeta^{\zeta_s} d\zeta'(\zeta'-\zeta)\frac{1}{\pi^2}[-2(V+\zeta')]^{1/2} \quad . \tag{3-19}$$

For small r, where $V \cong -Z/r$, the integrand in (19) is $\sim r^{-1/2}$, whereas it is $\sim r^{-5/2}$ in Eq.(1). It has been reduced by two orders of r. This is the first manifestation of the strong cancellations for $r \to o$ that are inherent in the structure of $E_{\zeta\zeta_s}$ in Eq.(16). As an immediate consequence, the contribution to $E_{\zeta\zeta_s}$ from $r \lesssim 1/Z$ is about $(\zeta_s-\zeta)^2/Z^2$ - with a numerical factor of order unity - which amount is small compared to Z^2 because of the relations (4). Indeed, the vicinity of the nucleus does not contribute significantly to $E_{\zeta\zeta_s}$. In other words: the splitting of E_1 into E_s and $E_{\zeta\zeta_s}$ successfully separates the strongly bound electrons from the rest of the atom.

We are now justified in evaluating $E_{\zeta\zeta_s}$ TF wise, i.e., by means of the semiclassical phase-space integral. In Eqs.(1-35) and (1-36) we find the relevant results,

$$N(\zeta_s) = \frac{2}{3}\nu_s^3 \tag{3-20}$$

and

$$-E_1(\zeta_s) + \zeta_s\, N(\zeta_s) = Z^2\, \nu_s \quad . \tag{3-21}$$

In combination with (10) the change in energy caused by our improved treatment of the strongly bound electrons is

$$\begin{aligned} \Delta_s E &= E_s - E_1(\zeta_s) + (\zeta_s - \zeta)N(\zeta_s) \\ &= Z^2(\nu_s - [\nu_s]) - \zeta(\tfrac{2}{3}\nu_s^3 - N_s) \quad . \end{aligned} \tag{3-22}$$

The continuous terms $Z^2\nu_s - \zeta\frac{2}{3}\nu_s^3$ come from the removal of the incorrect TF treatment of these innermost electrons, whereas the discontinuous terms $-Z^2[\nu_s]+\zeta N_s(=E_s)$ originate in the correct quantum mechanical description of these electrons which has been used to evaluate E_s.

The $\Delta_s E$ in (22) obviously depends on the particular value of ν_s, that is on the particular choice made for ζ_s. On the other hand, $E_1(\zeta)$ in (3) is clearly independent of ζ_s. What happened ? The result for E_s in (10) contains contributions from the shell structure of the corresponding Bohr atoms. But no shell effects are taken into account in computing $E_{\zeta\zeta_s}$. Consequently, in order to be consistent we must discard the shell structure in E_s, but retain all smooth (as a function of ν_s) parts.

Let us first look at the difference $\nu_s-[\nu_s]$, which supplies the neutral-atom value of $\Delta_s E$, when ζ=o. A plot of this indented function of ν_s is shown in Fig.2. It is visibly of the form

$$\nu_s - [\nu_s] = \frac{1}{2} + \text{oscillation} \quad . \tag{3-23}$$

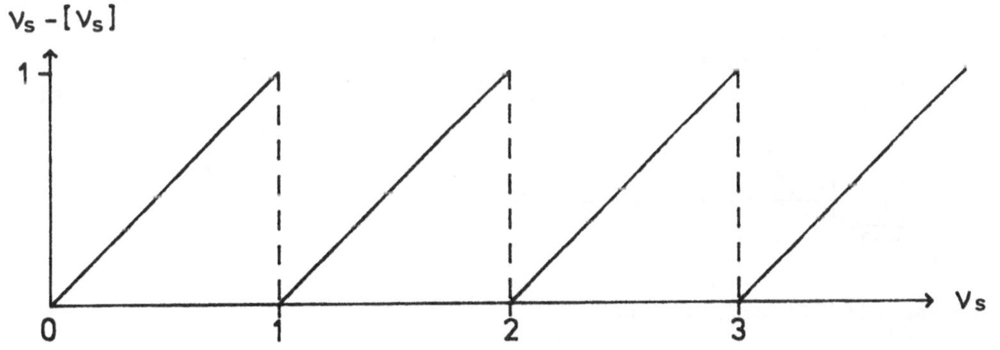

Fig.3-2. The difference $\nu_s-[\nu_s]$ *as a function of* ν_s.

As a matter of fact we know this "oscillation." It appeared early in the Introduction. According to Eq.(1-14) we have

$$\begin{aligned}\nu_s - [\nu_s] &= \frac{1}{2} + (\nu_s - \frac{1}{2}) - [(\nu_s - \frac{1}{2}) + \frac{1}{2}] \\ &= \frac{1}{2} + \langle \nu_s - \frac{1}{2} \rangle \quad ,\end{aligned} \tag{3-24}$$

or with Problem 1-2,

$$\nu_s - [\nu_s] = \frac{1}{2} - \sum_{m=1}^{\infty} \frac{1}{\pi m} \sin(2\pi m \nu_s) \quad . \tag{3-25}$$

Consequently, removing the Bohr-shell artifacts is done by the replacement

$$\nu_s - [\nu_s] \rightarrow \frac{1}{2} \quad . \tag{3-26}$$

In Eq.(22) this reproduces Scott's correction, which is the anticipated result. What remains to be shown is that the term proportional to ζ in (22) does not contribute, which means that it is entirely made of oscillations due to the Bohr shells and does not contain a smooth part.

Upon making use of (24), the cubic difference

$$\frac{2}{3}\nu_s^3 - N_s = \frac{2}{3}\nu_s^3 - \frac{2}{3}([\nu_s] + \frac{1}{2})^3 + \frac{1}{6}([\nu_s] + \frac{1}{2}) \tag{3-27}$$

appears as

$$\begin{aligned}\frac{2}{3}\nu_s^3 - N_s &= 2\nu_s^2 \langle \nu_s - \frac{1}{2} \rangle - 2\nu_s (\langle \nu_s - \frac{1}{2} \rangle^2 - \frac{1}{12}) \\ &\quad + \frac{2}{3} \langle \nu_s - \frac{1}{2} \rangle (\langle \nu_s - \frac{1}{2} \rangle^2 - \frac{1}{4}) \quad .\end{aligned} \tag{3-28}$$

These results of Problem 1-2:[4]

$$\begin{aligned}\sum_{m=1}^{\infty} \frac{(-1)^m}{(\pi m)^2} \cos(2\pi m y) &= \langle y \rangle^2 - \frac{1}{12} \quad , \\ \sum_{m=1}^{\infty} \frac{(-1)^m}{(\pi m)^3} \sin(2\pi m y) &= \frac{2}{3} \langle y \rangle (\langle y \rangle^2 - \frac{1}{4}) \quad ,\end{aligned} \tag{3-29}$$

are then used in arriving at

$$\frac{2}{3}\nu_s^3 - N_s = -2\nu_s^2 \sum_{m=1}^{\infty} \frac{1}{\pi m} \sin(2\pi m \nu_s) \quad -$$

$$- 2\nu_s \sum_{m=1}^{\infty} \left(\frac{1}{\pi m}\right)^2 \cos(2\pi m \nu_s) \qquad (3\text{-}30)$$

$$+ \sum_{m=1}^{\infty} \left(\frac{1}{\pi m}\right)^3 \sin(2\pi m \nu_s) \quad .$$

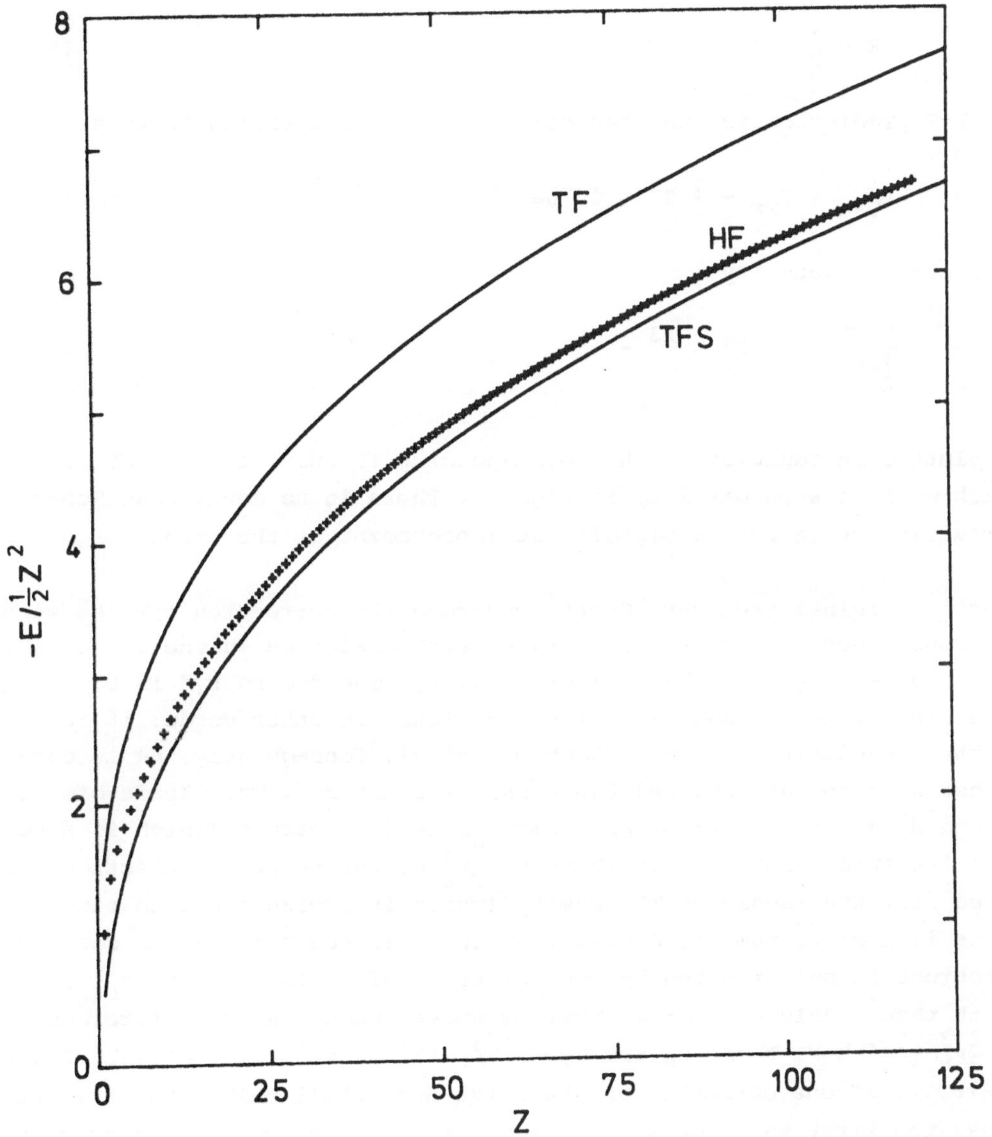

Fig. 3-3. Comparison of the TFS prediction (34) with the corresponding TF one, Eq. (2-160), and with HF binding energies (crosses); see also Fig. 2-2.

It is now obvious, that $\frac{2}{3}\nu_s^3 - N_s$ is, indeed, entirely composed of oscillations. Consequently, the analog of (26) is here the replacement

$$\frac{2}{3}\nu_s^3 - N_s \to 0 \ . \qquad (3\text{-}31)$$

Both this and (26) then turn Eq.(22) into Scott's result

$$\Delta_s E = \frac{1}{2} Z^2 \ . \qquad (3\text{-}32)$$

The TFS prediction for the neutral-atom binding energies is thus

$$-E_{TFS} = -E_{TF} - \frac{1}{2} Z^2 = 0.768745\ Z^{7/3} - \frac{1}{2} Z^2 \ . \qquad (3\text{-}33)$$

In Fig.3 the quantity

$$\frac{-E_{TFS}}{\frac{1}{2}Z^2} = 1.537\ Z^{1/3} - 1 \qquad (3\text{-}34)$$

is plotted in addition to the corresponding TF curve and the HF crosses, which we have seen earlier, in Fig.2-2. There is no doubt that Scott's correction represents a significant improvement of the theory.

Scott's original argument. Scott[1] regarded the correction for the strongly bound electrons as as "boundary effect" analogous to the decrease in particle density near the wall of a cavity. The "boundary" in Scott's reasoning is at the location of the nucleus; in other words: it is due to the singularity of the Coulomb potential. Consequently, he is concerned with the density and the count of electrons. This leads him to regard N_s as the number of specially treated electrons (which is a correct interpretation) and to think of $\frac{2}{3}\nu_s^3$ as the number of electrons removed from the incorrect TF density (which is a misconception since there is more to the TFS density, as we shall see below - the count of electrons is not affected by the splitting of E_1 into E_s and $E_{\zeta\zeta_s}$). Scott then concludes that ν_s must be chosen such that the difference $N_s - \frac{2}{3}\nu_s^3$ vanishes. These ν_s are $\nu_s = 3^{1/3}, 15^{1/3}, 42^{1/3}, \ldots = 1.442, 2.446, 3.476, \ldots$ if one corrects for the first Bohr shell only, the first two ones, the first three ones, ..., respectively. The corresponding results for the change in energy are $\Delta_s E/\frac{1}{2}Z^2 = 0.884, 0.932, 0.952, \ldots$ In general: if there are n_s Bohr shells in the atom, ν_s equals

$$\begin{aligned}\nu_s &= [(n_s+\tfrac{1}{2})^3 - \tfrac{1}{4}(n_s+\tfrac{1}{2})]^{1/3} \\ &\cong (n_s+\tfrac{1}{2}) - \tfrac{1}{12}/(n_s+\tfrac{1}{2}) \quad ,\end{aligned} \tag{3-35}$$

and $\Delta_s E$ is given by

$$\begin{aligned}\Delta_s E &= Z^2(\nu_s-[\nu_s]) = Z^2(\nu_s-n_s) \\ &\cong \tfrac{1}{2}\, Z^2\big(1-\tfrac{1}{6}/(n_s+\tfrac{1}{2})\big) \quad ,\end{aligned} \tag{3-36}$$

which approaches $\frac{1}{2}Z^2$ in the limit $n_s\to\infty$. Now, in the real atoms of the Periodic Table there are only very few Bohr shells (at most two), so that this limit is problematic. Certainly, the reliability of the numerical value of the factor $\frac{1}{2}$, as derived by Scott, is questionable. A typical reaction is that of March in his 1957 review article:[5] "it seems difficult to give a completely clearcut demonstration of the case." Just this was delivered - in the spirit of the treatment reported above - by Schwinger in 1980 (Ref.2).[6]

One reads occasionally that the first Bohr shell contributes 88%, the second another 5%, etc., to the Scott correction. Both this and the related remark, that there is a residual energy change of order $Z^{5/3}$ due to the strongly bound electrons, originate in Scott's reasoning which leads to Eq.(36). Now that it has been understood that Scott was simply paying too much attention to the oscillatory contributions from the Bohr shells, it is clear that these statements are wrong. The detailed shell effects in Eq.(22) have no physical significance. They are nothing more than a nuisance, inasmuch as we have to be particularly careful when extracting the smooth, non-oscillatory contribution.

<u>TFS energy functional.</u> So far we have been merely concerned with the change in energy resulting from the improved treatment of the strongly bound electrons. It is now time to study the corresponding changes in the effective potential and the density. The starting point is, as always, the stationary energy functional, which now will incorporate the Scott correction. For this purpose it is necessary to go beyond what we have done above because we must take into account the slight deviation of the effective potential from its limiting Coulomb shape. It will then be possible to demonstate that Scott's energy term $\frac{1}{2}Z^2$ is not affected by this deviation.

In evaluating the contribution to E_s from one of the strongly bound electrons, characterzied by its principal, its angular, and its magnetic quantum number (n, ℓ, and m, respectively), we can treat the difference between V(r) and -Z/r as small, so that the energy of said electron is given by

$$E_{n,\ell,m} = -\frac{Z^2}{2n^2} + \int(d\vec{r})\,(V+\frac{Z}{r})\,|\psi_{n,\ell,m}(\vec{r})|^2 \quad , \tag{3-37}$$

where $\psi_{n\ell m}$ is the corresponding wave function of the Bohr atom. Then, if ζ_s is such that there are n_s filled Bohr shells, we obtain for E_s (the factor of two is the spin multiplicity)

$$E_s = \sum_{n=1}^{n_s}\sum_{\ell=o}^{n-1}\sum_{m=-\ell}^{\ell} 2(E_{n,\ell,m}+\zeta) \quad . \tag{3-38}$$

When Eq.(37) is inserted, we meet the sum

$$\rho_s(r) = \sum_{n=1}^{n_s-}\sum_{\ell,m} 2|\psi_{n,\ell,m}(\vec{r})|^2 \quad , \tag{3-39}$$

which is, of course, the density of the specially treated strongly bound electrons. Because each closed ℓ-subshell is spherically symmetric, ρ_s depends only on r, the magnitude of the distance vector $\vec{r}$. The integral of $\rho_s(r)$ is equal to the total number N_s of Eq.(11),

$$\int(d\vec{r})\,\rho_s(\vec{r}) = \sum_{n=1}^{n_s-}\sum_{\ell,m} 2 = \sum_{n=1}^{n_s-} 2n^2 \tag{3-40}$$

$$= N_s = \frac{2}{3}(n_s+\frac{1}{2})^3 - \frac{1}{6}(n_s+\frac{1}{2}) \quad .$$

Also we know the integral of $\rho_s(r)$ times Z/r, since this is the negative of the potential energy of a Bohr atom with n_s filled shells. The virial theorem equates it to twice the binding energy, thus

$$\int(d\vec{r})\,\frac{Z}{r}\,\rho_s(r) = -2\sum_{n=1}^{n_s} 2n^2(-\frac{Z^2}{2n^2}) = 2\,Z^2\,n_s \quad . \tag{3-41}$$

Putting all these bits of information together, we arrive at

$$E_s = -Z^2\,n_s + \int(d\vec{r})\,(V+\frac{Z}{r})\,\rho_s + \zeta N_s = \tag{3-42}$$

$$= \int (d\vec{r}) V\rho_s + Z^2 n_s + \zeta N_s \quad . \tag{3-43}$$

This has to be supplemented by the TF evaluation of $E_{\zeta\zeta_s}$, obtained from Eq.(16) by inserting the TF expressions for $E_1(\zeta)$ and $N(\zeta)$, Eqs.(2-44) and (2-50), or equivalently by performing the ζ' integration in Eq.(19). Either way the result is

$$E_{\zeta\zeta_s} = \int (d\vec{r}) \{ -\frac{1}{15\pi^2}[-2(V+\zeta)]^{5/2} + \frac{1}{15\pi^2}[-2(V+\zeta_s)]^{5/2} + (\zeta_s-\zeta)\frac{1}{3\pi^2}[-2(V+\zeta_s)]^{3/2} \} \quad . \tag{3-44}$$

Consequently, at this stage the new expression for E_1 is

$$E_1 = \int (d\vec{r}) (-\frac{1}{15\pi^2}) [-2(V+\zeta)]^{5/2} + \int (d\vec{r}) \{ \frac{1}{15\pi^2}[-2(V+\zeta_s)]^{5/2} + (\zeta_s-\zeta)\frac{1}{3\pi^2}[-2(V+\zeta_s)]^{3/2} \} + \int (d\vec{r}) V\rho_s + Z^2 n_s + \zeta N_s \quad , \tag{3-45}$$

of which the first term is the previous TF result, the remaining ones its modification.

Since we are now taking into account that $V(r)$ deviates (slightly) from $-Z/r$, the relation (6) between ζ_s and n_s has to be reformulated appropriately. All electrons in the n_s-th shell have a binding energy larger than ζ_s, whereas those in the (n_s+1)-th shell have a smaller one:

$$\underset{\ell,m}{\mathrm{Min}}(-E_{n_s,\ell,m}) > \zeta_s > \underset{\ell,m}{\mathrm{Max}}(-E_{n_s+1,\ell,m}) \quad . \tag{3-46}$$

These energies refer to strongly bound electrons, so that there is no m dependence [because $V(\vec{r})$ is spherically symmetric in the vicinity of each nucleus of a molecule, the more so in our discussion concerning a single isolated atom], and little ℓ dependence is expected [otherwise it wouldn't be stronlgy bound electrons that we are talking about]. We shall therefore simplify matters by using averages over the shells instead of the maximum and minimum in (46). Thus

$$\zeta_1 > \zeta_s > \zeta_2 \quad , \tag{3-47}$$

with

$$\zeta_j = \frac{1}{2n_j^2} \sum_{\ell,m} (-2 E_{n_j,\ell,m}) \quad , \tag{3-48}$$

where

$$n_1 = n_s \quad , \quad n_2 = n_s + 1 \quad . \tag{3-49}$$

Upon utilizing the definition of $E_{n,\ell,m}$ in Eq. (37), the ζ_j's appear as

$$\zeta_j = \frac{Z^2}{2n_j^2} - \int (d\vec{r})\,(V + \frac{Z}{r})\,|\psi_{n_j}|^2_{av}(r) \tag{3-50}$$

where

$$|\psi_{n_j}|^2_{av}(r) \equiv \frac{1}{2n_j^2} \sum_{\ell,m} 2\,|\psi_{n_j,\ell,m}(\vec{r})|^2 \tag{3-51}$$

is the average single-electron density in the n_j-th shell. It is spherically symmetric for the same reason for which $\rho_s(r)$ of Eq. (39) has this property. So we can interpret (51) as taking the average over the angular dependences. Obviously, there is the connection

$$\rho_s(r) = \sum_{n=1}^{n_s} 2n^2\,|\psi_n|^2_{av}(r) \quad , \tag{3-52}$$

which emerges immediately when Eqs. (39) and (51) are combined. The well known Coulomb wave-functions are used in finding

$$|\psi_{n_j}|^2_{av}(r) = \frac{Z^3}{4\pi} \times \begin{cases} 4\,e^{-2Zr} \quad , \text{ for } n_j = 1, \\ \frac{4}{32}[1 - Zr + \frac{1}{2}(Zr)^2]\,e^{-Zr} \quad , \text{ for } n_j = 2 \;, \\ \frac{4}{243}[1 - \frac{4}{3}Zr + \frac{8}{9}(Zr)^2 - \frac{16}{81}(Zr)^3 + \frac{4}{243}(Zr)^4]\,e^{-2Zr/3}, \\ \qquad \text{for } n_j = 3 \;, \end{cases} \tag{3-53}$$

which illustrate the general structure of these averaged densities. As a consequence of their definition, they are normalized to unity

$$\int (d\vec{r})\,|\psi_{n_j}|^2_{av}(r) = 1 \quad , \tag{3-54}$$

which can easily be checked explicitly for the examples given above.

Clearly, the E_1 of Eq.(45) still contains all those spurious Bohr-shell effects that have to be removed in order to obtain correctly the Scott correction. For instance, if $V(r) \cong -Z/r$ is used to calculate the contribution to E_1 from the terms referring to the strongly bound electrons, the result (22) emerges without the replacements (26) and (31). It would be desirable to perform such an explicit deletion of the unphysical Bohr-shell artifacts in the expression for E_1 in (45) itself, so that all derived quantities would automatically be free of the spurious Bohr-shell oscillations. Unfortunately, it seems to be impossible to extract the smooth part out of E_1 of (45) without destroying the functional dependence on the effective potential V(r). Nevertheless, one can remove most of the unwanted Bohr-shell structure by performing a suitable average over ζ_s. Indeed, one easily imagines that the replacements (26) and (31) are the results of averaging over ν_s with an appropriately chosen weight function. To avoid a possible misunderstanding, let me emphasize that this averaging over ζ_s is not the essence of the TFS model; it is merely a technical procedure for eliminating the unphysical Bohr-shell effects. In doing so, however, use is made of the fact that ζ_s is not a physically uniquely defined quantity, but is [within the limits of Eq.(4)] quite arbitrary. Of course, one is free to employ any (reasonable) prescription for the averaging; depending on the particular application there may be one that is especially expedient. [Later in the development (Chapter Five) we shall arrive at a formulation which correctly incorporates the Scott correction without any reference to a separating binding energy like ζ_s.]

In Ref.7, the average over ζ_s is performed with uniform weight on the energy scale. This is natural since the energy is the fundamental quantity. Then, for a given number of specially treated Bohr shells n_s, the mean E_1 is just half the sum of its extreme values,

$$E_1 = \frac{1}{2}(E_{\zeta\zeta_1} + E_{\zeta\zeta_2}) + E_s \quad , \tag{3-55}$$

where ζ_1 and ζ_2 are the limiting values for ζ_s, Eq.(47). Applied to the structures appearing in Eq.(22) this averaging procedure produces

$$\nu_s - [\nu_s] \rightarrow \frac{1}{2}(n_1 - n_s) + \frac{1}{2}(n_2 - n_s) = \frac{1}{2} \tag{3-56}$$

and

$$\frac{2}{3}\nu_s^3 - N_s \rightarrow \frac{1}{2}(\frac{2}{3}n_1^3 - N_s) + \frac{1}{2}(\frac{2}{3}n_2^3 - N_s)$$
$$= \frac{2}{3}(n_s + \frac{1}{2}) \quad , \tag{3-57}$$

with n_1 and n_2 from (49), and N_s from (40). Obviously, the average (55) is good enough to simulate the replacement (22), but too simple to also reproduce (31), instead of which we now have (57). Some of the Bohr-shell effects are left. Nevertheless, the procedure (55) suffices for many applications of the TFS model.

It is advantageous to generalize Eq.(55) to

$$E_1 = \sum_{j=1}^{J} w_j \, E_{\zeta\zeta_j} + E_s \tag{3-58}$$

where ζ_1 and ζ_2 signify what they did before [Eqs.(47), (48), and (50)], whereas ζ_3, ζ_4, ..., ζ_J are intermediate values of ζ_s, for which n_j is a (non-integer) number between $n_1=n_s$ and $n_2=n_s+1$. The corresponding $|\psi_{n_j}|^2_{av}$ that appear in the generalization of Eq.(50) are appropriate (linear) averages of $|\psi_{n_1}|^2_{av}$ and $|\psi_{n_2}|^2_{av}$. The values of the ζ_j and their weights w_j are chosen such that in the application of interest all Bohr-shell oscillations are removed completely. Obviously, (58) reduces to (55) when J=2 and $w_1=w_2=\frac{1}{2}$. Another example is J=3 with

$$n_3 = n_s + \frac{1}{2} \tag{3-59}$$

and

$$w_1 = w_2 = -\frac{1}{6} \quad , \quad w_3 = \frac{4}{3} \quad , \tag{3-60}$$

which is the simplest average capable of simulating the replacements (26) and (31), see:

$$\nu_s - [\nu_s] \rightarrow -\frac{1}{6}(n_1-n_s) - \frac{1}{6}(n_2-n_s) + \frac{4}{3}(n_3-n_s)$$
$$= 0 - \frac{1}{6} + \frac{2}{3} = \frac{1}{2} \quad , \tag{3-61}$$

and

$$\frac{2}{3}\nu_s^3 - N_s \rightarrow -\frac{1}{6}(\frac{2}{3}n_1^3 - N_s) - \frac{1}{6}(\frac{2}{3}n_2^3 - N_s) + \frac{4}{3}(\frac{2}{3}(n_s+\frac{1}{2})^3 - N_s) =$$

$$= -\frac{1}{3}\times\frac{2}{3}(n_s+\tfrac{1}{2}) + \frac{4}{3}\times\frac{1}{6}(n_s+\tfrac{1}{2}) = 0 \quad , \tag{3-62}$$

indeed. Please note that the weights w_1 and w_2 are negative. This is disturbing but, unfortunately, unavoidable. Some additional discussion is contained in Problem 3. The occurence of negative weights requires particular caution to make sure that, for instance, the resulting density is positive.

Let us now imagine that we accept Eq. (58) and, without knowing of Eq. (22), use it to find the change in energy produced by the corrections for the strongly bound electrons. We insert $V \cong -Z/r$ into the terms referring to these innermost electrons and find

$$\Delta_s E = Z^2\Big(\sum_{j=1}^{J} w_j n_j - n_s\Big) - \zeta\Big(\sum_{j=1}^{J} w_j \frac{2}{3} n_j^3 - N_s\Big) \quad . \tag{3-63}$$

Can we give sense to this expression despite of the apparent ambiguities in choosing the values of n_j and w_j? Yes, of course, since (63) is clearly to be interpreted as the injunction to remove the Bohr-shell oscillations from the corresponding expression

$$\Delta_s E = Z^2(\nu_s - [\nu_s]) - \zeta\Big(\frac{2}{3}\nu_s^3 - \sum_{n=1}^{[\nu_s]} 2n^2\Big) \quad . \tag{3-64}$$

We are thus led back to Eq. (22), and identifying the smooth content gives Scott's correction (32), as we have seen above. In general terms: in an algebraic result, such as (63), the averaging over ζ_s is to be unterstood as the demand to construct the respective unaveraged expression in terms of ν_s and $[\nu_s]$, here: Eq. (64), and to remove the spurious Bohr-shell oscillations from it. We shall meet examples for this procedure as we proceed.

We are now set to finally write down the TFS energy functional. It is obtained from the TF functional (2-45) by replacing the TF version of E_1 by the TFS expression (58). So we have

$$E_{TFS} = \sum_{j=1}^{J} w_j \, E_{\zeta\zeta_j} + E_s + E_2 - \zeta N \quad , \tag{3-65}$$

which can be split into the TF energy functional plus its modification,

$$E_{TFS} = E_{TF} + \Delta_s E \quad , \tag{3-66}$$

with [this is Eq.(2-45)]

$$E_{TF} = \int (d\vec{r}) \left(-\frac{1}{15\pi^2}\right) [-2(V+\zeta)]^{5/2} - \frac{1}{8\pi}\int (d\vec{r}) [\vec{\nabla}(V+\frac{Z}{r})]^2 - \zeta N \qquad (3\text{-}67)$$

and

$$\Delta_s E = \sum_{j=1}^{J} w_j \int (d\vec{r}) \left\{\frac{1}{15\pi^2}[-2(V+\zeta_j)]^{5/2} + (\zeta_j-\zeta)\frac{1}{3\pi^2}[-2(V+\zeta_j)]^{3/2}\right\}$$
$$+ \int (d\vec{r}) V\rho_s + Z^2 n_s + \zeta N_s \quad . \qquad (3\text{-}68)$$

TFS density. As before we find the density by considering the response of E_1 to infinitesimal variations of the effective potential,

$$\delta_V E_1 = \int (d\vec{r}) \delta V(\vec{r}) n(\vec{r}) \quad , \qquad (3\text{-}69)$$

which is Eq.(2-14). The situation is different from the TF one, now, because in addition to the explicit dependence on V there is an implicit one, hidden in the ζ_j:

$$\delta_V \zeta_j = - \int (d\vec{r}) \delta V \, |\psi_{n_j}|^2_{av} \quad , \qquad (3\text{-}70)$$

which is a consequence of Eq.(50). This V dependence of the ζ_j gives rise to a contribution to the density

$$\sum_{j=1}^{J} w_j \frac{\partial E_{\zeta\zeta_j}}{\partial \zeta_j} (-|\psi_{n_j}|^2_{av}(r))$$
$$= \sum_{j=1}^{J} w_j \, Q_j \, |\psi_{n_j}|^2_{av}(r) \quad , \qquad (3\text{-}71)$$

where we have introduced [see Eq.(44)]

$$Q_j \equiv - \frac{\partial}{\partial \zeta_j} E_{\zeta\zeta_j} = (\zeta_j - \zeta) \int (d\vec{r}) \, \frac{1}{\pi^2} \, [-2(V+\zeta_j)]^{1/2} \quad . \qquad (3\text{-}72)$$

Equation (71) has to be supplemented by the part of the density that is obtained from the explicit dependence of E_1 on V,

$$\frac{1}{3\pi^2}[-2(V+\zeta)]^{3/2} - \sum_{j=1}^{J} w_j \{\frac{1}{3\pi^2}[-2(V+\zeta_j)]^{3/2} + (\zeta_j-\zeta)\frac{1}{\pi^2}[-2(V+\zeta_j)]^{1/2}\}$$

$$+ \rho_s \tag{3-73}$$

$$= \sum_{j=1}^{J} w_j \int_{\zeta}^{\zeta_j} d\zeta'(\zeta'-\zeta)\frac{1}{\pi^2}[-2(V+\zeta')]^{-1/2} + \rho_s .$$

The second version makes use of $E_{\zeta\zeta_j}$ as given in Eq. (19) and emphasizes the strong cancellations that occur for $r\to o$. The total electron density is the sum of (71) and (73). It is conveniently split into a density of the innermost electrons, n_{IME}, and the rest of the atom, $\tilde{n}$:

$$n = n_{IME} + \tilde{n} , \tag{3-74}$$

where

$$n_{IME} = \rho_s + \sum_{j=1}^{J} w_j \, Q_j \, |\psi_{n_j}|^2_{av} \tag{3-75}$$

and

$$\tilde{n} = \sum_{j=1}^{J} w_j \int_{\zeta}^{\zeta_j} d\zeta'(\zeta'-\zeta)\frac{1}{\pi^2}[-2(V+\zeta')]^{-1/2}$$

$$= \frac{1}{3\pi^2}[-2(V+\zeta)]^{3/2} - \sum_{j=1}^{J} w_j \{\frac{1}{3\pi^2}[-2(V+\zeta_j)]^{3/2} \tag{3-76}$$

$$+ (\zeta_j-\zeta)\frac{1}{\pi^2}[-2(V+\zeta_j)]^{1/2}\} .$$

Because of said cancellations $\tilde{n}$ is proportional to $r^{1/2}$ for $r\to o$, so that the density in the close proximity of the nucleus is entirely given by n_{IME}. In particular,

$$n(r=o) = n_{IME}(r=o)$$

$$= \frac{(2Z)^3}{4\pi}[\sum_{n'=1}^{n_s} \frac{1}{n'^3} + \sum_{j=1}^{J} w_j \, Q_j \, \frac{1}{2n_j^5}] , \tag{3-77}$$

which uses

$$|\psi_{n_j}|^2_{av}(r=o) = \frac{Z^3}{4\pi}\frac{4}{n_j^5} \quad , \tag{3-78}$$

as illustrated for $n_j=1,2,$ and 3 in Eq.(53). We shall have to say more about n(r=o), but a few checks of consistency are in order, first.

Consistency. The integrated density must equal the number of electrons. Is this so? The integrals over $(d\vec{r})$ of (75) and (76) are

$$\int(d\vec{r})n_{IME} = N_s + \sum_{j=1}^{J} w_j \, Q_j \tag{3-79}$$

and

$$\int(d\vec{r})\,\tilde{n} = \int(d\vec{r})\,\{\frac{1}{3\pi^2}[-2(V+\zeta)]^{3/2} - \sum_{j=1}^{J} w_j \frac{1}{3\pi^2}[-2(V+\zeta_j)]^{3/2}\}$$
$$- \sum_{j=1}^{J} w_j \, Q_j \quad , \tag{3-80}$$

respectively, where Eqs.(40) and (50) as well as the definition of Q_j in (72) have been made use of. Consequently,

$$\int(d\vec{r})n = \int(d\vec{r})\,\{\frac{1}{3\pi^2}[-2(V+\zeta)]^{3/2} - \sum_{j=1}^{J} w_j \frac{1}{3\pi^2}[-2(V+\zeta_j)]^{3/2}\} + N_s \, . \tag{3-81}$$

On the other hand, the count of electrons is, according to Eq. (2-12), given by

$$N = \frac{\partial}{\partial\zeta} E_1 \quad , \tag{3-82}$$

which for the TFS model reads

$$N = \int(d\vec{r})\,\frac{1}{3\pi^2}[-2(V+\zeta)]^{3/2} - \sum_{j=1}^{J} w_j \int(d\vec{r})\,\frac{1}{3\pi^2}[-2(V+\zeta_j)]^{3/2} + N_s \, . \tag{3-83}$$

It is, indeed, equal to the integrated density. Note, in particular, that this is true for any choice of values for the ζ_j and w_j. Neither does n_s, the number of shells of specially treated strongly bound electrons, matter. In other words: the statement

$$\int (d\vec{r})\, n = N \tag{3-84}$$

holds independent of the averaging procedure selected for the removal of the spurious Bohr-shell oscillations.

In Chapter Two, it was argued that Eq. (84) is equivalent to stating that E_1 does not depend on V and ζ individually, but only on their sum $V+\zeta$ [see after Eq. (2-23)]. Here, E_1 is the sum of its TF version and $\Delta_s E$ of Eq. (68),

$$E_1 = \int (d\vec{r}) \left(-\frac{1}{15\pi^2}\right) [-2(V+\zeta)]^{5/2} + \Delta_s E \quad , \tag{3-85}$$

so that the question is whether $\Delta_s E$ is a function(al) of $V+\zeta$. Since

$$\int (d\vec{r})\, V\rho_s + \zeta N_s = \int (d\vec{r})\,(V+\zeta)\,\rho_s \quad , \tag{3-86}$$

and

$$\begin{aligned} V+\zeta_j &= (V+\zeta) + (\zeta_j - \zeta) \\ &= (V+\zeta) + \frac{Z^2}{2n_j^2} - \int (d\vec{r})\,(V+\zeta)\,|\psi_{n_j}|^2_{av} \quad , \end{aligned} \tag{3-87}$$

the answer is affirmative.

Then there is the explicit dependence of $\Delta_s E$, and therefore of E_1, on Z, the nuclear charge. But E_1 must not make any reference to the external potential (here: $-Z/r$), it is solely determined by the effective potential (plus ζ, as we have just recalled). This is essential in relating the Z derivative of the energy to the electrostatic interaction energy of the nuclear charge and the electrons

$$Z\frac{\partial}{\partial Z} E = \int (d\vec{r}) \left(-\frac{Z}{r}\right) n(\vec{r}) \quad . \tag{3-88}$$

This is Eq. (1-96). We have already made use of it, repeatedly, not in the form of Eq. (88), but in one where the right-hand side is expressed in terms of the potential. It emerges from (88) when Poisson's equation is used and two partial integrations are performed:

$$Z \frac{\partial}{\partial Z} E = \int (d\vec{r}) \left(-\frac{Z}{r}\right) \left[-\frac{1}{4\pi}\nabla^2 \left(V + \frac{Z}{r}\right)\right]$$

$$= \int (d\vec{r}) \left[-\left(V + \frac{Z}{r}\right)\right] \left(-\frac{1}{4\pi}\nabla^2\right) \left(-\frac{Z}{r}\right) \tag{3-89}$$

$$= - Z \left(V + \frac{Z}{r}\right)\Big|_{r=o} \quad ;$$

the last step recognizes $Z\delta(\vec{r})$ as the source of the Coulomb potential $-Z/r$. If we, however, stop after the first partial integration, the result is

$$Z \frac{\partial}{\partial Z} E = - \frac{1}{4\pi} \int (d\vec{r}) \vec{\nabla} \left(V + \frac{Z}{r}\right) \cdot \vec{\nabla} \frac{Z}{r}$$

$$= Z \frac{\partial}{\partial Z} \left\{- \frac{1}{8\pi} \int (d\vec{r}) \left[\vec{\nabla} \left(V + \frac{Z}{r}\right)\right]^2\right\} \quad , \tag{3-90}$$

or, in view of Eqs. (66), (67), and (68), for the TFS energy,

$$Z \frac{\partial}{\partial Z} E_{TFS} = Z \frac{\partial}{\partial Z}(E_{TFS} - \Delta_s E) \quad , \tag{3-91}$$

which requires

$$Z \frac{\partial}{\partial Z} \Delta_s E = o \quad . \tag{3-92}$$

[Please do not miss that we have done nothing more than reverse the steps of Eq. (2-229).]

There are various Z dependences in ΔE_s, Eq. (68). Besides the explicit Z^2 terms in (68) and in (50), there is also the Z dependence of the Coulomb densities $|\psi_{n_j}|^2_{av}$ and ρ_s. They are of the structure Z^3 times a function of Zr, so that

$$Z \frac{\partial}{\partial Z} \begin{Bmatrix} |\psi_{n_j}|^2_{av}(r) \\ \rho_s(r) \end{Bmatrix} = (3 + \vec{r}\cdot\vec{\nabla}) \begin{Bmatrix} |\psi_{n_j}|^2_{av}(r) \\ \rho_s(r) \end{Bmatrix} . \tag{3-93}$$

This has the consequence

$$Z \frac{\partial}{\partial Z} \left\{\int (d\vec{r})\, V \rho_s + Z^2 n_s\right\} =$$

$$= \int (d\vec{r}) V (3+\vec{r}\cdot\vec{\nabla}) \rho_s + 2Z^2 n_s \quad . \tag{3-94}$$

A partial integration turns the integrand into

$$\begin{aligned} [3V - \vec{\nabla}\cdot(\vec{r}\, V)]\rho_s &= [-\vec{r}\cdot\vec{\nabla}V]\rho_s \\ &= [-r\frac{\partial}{\partial r} V]\rho_s = [V - \frac{\partial}{\partial r}(rV)]\rho_s \quad , \end{aligned} \tag{3-95}$$

which in conjunction with (41) produces

$$\begin{aligned} Z \frac{\partial}{\partial Z} \{\int (d\vec{r}) V\rho_s + Z^2 n_s\} \\ = \int (d\vec{r}) [V + \frac{Z}{r} - \frac{\partial}{\partial r}(rV)]\rho_s \quad . \end{aligned} \tag{3-96}$$

The analog of Eq.(41) for a single Bohr shell,

$$\int (d\vec{r}) \frac{Z}{r} |\psi_{n_j}|^2_{av} = (-2)(-\frac{Z^2}{2n_j^2}) = \frac{Z^2}{n_j^2} \quad , \tag{3-97}$$

can be employed in writing ζ_j as

$$\zeta_j = -\{\frac{Z^2}{2n_j^2} + \int (d\vec{r}) V |\psi_{n_j}|^2_{av}\} \quad , \tag{3-98}$$

after which the steps from (94) to (96) can be repeated with the necessary changes. The outcome is

$$Z \frac{\partial}{\partial Z} \zeta_j = -\int (d\vec{r}) [V + \frac{Z}{r} - \frac{\partial}{\partial r}(rV)] |\psi_{n_j}|^2_{av} \quad . \tag{3-99}$$

This, Eq.(96), and the definition of Q_j in (72) together produce

$$Z \frac{\partial}{\partial Z} \Delta_s E = \int (d\vec{r}) [V + \frac{Z}{r} - \frac{\partial}{\partial r}(rV)] [\rho_s + \sum_{j=1}^{J} w_j Q_j |\psi_{n_j}|^2_{av}] \quad , \tag{3-100}$$

or with (75),

$$Z \frac{\partial}{\partial Z} \Delta_s E = \int (d\vec{r}) [V + \frac{Z}{r} - \frac{\partial}{\partial r}(rV)] n_{IME} \quad . \tag{3-101}$$

This has to be reconciled, in some sense, with Eq. (92), otherwise the TFS model would be internally inconsistent.

Let us recall that, for a spherically symmetric density n(r), as is the situation for an isolated atom, Poisson's equation

$$- \frac{1}{4\pi} \nabla^2 (V + \frac{Z}{r}) = n(r) \tag{3-102}$$

is solved by

$$V + \frac{Z}{r} = \int (d\vec{r}\,') \frac{n(r')}{|\vec{r}-\vec{r}\,'|} = \int (d\vec{r}\,') \frac{n(r')}{r_>} \quad , \tag{3-103}$$

where $r_>$ denotes the larger one of r and r'; the latter identity is based upon the spherical symmetry of the density. The contents of the square brackets of (101) can be written as

$$V + \frac{Z}{r} - \frac{\partial}{\partial r}(rV) = - r \frac{\partial}{\partial r}(V + \frac{Z}{r}) \quad , \tag{3-104}$$

which, in connection with (103), draws our attention to

$$- r \frac{\partial}{\partial r} \frac{1}{r_>} = \begin{Bmatrix} 0 \, , & \text{for} \quad r < r' \\ \frac{1}{r} \, , & \text{for} \quad r > r' \end{Bmatrix} = \frac{1}{r} \eta(r-r') \quad . \tag{3-105}$$

Therefore,

$$Z \frac{\partial}{\partial Z} \Delta_s E = \int (d\vec{r}) (d\vec{r}\,') n_{IME}(r) n(r') \frac{1}{r} \eta(r-r') \quad . \tag{3-106}$$

Since the range of integration over r' is limited by r and a further integration over r weighted by $n_{IME}(r)$ is required, only those values of r' contribute for which $n_{IME}(r')$ is of significant size. Consequently, (106) is well approximated when $n(r') = n_{IME}(r') + \tilde{n}(r')$ is replaced by $n_{IME}(r')$, because $\tilde{n}$ is effectively zero where n_{IME} is large. After this replacement, the integrand can be symmetrized between r and r', so that

$$Z \frac{\partial}{\partial Z} \Delta_s E \cong \frac{1}{2} \int (d\vec{r}) (d\vec{r}\,') n_{IME}(r) n_{IME}(r') \frac{1}{r_>} =$$

$$= \frac{1}{2} \int (d\vec{r})\,(d\vec{r}\,')\; \frac{n_{IME}(r)\,n_{IME}(r')}{|\vec{r} - \vec{r}\,'|} \quad . \tag{3-107}$$

This is the electrostatic energy of the charge distribution due to the innermost electrons. Since $n_{IME}(r)$ equals a factor Z^3 times a function of Zr (just as ρ_s and $|\psi_{n_j}|^2_{av}$ do individually), we have

$$Z \frac{\partial}{\partial Z} \Delta_s E = Z \times \{\text{ a positive number }\} , \tag{3-108}$$

this "number" being composed of the particular values of n_s, w_j, and ζ_j. We have, thus, found that the implicit Z dependence of $\Delta_s E$ is associated with an energy of order Z - perfectly negligible on the scale set by $Z^{7/3}$ (TF) and Z^2 (Scott). In other words: Eq.(92) is obeyed within the accuracy of the TFS model; there is no internal inconsistency.

Fine, but didn't we just blow it? Certainly, $\Delta_s E$ is Scott's correction, it equals $\frac{1}{2}Z^2$; and, being independent of N, there is no difference between its partial and its total derivative with respect to Z. Shouldn't we, consequently, obtain

$$Z \frac{\partial}{\partial Z} \Delta_s E = Z^2 \quad ? \tag{3-109}$$

Or is, after all, Eq.(92) the correct statement? The puzzling answer is: all three equations - (92),(108), and (109) - are true.

This is so because the respective left-hand sides have different meanings. Equation (92) is derived in the framework of the general formalism: we start with a functional of V and ζ, which is specified by a given number of electrons, N, and a given external potential, here $-Z/r$; then we ask for the change in energy when this external potential is varied infinitesimally, here done by varying Z to $Z + \delta Z$; of course, there are induced changes of V and ζ [so that, e.g., the second integral in (67) still exists], but the stationary property of the energy functional under variations of V and ζ implies that these induced changes do not contribute to the change in energy to first order; the general answer

$$\cdot\, \delta_{V_{ext}} E = \int (d\vec{r})\, \delta V_{ext}\, n \tag{3-110}$$

[see Eq.(2-434)] appears as Eq.(88) in the present context; it leads to Eq.(92), which we now read correctly as the change in $\Delta_s E$ caused by varying nothing but the strength of the Coulomb potential of the nucleus. In the TFS energy functional this external potential $-Z/r$ occurs only in the second term of Eq.(67). Therefore, the Z, which is contained in (68) both explicitly and implicitly in the ζ_j and in the Bohr-atom density ρ_s, and with respect to which we differentiate in Eq.(108), must posses a different significance. It makes reference not to the external potential $-Z/r$, but to the effective potential V, in the sense of

$$Z = (-rV)\Big|_{r=0} = -\int(d\vec{r})\,r(V+\zeta)\delta(\vec{r}) \quad , \tag{3-111}$$

where we have added the (otherwise innocuous) constant ζ in order to emphasize the dependence of $\Delta_s E$ on the sum $V+\zeta$. Consequently, the differentiation in Eqs.(108), or (101), really means a variation of the (Coulomb part of the) effective potential, and not of the nuclear charge. Of course, for the actual V, the Z of (111) must equal the Z of the Coulomb potential of the nucleus, but not so for the trial potentials that we are free to use in $\Delta_s E$ of Eq.(68). Now that we have recognized that the Z in (68) changes when the effective potential is varied, we must also take into account the corresponding additional contribution to the density, labelled n_Z. It emerges from Eqs.(101) and (111), when combined into

$$\begin{aligned}\int(d\vec{r})\delta V\, n_Z &= \frac{\partial}{\partial Z}\Delta_s E\ \delta_V Z \\ &= \int(d\vec{r})\delta V\,(-r\delta(\vec{r}))\frac{\partial}{\partial Z}\Delta_s E \quad ,\end{aligned} \tag{3-112}$$

as

$$n_Z(r) = -\,r\delta(\vec{r})\,\frac{1}{Z}\int(d\vec{r}\,')n_{IME}(r')[V(r') + \frac{Z}{r'} - \frac{\partial}{\partial r'}(r'V(r'))] \quad , \tag{3-113}$$

and has almost no significance, because the product $r\delta(r)$ is effectively zero. Its only use consists in the possibility of evaluating the response of E_1 to an arbitrary variation of V in the standard way of Eq.(69), where the inclusion of n_Z into the density enables one to consider variations of the kind $\delta V = -\delta Z/r$. We then obtain from (69) the integrated version (101) that we know already. Note, in particular, that n_Z integrates to zero, and that its only contribution to the potential is an additive constant for r=o [see Eq.(103)], where V is infinite, anyhow. In short: as long as we remember that Eq.(101) must be taken into account when variations of the limiting Coulomb part of the effective potential are considered, we can forget about n_Z.

As to Eq. (109), we need only remark that it does not refer to the energy functional (68), but to its numerical value for $V=-Z/r$. Indeed, we have calculated Scott's correction by simply inserting the Coulomb potential into $\Delta_s E$. This raises the question, whether we can do better than that. How does one account for the deviation of V from its limiting Coulomb shape when evaluating the Scott term? The clue is Eq. (89) which relates the energy to the small-r form of the potential. In the TF model we exploited this equation in connection with the scaling properties of the TF energy functional. In the following section we shall do the analogous thing for the TFS model.

Scaling properties of the TFS model. As in Chapter Two, we consider scaling transformations that replace the actual V, ζ, and Z according to

$$\begin{aligned} V(r) &\to \mu^{\nu} V(\mu r) , \\ \zeta &\to \mu^{\nu} \zeta , \qquad (\mu>0) \\ Z &\to \mu^{\nu-1} Z , \end{aligned} \tag{3-114}$$

which repeat Eqs. (2-211), (2-213), and (2-214). Again, the stationary property of the energy functional, here: the TFS functional (66), implies that all first order changes must originate in the scaling of Z [the Z of Eqs. (88) and (89), to be precise]. Thus

$$\delta_{\mu} E^{\mu}_{TFS} = \delta\mu(\nu-1)\, Z \frac{\partial}{\partial Z} E_{TFS} , \tag{3-115}$$

as in Eq. (2-220), or with (89),[8]

$$\delta_{\mu} E^{\mu}_{TFS} = -\,\delta\mu(\nu-1)\, Z(V+\frac{Z}{r})\Big|_{r=0} , \tag{3-116}$$

where $\mu=1+\delta\mu$ with an infinitesimal $\delta\mu$ is understood. On the left-hand side of (116),

$$\delta_{\mu} E^{\mu}_{TFS} = \delta_{\mu} E^{\mu}_{TFS} + \delta_{\mu} \Delta_s E^{\mu} , \tag{3-117}$$

we already know $\delta_{\mu} E_{TF}$ from the earlier investigations,

$$\delta_{\mu} E^{\mu}_{TF} = \{ (\frac{5}{2}\nu-3) \int (d\vec{r}) (-\frac{1}{15\pi^2}) [-2(V+\zeta)]^{5/2} -$$

$$-(2\nu-1)\frac{1}{8\pi}\int(d\vec{r})\,[\vec{\nabla}(V+\frac{Z}{r})]^2 - \nu\zeta N\}\delta\mu \quad , \tag{3-118}$$

which is the left-hand side of Eq.(2-222). For the evaluation of $\delta_\mu \Delta_s E^\mu$, it is useful to prepare some tools first.

The scaling of Z [either the explicit statement in (114) or the (identical) result of inserting the scaled V into (111)] has an effect on the Bohr-shell densities $|\psi_{n_j}|^2_{av}$ and ρ_s, given by

$$\begin{pmatrix} \rho_s \\ |\psi_{n_j}|^2_{av} \end{pmatrix}(r) \rightarrow \mu^{3(\nu-1)} \begin{pmatrix} \rho_s \\ |\psi_{n_j}|^2_{av} \end{pmatrix}(\mu^{\nu-1}\, r) \quad . \tag{3-119}$$

A consequence thereof is

$$\begin{aligned} \{\int(d\vec{r})V(r)\;\rho_s(r)\}^\mu &= \mu^{4\nu-3}\int(d\vec{r})V(\mu r)\;\rho_s(\mu^{\nu-1}\,r) \\ &= \mu^{4\nu-3}\,\mu^{-3(\nu-1)}\int(d\mu^{\nu-1}\vec{r})V(\mu r)\rho_s(\mu^{\nu-1}r) \\ &= \mu^\nu\int(d\vec{r})V(\mu^{2-\nu}r)\rho_s(r) \end{aligned} \tag{3-120}$$

so that

$$\begin{aligned} \delta_\mu\{\int(d\vec{r})V\rho_s\}^\mu &= \delta\mu\{\nu\int(d\vec{r})V\rho_s + (2-\nu)\int(d\vec{r})\,(r\frac{\partial}{\partial r}V)\,\rho_s\} \\ &= \delta\mu\{2(\nu-1)\int(d\vec{r})V\rho_s + (2-\nu)\int(d\vec{r})\frac{\partial}{\partial r}(rV)\,\rho_s\} \quad . \end{aligned} \tag{3-121}$$

An analogous statement holds for the integral appearing in Eq.(98), implying

$$\delta_\mu\;\zeta_j^\mu = \delta\mu\{2(\nu-1)\zeta_j - (2-\nu)\int(d\vec{r})\frac{\partial}{\partial r}(rV)\,|\psi_{n_j}|^2_{av}\} \quad , \tag{3-122}$$

or, more conveniently for the sequel,

$$\delta_\mu(\mu^{-\nu}\zeta_j^\mu) = \delta\mu(\nu-2)\{\zeta_j + \int(d\vec{r})\frac{\partial}{\partial r}(rV)\,|\psi_{n_j}|^2_{av}\} \quad . \tag{3-123}$$

It is used in exhibiting the scaling behavior of $E_{\zeta\zeta_j}$, preferably studied in the concise form (19):

$$\delta_\mu E^\mu_{\zeta\zeta_j} = \delta_\mu\{-\int(d\vec{r})\int_{\mu^\nu\zeta}^{\zeta_j^\mu} d\zeta'(\zeta'-\mu^\nu\zeta)\frac{1}{\pi^2}[-2(\mu^\nu V(\mu r)+\mu^\nu\zeta)]^{1/2}\}$$

$$= \delta_\mu\{-\mu^{\frac{5}{2}\nu-3}\int(d\vec{r})\int_\zeta^{\mu^{-\nu}\zeta_j^\mu} d\zeta'(\zeta'-\zeta)[-2(V(r)+\zeta)]^{1/2}\} \tag{3-124}$$

$$= \delta\mu(\frac{5}{2}\nu-3)E_{\zeta\zeta_j} + \frac{\partial E_{\zeta\zeta_j}}{\partial\zeta_j}\delta_\mu(\mu^{-\nu}\zeta_j^\mu) \quad .$$

With Eq.(72) and (123) this supplies

$$\delta_\mu E^\mu_{\zeta\zeta_j} = \delta\mu\{(\frac{5}{2}\nu-3)E_{\zeta\zeta_j} - (\nu-2)Q_j[\zeta_j + \int(d\vec{r})\frac{\partial}{\partial r}(rV)|\psi_{n_j}|^2_{av}]\} \quad . \tag{3-125}$$

This combines with both Eq.(121) and

$$\delta_\mu\{Z^2 n_s+\zeta N_s\}^\mu = \delta\mu\{2(\nu-1)Z^2 n_s + \nu\zeta N_s\} \tag{3-126}$$

to

$$\delta_\mu\ \Delta_s E^\mu = \delta\mu\{(\frac{5}{2}\nu-3)\Delta_s E - (\nu-2)[\sum_j w_j Q_j\zeta_j + \frac{1}{2}\int(d\vec{r})V\rho_s$$

$$+\int(d\vec{r})n_{IME}\frac{\partial}{\partial r}(rV) + \frac{1}{2}Z^2 n_s + \frac{3}{2}\zeta N_s]\} \quad . \tag{3-127}$$

Together with Eq.(118), we then conclude from Eqs.(116) and (117) that

$$(\frac{5}{2}\nu-3)E_{TFS} + \frac{3}{2}(\nu-2)\zeta N + (\nu-1)Z(V+\frac{Z}{r})\Big|_{r=0}$$

$$+ (\frac{1}{2}\nu-2)\frac{1}{8\pi}\int(d\vec{r})[\vec{\nabla}(V+\frac{Z}{r})]^2 \tag{3-128}$$

$$= (\nu-2)[\sum_{j=1}^{J} w_j Q_j\zeta_j + \frac{1}{2}\int(d\vec{r})V\rho_s + \int(d\vec{r})n_{IME}\frac{\partial}{\partial r}(rV) + \frac{1}{2}Z^2 n_s + \frac{3}{2}\zeta N_s] \quad ,$$

which, finally, states the scaling behavior of the TFS model.

Please observe that the terms on the right-hand side of Eq.(128) all refer to the specially treated strongly bound electrons. Therefore, replacing E_{TFS} by E_{TF} and setting the right-hand side equal to zero should reproduce the corresponding statement about the TF model. Indeed, what is obtained is (a rewritten version of) Eq.(2-222). There it was noticed that the most useful choice for ν is $\nu=4$. This is still true. After employing

$$\zeta(Z,N) = -\frac{\partial}{\partial N} E_{TFS}(Z,N) \tag{3-129}$$

[this has appeard earlier as Eq.(2-225)] and Eq.(89), Eq.(128) reads, for $\nu=4$,

$$[7 - 3(Z\frac{\partial}{\partial Z} + N\frac{\partial}{\partial N})]\ E_{TFS}(Z,N) \tag{3-130}$$

$$= 2[\sum_{j=1}^{J} w_j Q_j \zeta_j + \frac{1}{2}\int(d\vec{r})V\rho_s + \int(d\vec{r})n_{IME}\frac{\partial}{\partial r}(rV) + \frac{1}{2}Z^2 n_s - \frac{3}{2}N_s\frac{\partial}{\partial N}E_{TFS}(Z,N)],$$

where it is made explicit that we are now interested in the dependence of E_{TFS} on Z and N. This generalizes Eq.(2-226).

Second quantitative derivation of Scott's correction. Until now we have always been content with the approximation $V\cong -Z/r$ when evaluating the Scott correction. It is time to pay attention to the difference between V and its small-r Coulomb part. In general, V(r) is given by Eq.(103). We rewrite it by using the identity

$$\frac{1}{r_>} = \frac{1}{r'}\eta(r'-r) + \frac{1}{r}\eta(r-r') = \frac{1}{r'} - (\frac{1}{r'} - \frac{1}{r})\eta(r-r') , \tag{3-131}$$

obtaining

$$V(r) = -\frac{Z}{r} + \int(d\vec{r}')\,\frac{n(r')}{r'} - \int(d\vec{r}')n(r')(\frac{1}{r'} - \frac{1}{r})\eta(r-r') . \tag{3-132}$$

The first integral appeared in Eq.(99), it equals $-\frac{\partial}{\partial Z}E_{TFS}(Z,N)$. For the second one we write v(r),

$$v(r) = \int(d\vec{r}')n(r')(\frac{1}{r'} - \frac{1}{r})\eta(r-r') , \tag{3-133}$$

so that

$$V(r) = -\frac{Z}{r} - \frac{\partial E_{TFS}}{\partial Z} - v(r) . \tag{3-134}$$

In this form, we shall insert it into the right-hand side of Eq.(130). It is thereby not necessary to keep track of more than the first order in $V+Z/r$, since the TFS model is based on the physical argument that this is a small quantity for the strongly bound electrons [see Eq.(37)].

From Eq. (50) we get

$$\zeta_j = \frac{Z^2}{2n_j^2} + \frac{\partial E_{TFS}}{\partial Z} + \bar{v}_{n_j} \quad , \tag{3-135}$$

with

$$\bar{v}_{n_j} \equiv \int (d\vec{r}) v(r) |\psi_{n_j}|^2_{av}(r) \tag{3-136}$$

being the average of $v(r)$ over the n_j-th Bohr shell. When evaluating the integral, that gives Q_j in (72), to first order in v, another average of $v(r)$ is also met,

$$\bar{v}'_{n_j} \equiv \int (d\vec{r}) v(r) \frac{4}{5\pi^2} \left(\frac{Z}{2n_j^2}\right)^3 \left(\frac{2n_j^2}{Zr} - 1\right)^{-1/2} \quad , \tag{3-137}$$

where the range of integration is $r<2n_j^2/Z$, of course. Then

$$Q_j = (\zeta_j - \zeta) \frac{2n_j^5}{Z^2} [1 - 5\frac{n_j^2}{Z^2}(\bar{v}_{n_j} - \bar{v}'_{n_j})] \quad , \tag{3-138}$$

which in conjunction with (135) and (129) produces

$$\begin{aligned} Q_j = n_j^3 - \frac{n_j^5}{Z^2}[3\bar{v}_{n_j} - 5\bar{v}'_{n_j}] + \frac{\partial E_{TFS}}{\partial Z} \frac{2n_j^5}{Z^2} \\ + \frac{\partial E_{TFS}}{\partial N} \frac{2n_j^5}{Z^2}[1 - \frac{5n_j^2}{Z^2}(\bar{v}_{n_j} - \bar{v}'_{n_j})] \quad , \end{aligned} \tag{3-139}$$

to first order in v and $\partial E_{TFS}/\partial Z$. Further we have

$$\int (d\vec{r}) V \rho_s = - 2Z^2 n_s - \frac{\partial E_{TFS}}{\partial Z} N_s - \bar{v}_s \quad , \tag{3-140}$$

where

$$\bar{v}_s \equiv \int (d\vec{r}) v(r) \rho_s(r) = \sum_{n'=1}^{n_s-} 2n'^2 \, \bar{v}_{n'} \quad . \tag{3-141}$$

Then there is the quantity

$$\int (d\vec{r}) n_{IME}(r) \frac{\partial}{\partial r} \left(rV(r)\right) =$$

$$= \int (d\vec{r}) [\rho_s + \sum_{j=1}^{J} w_j Q_j |\psi_{n_j}|^2_{av}] [-\frac{\partial E_{TFS}}{\partial Z} - \frac{\partial}{\partial r}(rv)]$$

$$= -\frac{\partial E_{TFS}}{\partial Z}(N_s + \sum_{j=1}^{J} w_j Q_j) - \overline{\partial (rv)/\partial r}_s \tag{3-142}$$

$$- \sum_{j=1}^{J} w_j Q_j \overline{\partial (rv)/\partial r}_{n_j} \quad ,$$

which to first order in V+Z/r is given by

$$\int (d\vec{r}) n_{IME} \frac{\partial}{\partial r}(rV) = -\frac{\partial E_{TFS}}{\partial Z} - \overline{\partial (rv)/\partial r}_s$$

$$-\sum_{j=1}^{J} w_j (n_j^3 + \frac{2n_j^5}{Z^2} \frac{\partial E_{TFS}}{\partial N}) (\frac{\partial E_{TFS}}{\partial Z} + \overline{\partial (rv)/\partial r}_{n_j}) . \tag{3-143}$$

We are now prepared to evaluate the right-hand side of (130). The outcome is

$$[7 - 3(Z\frac{\partial}{\partial Z} + N\frac{\partial}{\partial N})] E_{TFS}(Z,N)$$

$$= Z^2 \{\sum_{j=1}^{J} w_j n_j - n_s\} + 2(\frac{\partial}{\partial Z} + \frac{\partial}{\partial N}) E_{TFS}(Z,N) \{\sum_{j=1}^{J} w_j n_j^3 - \frac{3}{2} N_s\}$$

$$- \{\sum_{j=1}^{J} w_j n_j^3 (\bar{v}_{n_j} - 5\bar{v}'_{n_j} + 2\overline{\partial (rv)/\partial r}_{n_j}) \tag{3-144}$$

$$+ \bar{v}_s + 2\overline{\partial (rv)/\partial r}_s\}$$

$$- \frac{2}{Z^2} \frac{\partial E_{TFS}(Z,N)}{\partial N} \{\sum_{j=1}^{J} w_j n_j^5 (3\bar{v}_{n_j} - 5\bar{v}'_{n_j} + 2 \overline{\partial (rv)/\partial r}_{n_j})\}$$

where the various curly brackets have to be replaced by their "smooth parts" according to our general recipy for removing the spurious Bohr-shell oscillations. The first two such expressions in (144) are the familiar ones of Eq.(63) - we know that they are replaced by $\frac{1}{2}$ and 0, respectively. We do not know the corresponding numbers for the two expressions referring to v; fortunately, we do not need them, because the averages of v are essentially equal to Z times a number.

To make this point let us look back at the definition of $v(r)$ in Eq.(133) and insert $n = n_{IME} + \tilde{n}$ [Eq.(74)] to split $v(r)$ into v_{IME} and $\tilde{v}$, correspondingly. As stated repeatedly, n_{IME} has the structure: Z^3 times a function of Zr. This implies immediately that v_{IME} equals Z times a function of Zr. Concerning $\tilde{v}$, we first remark that the small-r behavior of $\tilde{n}$, which is relevant here, is given by

$$\begin{aligned}\tilde{n}(r) &\cong \sum_{j=1}^{J} w_j \frac{1}{2}(\zeta_j - \zeta)^2 \frac{1}{\pi^2}\left(2\frac{Z}{r}\right)^{-1/2} \\ &\cong \frac{1}{2\pi^2}\sum_{j=1}^{J} w_j \left(\frac{Z^2}{2n_j^2}\right)^2 \frac{1}{Z}\sqrt{Zr/2} \\ &= \frac{1}{8\pi^2}\left[\sum_{j=1}^{J} w_j/n_j^4\right] Z^3 \sqrt{Zr/2} \quad ,\end{aligned} \tag{3-145}$$

also of the form: Z^3 times a function of Zr, so that $\tilde{v}$, like v_{IME}, equals Z times a function of Zr, at least for the small r of importance. Since the essential measure of distance is Zr in both (136) and (137), we have, as announced above,

$$\begin{aligned}&\{\bar{v}_{n_j};\ \bar{v}'_{n_j};\ \bar{v}_s;\ \overline{\partial(rv)/\partial r}_{n_j};\ \overline{\partial(rv)/\partial r}_s\} \\ &= Z \times \{\text{ a corresponding number }\} \quad .\end{aligned} \tag{3-146}$$

Consequently, the contents of the third and fourth curly brackets in (144) are to be replaced by $a_1 Z$ and $a_2 Z$, respectively, with a_1 and a_2 being numbers that are (practically) independent of Z (if they don't vanish to begin with).

We then arrive at

$$\begin{aligned}&\left[7 - 3\left(Z\frac{\partial}{\partial Z} + N\frac{\partial}{\partial N}\right)\right] E_{TFS}(Z,N) \\ &= \frac{1}{2}Z^2 - a_1 Z - a_2 \frac{1}{Z}\frac{\partial}{\partial N} E_{TFS}(Z,N) \quad .\end{aligned} \tag{3-147}$$

This extends Eq.(2-226), the corresponding equation obeyed by $E_{TF}(Z,N)$, for which the right-hand side in (147) is zero. Upon inserting the ansatz

$$E_{TFS}(Z,N) = E_{TF}(Z,N) + Z^{7/3} \sum_{k=1}^{\infty} e_k(N/Z)\ Z^{-k/3} \tag{3-148}$$

into (147), we find[9]

$$\begin{aligned} E_{TFS}(Z,N) &= E_{TF}(Z,N) + \frac{1}{2} Z^2 - \frac{1}{4} a_1 Z \\ &\quad - \frac{1}{6} a_2 \frac{1}{Z} \frac{\partial}{\partial N} E_{TF}(Z,N) + \ldots \qquad (3\text{-}149) \\ &= E_{TF}(Z,N) + \frac{1}{2} Z^2 + O(Z) \quad . \end{aligned}$$

Indeed, here is Scott's term again; and the contribution to E_{TFS} of order Z is utterly insignificant, because our model contains physical approximations already at the order $Z^{5/3}$ (of which the lion's share belongs to the exchange energy; see Chapter Four). Thus, there is not the slightest doubt left about Scott's correction to the energy; we have checked, in detail, that v(r), the deviation of the effective potential from its limiting -Z/r shape, does not contribute to the energy above the order $Z^{3/3}$ (notwithstanding the likely possibility that a_1 and a_2 are both equal to zero).

Some implications concerning energy. The Scott term is independent of N, with the consequence

$$\frac{\partial}{\partial N} E_{TFS}(Z,N) = \frac{\partial}{\partial N} E_{TF}(Z,N) \quad , \tag{3-150}$$

or after using Eq.(129),

$$\zeta_{TFS}(Z,N) = \zeta_{TF}(Z,N) \quad . \tag{3-151}$$

This is to say that the description of the outer reaches of the atom has not been altered. Another way of looking at the same thing is to state that the ionization energy required to strip Z-N electrons off the neutral atom, E(Z,N)-E(Z,Z), is the same for an TF atom and an TFS atom:

$$E_{TF}(Z,N) - E_{TF}(Z,Z) = E_{TFS}(Z,N) - E_{TFS}(Z,Z) \ . \tag{3-152}$$

This is quite reasonable since the modifications that distinguish the TFS model from the TF model refer exclusively to the deep interior of the atom.

Next, let us check if the virial theorem for Coulomb systems

$$2E_{kin} = - E_{pot} = - (E_{ee} + E_{Ne}) \tag{3-153}$$

holds in the TFS model, as it should. This is another consistency test. For this purpose, we return to Eq. (128) and set $\nu=2$ [recall that in a general theory this is the only reasonable value for ν, as pointed out around Eq. (2-477)]. This produces

$$2E_{TFS} + Z(V+\frac{Z}{r})\Big|_{r=0} - \frac{1}{8\pi}\int(d\vec{r})\,[\vec{\nabla}(V+\frac{Z}{r})]^2 = 0 \ . \tag{3-154}$$

The second term is equal to the negative of the interaction energy of the electrons with the nucleus, E_{Ne}, whereas the third is the negative of the electron-electron interaction energy, E_{ee}, so that

$$2E_{TFS} = E_{Ne} + E_{ee} = E_{pot} \ , \tag{3-155}$$

which, in combination with $E_{TFS} = E_{kin} + E_{pot}$, immediately implies Eq. (153). Everything is alright.

We are now justified in writing

$$E_{kin}(Z,N) = - E_{TFS}(Z,N) = - E_{TF}(Z,N) - \frac{1}{2} Z^2 \tag{3-156}$$

and

$$E_{Ne}(Z,N) = Z\frac{\partial}{\partial Z} E_{TFS}(Z,N) = Z\frac{\partial}{\partial Z} E_{TF}(Z,N) + Z^2 \ , \tag{3-157}$$

so that

$$\begin{aligned} E_{ee}(Z,N) &= E_{TFS}(Z,N) - E_{kin}(Z,N) - E_{Ne}(Z,N) \\ &= 2E_{TF}(Z,N) - Z\frac{\partial}{\partial Z} E_{TF}(Z,N) \ . \end{aligned} \tag{3-158}$$

It turns out, that both E_{kin} and E_{Ne} differ from their TF values by an amount proportional to Z^2; in contrast, E_{ee} is the same in the TF and the TFS model. In other words: the electrostatic energy of the electron cloud remains unchanged by the Scott correction; what is altered is the kinetic energy of the electrons and the interaction energy of the nucleus with the electronic atmosphere of the atom.

For a neutral atom, N=Z, we have $E_{TF}(Z,Z)=-(3/7)(B/a)Z^{7/3}$, so that the relative sizes of these various energies are

$$\begin{aligned} E_{ee} : E_{kin} : (-E_{Ne}) \\ &= 1 : 3(1-\frac{7}{6}\frac{a}{B}Z^{-1/3}) : 7(1-\frac{a}{B}Z^{-1/3}) \\ &= 1 : 3(1-0.65/Z^{1/3}) : 7(1-0.56/Z^{1/3}) . \end{aligned} \tag{3-159}$$

These proportions approach the TF limit of 1:3:7 [Eq.(2-238)] rather slowly; for Z in the range of the Periodic Table, i.e. $Z^{1/3}\lesssim 5$, the deviation from this limit is significant. For instance, in mercury (Z=80), the proportions are

$$E_{ee} : E_{kin} : (-E_{Ne}) = 1 : 2.55 : 6.09 \quad . \tag{3-160}$$

Electron density at the site of the nucleus. It has been said already that the density of the electrons at r=o is finite in the TFS model, whereas it grows infinite (proportional to $r^{-3/2}$) in the TF model. Let us now make sure that the TFS prediction for n(r=o) is not only finite, but gives the correct numerical amount.

Upon inserting Eq.(139) into Eq.(77), we have

$$\begin{aligned} n_o \equiv n(r=o) = \frac{(2Z)^3}{4\pi}[\sum_{n'=1}^{n_{s-}}\frac{1}{n'^3} + \sum_{j=1}^{J} w_j \frac{1}{2n_j^2} + \frac{1}{Z^2}(\frac{\partial}{\partial Z}+\frac{\partial}{\partial N})E_{TFS}(Z,N) \\ + \frac{3/2}{Z^2}\sum_{j=1}^{J} w_j(\frac{5}{3}\bar{v}'_{n_j} - \bar{v}_{n_j}) \\ + \frac{5}{Z^2}\frac{\partial E_{TFS}(Z,N)}{\partial N}\sum_{j=1}^{J} w_j n_j^2(\bar{v}'_{n_j} - \bar{v}_{n_j})] \quad . \end{aligned} \tag{3-161}$$

We confine ourselves to the situation of a neutral atom, when $\partial E_{TFS}/\partial Z = -(B/a)Z^{4/3} + Z$ and $\partial E_{TFS}/\partial N=0$, for which

$$n_o/\frac{(2Z)^3}{4\pi} = \{\sum_{n'=1}^{n_{s-}}\frac{1}{n'^3} + \sum_{j=1}^{J} w_j\frac{1}{2n_j^2}\} - \frac{B}{a}Z^{-2/3} + \frac{1}{Z} +$$

$$+ \frac{3/2}{Z^2} \{\sum_{j=1}^{J} w_j (\frac{5}{3} \bar{v}'_{n_j} - \bar{v}_{n_j})\} \quad , \tag{3-162}$$

where the deletion of the unphysical Bohr-shell oscillations in the two curly-bracket terms is called for. In the first expression this is done without much effort. According to the general recipy [explained at the examples of Eqs. (63) and (64)], we write

$$\sum_{n'=1}^{n_s} \frac{1}{n'^3} + \sum_{j=1}^{J} w_j \frac{1}{2n_j^2} \;\hat{=}\; \sum_{n'=1}^{[\nu_s]} \frac{1}{n'^3} + \frac{1}{2\nu_s^2}$$

$$= \sum_{n'=1}^{\infty} \frac{1}{n'^3} + \frac{1}{2\nu_s^2} - \sum_{n'=[\nu_s]+1}^{\infty} \frac{1}{n'^3} \tag{3-163}$$

$$= \sum_{n'=1}^{\infty} \frac{1}{n'^3} + \text{oscillation} \quad .$$

The last equality makes use of

$$\sum_{n'=[\nu_s]+1}^{\infty} \frac{1}{n'^3} = \frac{1}{2\nu_s^2} - \frac{1}{\nu_s^3} \sum_{k=1}^{\infty} \frac{\sin(2\pi k \nu_s)}{\pi k} + \frac{3/2}{\nu_s^4} \sum_{k=1}^{\infty} \frac{\cos(2\pi k \nu_s)}{(\pi k)^2}$$

$$+ \frac{3}{\nu_s^5} \sum_{k=1}^{\infty} \frac{\sin(2\pi k \nu_s)}{(\pi k)^3} + \ldots \quad , \tag{3-164}$$

the derivation of which is presented as Problem 4. Consequently, the contents of the first pair of curly brackets in (162) are to be replaced by

$$\sum_{n'=1}^{\infty} \frac{1}{n'^3} = 1.2020569\ldots \quad . \tag{3-165}$$

[In terms of Riemann's Zeta function this sum is $\zeta(3)$.] At this stage, we have[10]

$$n_o / \frac{(2Z)^3}{4\pi} = 1.2021 - 1.7937\, Z^{-2/3} + \ldots \quad , \tag{3-166}$$

where the ellipsis indicates the terms of order Z^{-1} in (162) [recall that, as stated in (146), $\bar{v}_{n_j}$ and $\bar{v}'_{n_j}$ are proportional to Z]. Note that the coefficient of the $Z^{-2/3}$ term is almost five times as large as the corresponding one for a Bohr atom, which, according to Problem 1-5, equals $(1/2)(2/3)^{2/3} = 0.38$. The mutual repulsion of the electrons makes the density decrease near the nucleus.

In Table 1 the HF predictions[11] for n_o are compared to the asymptotic value, the first term on the right-hand side of (166), and to the TFS result (166). We observe that the asymptotic value is ap-

Table 3-1. HF and TFS predictions for $4\pi n_o/(2Z)^3$ for Z = 17,34,... ..,102. The columns DAV and DTFS give the deviations, in percent, of the asymptotic value and the TFS one from the HF number. The difference between the HF and the TFS results is listed in the column $O(1/Z)$.

Z	HF	DAV	TFS	DTFS	$O(1/Z)$
17	1.0291	16.8	0.9308	-9.6	1.672/Z
34	1.0816	11.1	1.0311	-4.7	1.715/Z
51	1.1063	8.7	1.0716	-3.1	1.766/Z
68	1.1205	7.3	1.0944	-2.3	1.776/Z
85	1.1303	6.3	1.1093	-1.9	1.784/Z
102	1.1375	5.7	1.1199	-1.5_5	1.797/Z

proached rather slowly, and that the difference between the HF and the TFS numbers is only a few percent for atoms that are not too small. In the last column of Table 1 this difference is recognized as equaling about 1.8/Z for large values of Z. This is a reassurance that Eq.(166) does display the correct constant and $Z^{-2/3}$ term, indeed.

Looking back at Eq.(162) we thus conclude that the term 1/Z, which originates in the Scott correction of $\frac{1}{2}Z^2$ to the energy, only accounts for about half of the 1/Z term supplementing Eq.(166). The remaining contribution, however, cannot be produced correctly by the second curly-bracket term in (162), because there are additional corrections on this level of approximation. In particular, it is necessary to include into the description the changes of the wave functions to first order in v(r), which means that E_s, the energy of the strongly bound electrons, has to be evaluated to second order in the difference $V-(-Z/r)$.[12] For the energy considerations it was sufficient to use the first-order expression (42). Another contribution arises from modifications due to the inclusion of the exchange energy (to be described in Chapter Four), which causes a change of the potential of relative size 1/Z at small distances. In short: at the present stage we are unable to predict the 1/Z supplement to the TFS prediciton (166) accurately. What we can do is make a numerical estimate. For example, when

$$\frac{1}{Z} + \frac{0.82}{Z}\,(1 - Z^{-2/3}) \qquad (3\text{-}167)$$

is added to the right-hand side of (166), the agreement with the HF predictions is better than 1.0; 0.5; 0.1 percent for Z larger than 3; 11; 18, respectively.

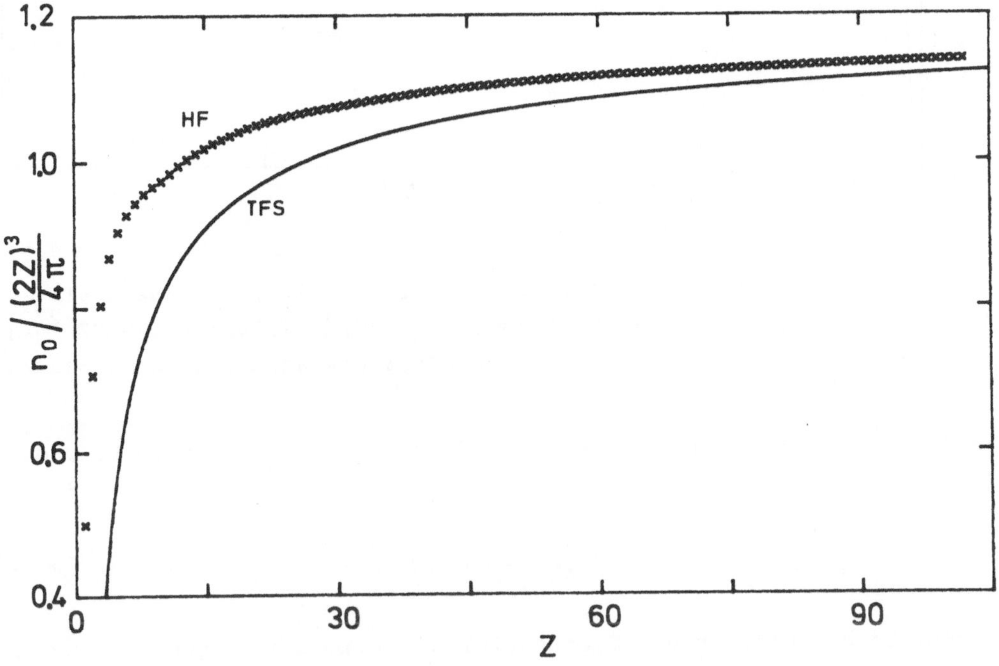

Fig. 3-4. HF prediction (crosses; for Z = 1,2,...,102) and TFS result (smooth curve) for $4\pi n_o/(2Z)^3$.

For illustration, Fig. 4 shows the HF and TFS predictions for $4\pi n_o/(2Z)^3$, and Fig.5 displays the HF results for the order-of-Z^{-1} contribution along with the smooth curve corresponding to the interpolation (167). Please note that Fig.5 indicates that n_o contains an oscillatory part. This is the first time during the development that we are confronted with a manifestation of the atomic shell structure.

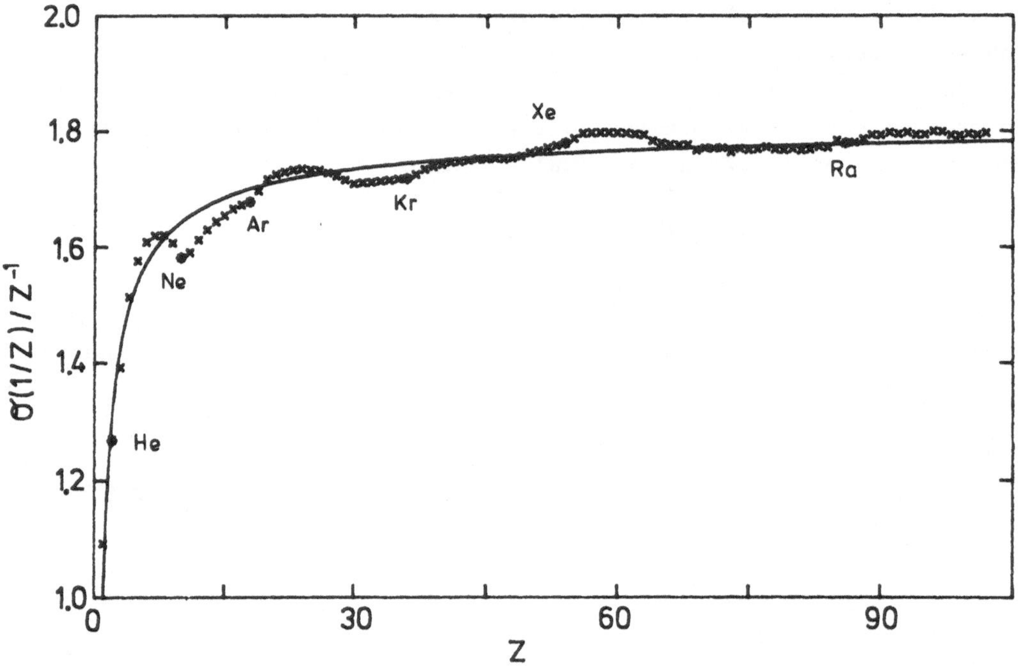

Fig. 3-5. HF prediction (crosses; for Z = 1,2,...,102) and interpolation (167) (smooth curve) for $O(1/Z)/Z^{-1} = [4\pi n_o/(2Z)^3 - 1.2021 + 1.7937Z^{-2/3}]/Z^{-1}$. The HF predictions for inert gas atoms are represented by stars.

Numerical procedure. The TFS density of Eqs. (74) - (76) is to be used in Poisson's equation,

$$-\frac{1}{4\pi}\nabla^2(V+\frac{Z}{r}) = n = n_{IME} + \tilde{n} \quad , \tag{3-168}$$

in order to calculate the TFS potential, V. Inasmuch as the density involves not only the potential and the minimum binding energy, ζ, but also the parameters ζ_j and Q_j, solving Eq. (168) under the boundary conditions

$$r\,V(r) \rightarrow \begin{cases} -Z\ , & \text{for } r \rightarrow o\ , \quad \text{(a)} \\ -(Z-N)\ , & \text{for } r \rightarrow \infty\ , \quad \text{(b)} \end{cases} \tag{3-169}$$

is a considerably more complex numerical task than it is in the TF model of the preceding Chapter. The principal complication is that the ζ_j and Q_j are not independent quantities, but are given in terms of integrals involving V (and ζ); these are Eqs. (50) and (72).

Before any calculation we must decide upon the number of shells of strongly bound electrons to be corrected for, that is: we choose a value for n_s. Then there is the question of how to average over ζ_s so as to remove - to the desired extent - the unphysical Bohr-shell artifacts, that is: we make a choice for the number J of representative values for ζ_s, which are called ζ_j ($j=1,2,\ldots,J\geq 2$). Along with J we select appropriate weights w_j to be used for the averaging. To each j there corresponds, in Eq.(50), a certain principal quantum number n_j and its Bohr-shell density $|\psi_{n_j}|^2_{av}$. For $n=n_s$, and $n_2=n_s+1$, which are integers, these can be found in Eq.(53); for the non-integer numbers n_3.. ..n_J we use appropriate (linear) averages of $|\psi_{n_1}|^2_{av}$ and $|\psi_{n_2}|^2_{av}$, as remarked after Eq.(58). "Appropriate" means, of course, fitting for the purpose for which the computation is made; in the typical situation the choice for J, w_j, n_j, and $|\psi_{n_j}|^2_{av}$ will be dictated by the particular Bohr-shell oscillations that one wants to remove. These decisions being made the numerical procedure is the following.

For positive ions, N<Z, the search for V begins with a reasonable guess for ζ as well as for the ζ_j and Q_j ($j=1,\ldots,J$). For instance, one can use for ζ the corresponding TF value, and for the ζ_j and Q_j the numbers obtained from Eqs.(135) and (139) when the various averages of v(r) are neglected. Then, with these guessed ζ,ζ_j, and Q_j, starting at a sufficiently large distance (where n=o) with the known asymptotic form of V, Eq.(169b), one integrates the differential equation (168) inwards and compares the evaluations of the integrals for the ζ_j and the Q_j with the initial guesses, thereby obtaining improved values of these parameters. Further, one checks if (169a) is obeyed, and the outcome of this test leads to an improved ζ. Then one tries again with the new parameters. For an initial guess not too bad, this scheme is rapidly converging.

For a neutral atom, N=Z, we know that ζ is zero as it is in the TF model. But we have less knowledge about the asymptotic form of V. For large r, which means outside the region of strongly bound electrons, the density has the TF form (with ζ=o), given by the first term in the second version of (76), and the potential now satisfies the TF equation

$$-\frac{1}{4\pi}\nabla^2 V = \frac{1}{3\pi^2}(-2V)^{3/2} \quad \text{for} \quad r \text{ large} . \tag{3-170}$$

Thus, asymptotically V must be equal to a rescaled TF potential,

$$V(r) = \mu^4 V_{TF}(\mu r) \quad , \tag{3-171}$$

with μ close to unity. Again, for given parameters - now they are μ, ζ_j, and Q_j - one integrates the differential equation for V inwards and by iteration improves their values.

Numerical results for neutral mercury. For illustration, such a calculation has been performed for neutral mercury, for which N=Z=80. For the sake of simplicity, the simplest averaging procedure was chosen, the one with J=2 and $w_1=w_2=1/2$. The initial guesses for the various quantities, as they are obtained from Eqs.(135) and (139) are compared with their actual values in Table 2. For this choice of w_j's, the second curly-bracket term on the right-hand side of Eq.(144) does not vanish [look back at Eq.(57)] which has the consequence that, upon neglecting the terms containing averages of v(r), the energy emerges as[13]

$$E_{TFS} \cong -\frac{3}{7}\frac{B}{a}Z^{7/3}\left(1+\frac{2n_s+1}{3Z}\right)^{7/3} + Z^2\left(\frac{1}{2}+\frac{2n_s+1}{4Z}+\frac{(2n_s+1)^2}{28Z^2}\right) = -\frac{3}{7}\frac{B}{a}Z^{7/3}+\frac{1}{2}Z^2+O(Z^{4/3}) \quad . \tag{3-172}$$

Accordingly, the energy derivative needed in (135) and (139) is

$$\frac{\partial E_{TFS}}{\partial Z} \cong -\frac{B}{a}Z^{4/3}\left(1+\frac{2n_s+1}{3Z}\right)^{4/3} + Z + \frac{1}{4}(2n_s+1) \quad . \tag{3-173}$$

For the scaling parameter μ, the natural initial guess is $\mu=1$. As we see in Table 2, the initial guesses for $\partial E/\partial Z, E, \mu$, as well as ζ_1 and Q_1 dif-

Table 3-2. Comparison of initial guesses (IG) with actual values (AV) and their deviation (DEV) in percent; for N=Z=8o, $n_s=1$, J=2, $w_1=w_2=1/2$.

Quantity	IG	AV	DEV
$-\partial E/\partial Z$	547.9	550.5	-0.47
$-E$	18560	18340	+1.2
μ	1.0000	1.0045	-0.45
ζ_1	2652.0	2788.0	-2.8
ζ_2	252.1	443.6	-43.0
Q_1	0.8828	0.8667	-4.4
Q_2	2.521	5.115	-51.0

fer only by a small amount from the actual values, whereas the agreement is much poorer for ζ_2 and Q_2. This could have been anticipated, because the integrations to be performed for ζ_2 and Q_2 cover a much larger range of r than the ones for ζ_1 and Q_1. In this larger range, neglecting v(r) causes a significant error, which it does not for the small r associated with ζ_1 and Q_1. This situation is improved, however, when the calculation is repeated for a larger value of Z. As noted in Ref.7, for a Z ten times larger, the percentage deviation is smaller by a factor of 5 (for E) to 15 (for ζ_2).

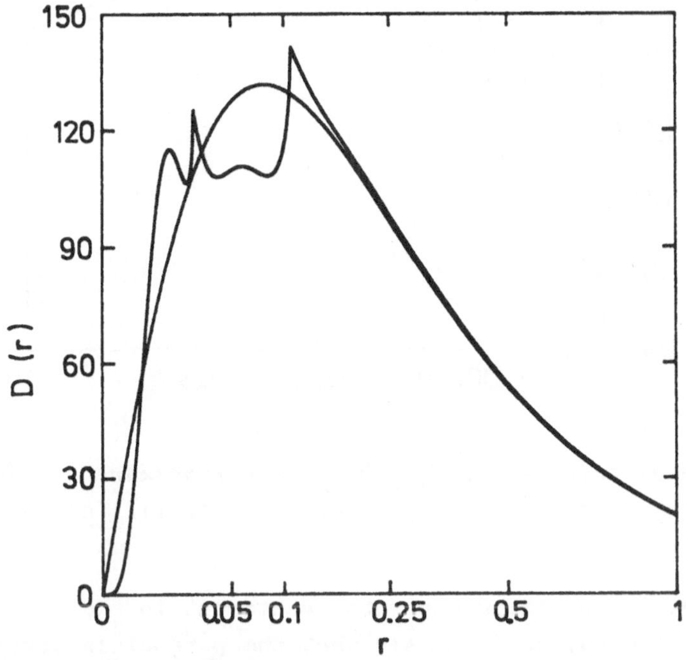

Fig.3-6. Comparison of radial densities $D=4\pi r^2 n$ *for neutral mercury. Smooth curve TF; curve with structure: TFS (with the parameters of Table 2). The abscissa is linear in the square root of* r.

This computation for neutral mercury also supplies a TFS density, which is compared to the corresponding TF density in Fig.6. In order to stretch the small-r region where the interesting structure is located, the abscissa in this plot is chosen linear in the square root of r. The two radial densities differ significantly for $r \lesssim 0.2$. Please note in particular that the TF density is much larger in the immediate vicinity of the nucleus at r=o. Of course, one must not take this TFS density too serious in the intermediate region, where we see two sharp peaks. These originate in the sum-over-j term in the second version of

Eq.(76), and are consequently artifacts of the typical TF discontinuity associated with the square root. Another averaging scheme will naturally result in a TFS density that looks different in this intermediate region.

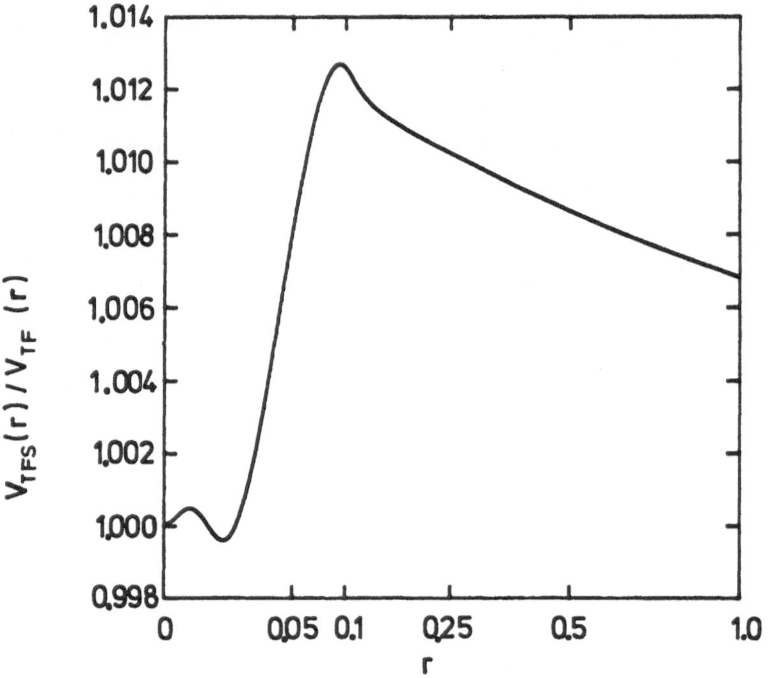

Fig.3-7. Ratio of potentials, V_{TFS}/V_{TF}, *as a function of* r *for neutral mercury (the TFS parameters being those of Table 2). The abscissa is linear in the square root of* r.

Finally, let us see whether the statement in the paragraph after Eq.(2-408) is, indeed, true, namely that the potentials obtained in the extensions of the TF model do not differ much from the TF potential itself. Here we have the first example of such an extension: the TFS model. In Fig.7, a plot is presented not of the TFS and the TF potential - they would be indiscernable - but of their ratio. We observe that this ratio is close to unity, the maximal deviation being hardly more than one percent. In contrast, the respective densities differ by an enormous amount for $r \lesssim 0.1$ as illustrated in Fig.6. Thus we are right in preferring the potential over the density as the fundamental quantity. As a matter of fact, up to date all attempts of deriving Scott's correction in the framework of "density functional theory" have been unsuccessful (unless some ad-hoc modifications of the theory are introduced - a strategy that is hardly acceptable).

Problems

3-1. To prove Eq. (25) without making use of Problem 1-2, first note that one can express $[\nu_s]$ as an integral involving Dirac's Delta function:

$$[\nu_s] = \sum_{n'=1}^{[\nu_s]} 1 = \int_{\nu_o}^{\nu_s} d\nu \sum_{n'=-\infty}^{\infty} \delta(n'-\nu) \quad ,$$

with $0<\nu_o<1$. Then employ Poisson's identity

$$\sum_{n'=-\infty}^{\infty} \delta(n'-\nu) = \sum_{m=-\infty}^{\infty} e^{i2\pi m\nu} = 1 + 2\sum_{m=1}^{\infty} \cos(2\pi m\nu) \quad ,$$

perform the ν integration, and arrive at Eq. (25). Repeat this procedure to evaluate

$$\sum_{n'=1}^{[\nu_s]} n'^2$$

and derive Eq. (30).

3-2. As an illustration of Scott's "boundary effect" argument, consider N non-interacting particles, restricted to the one-dimensional motion along the x-axis, and confined to the range $0 \leq x \leq a$. These particles occupy the N states with least energy, one per state. Compare the TF approximation to the density and the energy with the exact results. Note, in particular, that the requirement of vanishing wave functions at x=0 and x=a cause the exact energy to be larger than the TF result. Repeat for N particles confined to the interior of a three-dimensional sphere, and observe that there is much less structure in the density than in the previous one-dimensional situation. Why?

3-3. Simulate the replacement (31) with the aid of a weight function $w(\nu_s)$,

$$\int d\nu_s \, w(\nu_s) \left(\tfrac{2}{3}\nu_s^3 - N_s\right) = 0.$$

Since N_s is constant for $n_s<\nu_s<n_s+1$ ($n_s=1,2,\ldots$), it is natural to choose $w(\nu_s)$ periodic in ν_s:

$$w(\nu_s) = f(\nu_s - [\nu_s]) \equiv f(\mu) \quad .$$

Then the range of integration covers one or more periods of w. Conclude that $f(\mu)$ must obey

$$\int_0^1 d\mu \; f(\mu)(n_s+\mu)^3 = (n_s+\tfrac{1}{2})^3 - \tfrac{1}{4}(n_s+\tfrac{1}{2})$$

for $n_s=1,2,3,\ldots$. Show that any such $f(\mu)$ also simulates the replacement (26). Evaluate

$$\int_0^1 d\mu \; f(\mu)(1-2\mu)^2 \quad ,$$

and conclude that $f(\mu)$ cannot be non-negative for all μ.

3-4. Use a procedure similar to the one of Problem 1 to derive

$$\sum_{n'=[\nu_s]+1}^{\infty} f(n') = \int_{\nu_s}^{\infty} d\nu f(\nu) + 2\sum_{m=1}^{\infty} \int_{\nu_s}^{\infty} d\nu f(\nu) \cos(2\pi m\nu) \quad .$$

Then integrate by parts repeatedly to find

$$\sum_{n'=[\nu_s]+1}^{\infty} f(n') = \int_{\nu_s}^{\infty} d\nu f(\nu) - f(\nu_s)\sum_{m=1}^{\infty} \frac{\sin(2\pi m\nu_s)}{\pi m}$$

$$- \frac{1}{2} f'(\nu_s)\sum_{m=1}^{\infty} \frac{\cos(2\pi m\nu_s)}{(\pi m)^2}$$

$$+ \frac{1}{4} f''(\nu_s)\sum_{m=1}^{\infty} \frac{\sin(2\pi m\nu_s)}{(\pi m)^3} + \ldots$$

$$= \int_{\nu_s}^{\infty} d\nu f(\nu) + \text{oscillation.}$$

Specify $f(\nu) = \frac{1}{\nu^3}$, and arrive at Eqs. (164) and (163).

Chapter Four

QUANTUM CORRECTIONS AND EXCHANGE

In Chapter Two we learned that the TF energy of an atom is proportional to $Z^{7/3}$; in Chapter Three it was established that the leading correction to this TF energy is proportional to $Z^2 = Z^{7/3}/Z^{1/3}$. In this Chapter we shall be concerned with the second correction which, not surprisingly, supplies a term of order $Z^{7/3}/Z^{2/3} = Z^{5/3}$ to the binding energy of atoms. It will account for the difference between the integer-Z HF crosses and the continuous TFS curve in Fig.3-3.

There are two different contributions to this $Z^{5/3}$ term. The first originates in what we called "quantum corrections" when discussing the relation between quantum mechanical traces and semiclassical phase space integrals [see after Eq.(1-43)]. It thus means an improved evaluation of the trace in

$$E_1(V+\zeta) = \operatorname{tr}(\tfrac{1}{2}p^2 + V + \zeta)\eta(-\tfrac{1}{2}p^2 - V - \zeta) \quad . \tag{4-1}$$

This E_1 is, however, only part of the energy functional (2-434),

$$E(V,n,\zeta) = E_1(V+\zeta) - \int(d\vec{r}\,')(V - V_{ext})n + E_{ee}(n) - \zeta N, \tag{4-2}$$

in which the electron-electron interaction energy E_{ee} is also the object of approximations. So far it was sufficient to be content with the Coulomb energy

$$E_{ee}(n) \cong \frac{1}{2}\int(d\vec{r})(d\vec{r}\,')\,\frac{n(\vec{r})n(\vec{r}\,')}{|\vec{r}-\vec{r}\,'|} \quad , \tag{4-3}$$

but now it will be necessary to include the exchange energy as well (in an appropriate approximate way). This is the second contribution to the $Z^{5/3}$ term in the binding energy.

Since both $\Delta_{qu}E$, the change in energy due to the quantum corrections, and E_{ex}, the exchange energy, are of the same order, namely $Z^{5/3}$, consistent models must not prefer one over the other. We shall therefore refrain from considering those extensions of the TF model which include either only $\Delta_{qu}E$ (the "Thomas-Fermi-von Weizsäcker model") or E_{ex} (the "Thomas-Fermi-Dirac model"). Instead we shall aim at a description in which the TFS model is supplemented by both the quantum corrections and exchange.

<u>Qualitative arguments.</u> In order to justify the remark that both $\Delta_{qu}E$ and E_{ex} are proportional to $Z^{5/3}$ let us briefly discuss the situation in a qualitative way.

The error in Eq.(1-43) is due to the noncommutativity of $\vec{r}$ and $\vec{p}$, which appear in the Hamilton operator in the potential energy $V(\vec{r})$ and the kinetic energy $\frac{1}{2}p^2$. So we are confronted with corrections that are associated with the finiteness of $\vec{\nabla}V$. (Accordingly, these are called "gradient corrections" or "inhomogeneity corrections" by other authors; we shall stick to the name "quantum corrections.") The relevant measure of the size of $\vec{\nabla}V$ is the one of Eq.(2-400), namely $|\lambda\vec{\nabla}V|/|V|$. In view of $r \sim Z^{-1/3}$, $\vec{\nabla} \sim Z^{1/3}$, and $\lambda \sim |V|^{-1/2} \sim Z^{-2/3}$, this quantity is of order $Z^{-1/3}$. And since $\vec{\nabla}V$ is a vector, the corresponding energy correction, which is a scalar, is (to first order) proportional to the square of $\vec{\nabla}V$, so that it is smaller than the leading energy term by two factors of $Z^{-1/3}$. Therefore,

$$\frac{\Delta_{qu}E}{E_{TF}} \sim Z^{-2/3} \tag{4-4}$$

or with $E_{TF} \sim Z^{7/3}$,

$$\Delta_{qu} E \sim Z^{5/3} \quad , \tag{4-5}$$

indeed. Incidentally, we shall see below that consistency requires to include a contribution from the second derivative of the potential; it also leads to a $Z^{5/3}$ term in the energy.

We turn to the exchange energy now. The electrostatic energy of each electron with the other electrons, constituting Z electrons at a distance $\sim Z^{-1/3}$, is of order $Z/Z^{-1/3} = Z^{4/3}$. Consequently, the total electrostatic energy is proportional to $Z \times Z^{4/3} = Z^{7/3}$, a result familiar to us since the discussion of the TF model in Chapter Two. In contrast with this electrostatic energy, exchange is limited to electrons with overlapping wave functions at a distance $\sim\lambda\sim Z^{-2/3}$; thus the exchange energy of each electron is of the order $1/Z^{-2/3} = Z^{2/3}$, that of all Z electrons being $\sim Z \times Z^{2/3} = Z^{5/3}$. Indeed, we have

$$E_{ex} \sim Z^{5/3} \quad , \tag{4-6}$$

as stated above.

Quantum corrections I (time transformation function). The quantum corrections concern the term $E_{\zeta\zeta_s}$ of Eq.(3-3), since the contribution from the strongly bound electrons has already been taken care of. According to Eq.(3-18), this quantity is given by

$$E_{\zeta\zeta_s} = \int_{\zeta}^{\zeta_s} d\zeta'(\zeta'-\zeta) \frac{d}{d\zeta'} N(\zeta') \quad , \tag{4-7}$$

which combined with Eq.(2-10),

$$N(\zeta') = \mathrm{tr}\, \eta(-H-\zeta') \quad , \tag{4-8}$$

reads

$$E_{\zeta\zeta_s} = \int_{\zeta}^{\zeta_s} d\zeta'(\zeta'-\zeta)\, \mathrm{tr} \frac{d}{d\zeta'} \eta(-H-\zeta') \tag{4-9}$$

We remember that in these equations H denotes the independent-particle Hamilton operator,

$$H = \frac{1}{2}p^2 + V(\vec{r}) \quad , \tag{4-10}$$

V being the effective potential.

The result of differentiating Heaviside's unit step function $\eta(x)$ is Dirac's Delta function $\delta(x)$,

$$\frac{d}{dx} \eta(x) = \delta(x) \quad , \tag{4-11}$$

the Fourier transform of which is

$$\delta(x) = \int_{-\infty}^{\infty} \frac{dt}{2\pi} e^{ixt} \quad . \tag{4-12}$$

Consequently, the trace in Eq.(9) can be written as

$$\begin{aligned} \mathrm{tr} \frac{d}{d\zeta'} \eta(-H-\zeta') &= -\, \mathrm{tr}\, \delta(-H-\zeta') \\ &= -\, \mathrm{tr} \int \frac{dt}{2\pi} e^{-i(H+\zeta')t} \quad . \end{aligned} \tag{4-13}$$

Upon evaluating the trace as the diagonal sum in configuration space, the last equality is

$$\mathrm{tr}\, \delta(-H-\zeta') = 2\int (d\vec{r}\,') \int \frac{dt}{2\pi} \langle \vec{r}\,' | e^{-i(H+\zeta')t} | \vec{r}\,' \rangle \quad . \tag{4-14}$$

We meet here the time transformation function

$$\langle\vec{r}',t|\vec{r}'',o\rangle = \langle\vec{r}'|e^{-iHt}|\vec{r}''\rangle \quad , \tag{4-15}$$

needed for $\vec{r}'=\vec{r}''$. With Eq.(15), Eqs.(14) and (9) appear as

$$\mathrm{tr}\,\delta(-H-\zeta') = 2\int(d\vec{r}')\int\frac{dt}{2\pi}\,e^{-i\zeta't}\langle\vec{r}',t|\vec{r}',o\rangle \quad , \tag{4-16}$$

and

$$E_{\zeta\zeta_s} = -2\int_{\zeta}^{\zeta_s} d\zeta'(\zeta'-\zeta)\int(d\vec{r}')\int\frac{dt}{2\pi}\,e^{-i\zeta't}\langle\vec{r}',t|\vec{r}',o\rangle \ , \tag{4-17}$$

respectively.

So far we have been approximating traces by the corresponding phase space integrals [see Eq.(1-43)], which gives

$$\begin{aligned}\mathrm{tr}\,\delta(-H-\zeta') &\cong 2\int\frac{(d\vec{r}')(d\vec{p}')}{(2\pi)^3}\,\delta(-\tfrac{1}{2}p'^2-V(\vec{r}')-\zeta') \\ &= 2\int(d\vec{r}')\int\frac{dt}{2\pi}\int\frac{(d\vec{p}')}{(2\pi)^3}\,e^{-i\left(\frac{1}{2}p'^2+V(\vec{r}')+\zeta'\right)t}\end{aligned} \tag{4-18}$$

when applied to the left-hand side of Eq.(16). The comparison with the right-hand side of this equation shows that this semiclassical approximation can be regarded as

$$\begin{aligned}\langle\vec{r}',t|\vec{r}',o\rangle &= \langle\vec{r}'|e^{-iHt}|\vec{r}'\rangle \\ &\cong \int\frac{(d\vec{p}')}{(2\pi)^3}\,e^{-i\left(\frac{1}{2}p'^2+V(\vec{r}')\right)t} \\ &= \left(\frac{1}{2\pi it}\right)^{3/2} e^{-iV(\vec{r}')t} \quad .\end{aligned} \tag{4-19}$$

This relation is exact for a spatially constant potential. It is a good approximation for $\vec{r}$-dependent potentials if the considered time t is small, since then the particle has not enough time to propagate far enough to become aware of changes in the potential.

As a first step towards improving (19) by including effects of the derivatives of the potential, let us consider a linear potential describing a constant force $\vec{F}$,

$$V_1(\vec{r}) = V_o - \vec{F}\cdot\vec{r} \quad . \tag{4-20}$$

According to Eq.(1-42), the ordered version of

$$e^{-iHt} = e^{-i\left(\frac{1}{2}p^2+V_0-\vec{F}\cdot\vec{r}\right)t} \tag{4-21}$$

is

$$e^{-iHt} = e^{-i(V_0-\vec{F}\cdot\vec{r})t}\; e^{-i\,\frac{1}{2}(\vec{p}+\frac{1}{2}\vec{F}t)^2t}\; e^{-iF^2t^3/24} \quad , \tag{4-22}$$

so that Eq.(1-41) implies

$$\begin{aligned} \mathrm{tr}\, e^{-iHt} &= 2\int (d\vec{r}\,')\langle \vec{r}\,',t|\vec{r}\,',o\rangle \\ &= 2\int (d\vec{r}\,')\int \frac{(d\vec{p}\,')}{(2\pi)^3}\, e^{-i(V_0-\vec{F}\cdot\vec{r}\,')t}\; e^{-i\frac{1}{2}(\vec{p}\,'+\frac{1}{2}\vec{F}t)^2t}\; e^{-iF^2t^3/24}. \end{aligned} \tag{4-23}$$

Thus, after translating the origin in $\vec{p}\,'$ space,

$$\begin{aligned} \langle \vec{r}\,',t|\vec{r}\,',o\rangle &= \int \frac{(d\vec{p}\,')}{(2\pi)^3}\, e^{-iV_1(\vec{r}\,')t}\; e^{-i\frac{1}{2}p'^2t}\; e^{-iF^2t^3/24} \\ &= \left(\frac{1}{2\pi it}\right)^{3/2} e^{-i[V_1(\vec{r}\,')t+(\vec{\nabla}'V_1(\vec{r}\,'))^2t^3/24]} \quad , \end{aligned} \tag{4-24}$$

which, for the potential (20) is, indeed, the correct result.

It is tempting to use the expression (24) as an improved approximation to $\langle \vec{r}\,',t|\vec{r}\,',o\rangle$ for potentials with small second derivatives, just as (19) is employed when the gradient, the first derivative of the potential, is small. Doing this would, indeed, result in a correction of relative order $Z^{-2/3}$, because $V \sim Z^{4/3}$ implies that the relevant values of t are of order $Z^{-4/3}$, which combined with $\vec{\nabla}' \sim \frac{1}{r'} \sim Z^{1/3}$ shows that

$$(\vec{\nabla}'V(\vec{r}\,'))^2\, t^3 \sim (Z^{1/3}\, Z^{4/3})^2 (Z^{-4/3})^3 = Z^{-2/3} \quad . \tag{4-25}$$

This much is fine; what is wrong, however, with the approximation (24) is that it does not contain all corrections of relative order $Z^{-2/3}$.

To illustrate this point, consider a quadratic potential

$$V_2(\vec{r}) = V_0 + \vec{k}\cdot\vec{r} + \frac{1}{2}\,\vec{r}\cdot\overleftrightarrow{\omega^2}\cdot\vec{r} \quad , \tag{4-26}$$

which is a second order approximation to any given potential $V(\vec{r})$ around $\vec{r} = \vec{r}\,'$, if the constants $V_0, \vec{k}$, and $\overleftrightarrow{\omega^2}$ are such that

$$V(\vec{r}\,') = V_0 + \vec{k}\cdot\vec{r}\,' + \frac{1}{2}\,\vec{r}\,'\cdot\overleftrightarrow{\omega^2}\cdot\vec{r}\,' = V_2(\vec{r}\,') \quad ,$$

$$\vec{\nabla}' V(\vec{r}') = \vec{k} + \overleftrightarrow{\omega^2}\cdot\vec{r}' = \vec{\nabla}' V_2(\vec{r}') \quad , \tag{4-27}$$

$$\vec{\nabla}'\vec{\nabla}' V(\vec{r}') = \overleftrightarrow{\omega^2} = \vec{\nabla}'\vec{\nabla}' V_2(\vec{r}') \quad .$$

These constants given, it is always possible to adopt a coordinate system in which the (symmetric) dyadic $\overleftrightarrow{\omega^2}$ is diagonal:

$$\frac{1}{2}\,\vec{r}\cdot\overleftrightarrow{\omega^2}\cdot\vec{r} = \frac{1}{2}(\omega_x^2\,x^2 + \omega_y^2\,y^2 + \omega_z^2\,z^2) \quad , \tag{4-28}$$

with the consequence that the dynamics in the three perpendicular directions of x, y, and z is independent. We can thus simplify matters, for a start, by considering the one-dimensional motion along, say, the x-axis, governed by the Hamilton operator (we choose to distribute the constant V_o equally among x,y, and z)

$$H_x = \frac{1}{2}\,p_x^2 + \frac{1}{3}\,V_o + k_x\,x + \frac{1}{2}\,\omega_x^2\,x^2 \quad . \tag{4-29}$$

The time transformation function $\langle x',t|x'',o\rangle$ for such a one-dimensional harmonic oscillator is well known,

$$\langle x',t|x'',o\rangle = \left(\frac{1}{2\pi i T_x}\right)^{1/2} e^{-i\Phi_x} \quad , \tag{4-30}$$

where the phase Φ_x is given by

$$\begin{aligned}\Phi_x = {} & \frac{1}{3}\,V_o t - \frac{k_x^2}{\omega_x^3}\left(\frac{\omega_x t}{2} - \tan\left(\frac{\omega_x t}{2}\right)\right) \\ & + \frac{k_x}{\omega_x}\,(x'+x'')\,\tan\left(\frac{\omega_x t}{2}\right) \\ & + \frac{1}{4}\,(x'+x'')^2\,\omega_x\,\tan\left(\frac{\omega_x t}{2}\right) \\ & - \frac{1}{4}\,(x'-x'')^2\,\omega_x\,\cot\left(\frac{\omega_x t}{2}\right) \quad ,\end{aligned} \tag{4-31}$$

and the "tyme" T_x by

$$T_x = \frac{1}{\omega_x}\,\sin(\omega_x t) \quad . \tag{4-32}$$

(The dependence on ω_x is even, so that these equations hold both for $\omega_x^2 > o$ and for $\omega_x^2 < o$.) Equations (30) to (32) can be easily produced by a variety of techniques;[1] at worst, one verifies that the Schrödinger equation

$$i \frac{\partial}{\partial t} \langle x',t|x'',o\rangle = \langle x',t|H_x|x'',o\rangle \tag{4-33}$$

is obeyed, as well as the initial condition

$$\langle x',t|x'',o\rangle \to \delta(x'-x'') \ , \quad \text{for} \quad t \to o \ . \tag{4-34}$$

Please note that Φ_x splits into two parts depending on the sum and difference of x' and x", respectively. Thus, if we now denote their difference by s_x and half their sum by x', we have

$$\begin{aligned} &\Phi_x(x',s_x,t) \\ &\qquad = \Phi_x(x',o,t) - \frac{1}{4} s_x^2 \, \omega_x \cot\left(\frac{\omega_x t}{2}\right) \ , \end{aligned} \tag{4-35}$$

where

$$\begin{aligned} \Phi_x(x',o,t) &= \frac{1}{3} V_o t - \frac{k_x^2}{\omega_x^3} \left(\frac{\omega_x t}{2} - \tan\left(\frac{\omega_x t}{2}\right)\right) \\ &\quad + (2k_x x' + \omega_x^2 x'^2) \frac{1}{\omega_x} \tan\left(\frac{\omega_x t}{2}\right) \ . \end{aligned} \tag{4-36}$$

Let us now find out of which order in $Z^{-1/3}$ the various terms are in Eqs.(32), (35), and (36). As in the discussion of (24) we have

$$\begin{aligned} &r' \sim x' \sim Z^{-1/3} \ , \quad \vec{\nabla}' \sim Z^{1/3} \\ &V(r') \sim Z^{4/3} \ , \quad t \sim Z^{-4/3} \end{aligned} \tag{4-37}$$

so that [Eq.(27)]

$$\begin{aligned} &\overleftrightarrow{\omega}^2 \sim \omega_x^2 \sim Z^{6/3} \ , \quad \omega_x \sim Z^{3/3} \\ &\vec{k} \sim k_x \sim Z^{5/3} \ , \\ &V_o \sim Z^{4/3} \ . \end{aligned} \tag{4-38}$$

Further, s_x signifies the distance between two x-coordinates, which is relevant only when overlap integrals are evaluated. Therefore, just as in our qualitative discusssion of the exchange energy, $\vec{s} = (s_x, s_y, s_z)$ is of order of the electrons' de Broglie wave length,

$$\vec{s} \sim s_x \sim \lambda \sim | V |^{-1/2} \sim Z^{-2/3} \quad . \tag{4-39}$$

As a consequence of (37) and (38), we observe

$$\omega_x t \sim Z^{-1/3} \quad , \tag{4-40}$$

so that

$$\tan(\frac{\omega_x t}{2}) = \frac{\omega_x t}{2} + \frac{1}{3}(\frac{\omega_x t}{2})^3 + \ldots \tag{4-41}$$

is an expansion in powers of $Z^{-1/3}$. Inserted into (36), this produces

$$\begin{aligned} \Phi_x(x',0,t) &= \frac{1}{3} V_o t + \frac{1}{24} k_x^2 t^3 \\ &+ (2k_x x' + \omega_x^2 x'^2)(\frac{t}{2} + \frac{1}{24} \omega_x^2 t^3) \\ &+ O(Z^{-4/3}) \quad . \end{aligned} \tag{4-42}$$

Up to order $Z^{-2/3}$ we thus have

$$\begin{aligned} \Phi_x(x',0,t) &\cong (\frac{1}{3} V_o + k_x x' + \frac{1}{2}\omega_x^2 x'^2)t \\ &+ \frac{1}{24} (k_x + \omega_x^2 x')^2 t^3 \quad . \end{aligned} \tag{4-43}$$

Analogous expressions are obtained for Φ_y and Φ_z, their sum being

$$\begin{aligned} \Phi(\vec{r}',0,t) &= \Phi_x(x',0,t) + \Phi_y(y',0,t) + \Phi_z(z',0,t) \\ &\cong (V_o + \vec{k}\cdot\vec{r}' + \frac{1}{2} \vec{r}'\cdot\overleftrightarrow{\omega^2}\cdot\vec{r}')t \\ &+ \frac{1}{24} (\vec{k} + \overleftrightarrow{\omega^2}\cdot\vec{r}')^2 t^3 \quad , \end{aligned} \tag{4-44}$$

or with Eqs.(26) and (27)

$$\Phi(\vec{r}',0,t) \cong V(\vec{r}')\ t + \frac{1}{24}[\vec{\nabla}'V(\vec{r}')]^2\ t^3 \quad . \tag{4-45}$$

Likewise we find for the s_x term in (35)

$$\frac{1}{4}s_x^2\ \omega_x \cot(\frac{\omega_x t}{2}) \cong \frac{s_x^2}{2t} - \frac{1}{24}\ s_x^2\ \omega_x^2\ t\ , \tag{4-46}$$

which then leads to the three dimensional result

$$\Phi(\vec{r}',\vec{s},t)$$
$$\cong \Phi(\vec{r}',0,t) - \frac{s^2}{2t} + \frac{t}{24}(\vec{s}\cdot\vec{\nabla}')^2\ V(\vec{r}') \quad . \tag{4-47}$$

The corresponding approximation for the tyme T_x of (32) is

$$T_x \cong t(1 - \frac{1}{6}\ \omega_x^2\ t^2) \quad , \tag{4-48}$$

with this implication for three dimensional T:

$$T \equiv (T_xT_yT_z)^{1/3} \cong t\left(1 - \frac{1}{18}(\omega_x^2 + \omega_y^2 + \omega_z^2)\ t^2\right) \quad , \tag{4-49}$$

or

$$T(\vec{r}',\vec{s},t) \cong t\left(1 - \frac{t^2}{18}\ \nabla'^2\ V(\vec{r}')\right) \quad . \tag{4-50}$$

The tyme T does not depend upon $\vec{s}$ up to the order considered here, that is up to corrections of relative size $Z^{-2/3}$.

In summing up, we state that our new approximation for the time transformation function needed in Eqs. (16) and (17) is

$$\langle\vec{r}' + \frac{1}{2}\ \vec{s},\ t|\vec{r}' - \frac{1}{2}\ \vec{s},\ 0\rangle = (\frac{1}{2\pi i T})^{3/2}\ e^{-i\Phi} \quad , \tag{4-51}$$

where the phase $\Phi(\vec{r}',\vec{s},t)$ and the tyme $T(\vec{r}',\vec{s},t)$ are given by Eqs. (47), (45), and (50), which are correct up to order $Z^{-2/3}$. One checks immediately that Eqs. (24) and (19) are reproduced in the situation of a linear or a constant potential, respectively.

After arriving at Eq. (24) we resisted the temptation of using this expression as the basis of improved approximations because the corresponding Φ and T did not contain all terms of relative order $Z^{-2/3}$. While the "local oscillator approximation" of Eqs. (26) and (27) does, indeed, produce all the terms missing in (24), we have, so far, no way of knowing that nothing has been left out. An independent count of the powers of $Z^{-1/3}$ is asked for. It is supplied by the Schrödinger equation obeyed by $<\vec{r}',t|\vec{r}'',0>$, of which the one dimensional version is written as Eq. (33). With the Hamilton operator (10) it reads in three dimensions:

$$i \frac{\partial}{\partial t} <\vec{r}_1',t|\vec{r}_2',0> = \left(-\frac{1}{2}\nabla_1'^2 + V(\vec{r}_1')\right) <\vec{r}_1',t|\vec{r}_2',0>$$

$$= \left(-\frac{1}{2}\nabla_2'^2 + V(\vec{r}_2')\right) <\vec{r}_1',t|\vec{r}_2',0> \quad . \tag{4-52}$$

Upon setting

$$\vec{r}' = \frac{1}{2}(\vec{r}_1' + \vec{r}_2') \quad , \quad \vec{s} = \vec{r}_1' - \vec{r}_2' \quad , \tag{4-53}$$

accompanied by

$$\vec{\nabla}_1' = \frac{1}{2}\vec{\nabla}' + \vec{\nabla}_s \quad , \quad \vec{\nabla}_2' = \frac{1}{2}\vec{\nabla}' - \vec{\nabla}_s \quad , \tag{4-54}$$

the sum and difference of the two versions of (52) appear as

$$\left[i\frac{\partial}{\partial t} + \frac{1}{8}\nabla'^2 + \frac{1}{2}\nabla_s^2 - \frac{1}{2}\left(V(\vec{r}' + \frac{1}{2}\vec{s}) + V(\vec{r}' - \frac{1}{2}\vec{s})\right)\right]\left(\frac{1}{2\pi i T}\right)^{3/2} e^{-i\Phi} = 0 \, , \tag{4-55}$$

and

$$\left[\vec{\nabla}'\cdot\vec{\nabla}_s - \left(V(\vec{r}' + \frac{1}{2}\vec{s}) - V(\vec{r}' - \frac{1}{2}\vec{s})\right)\right]\left(\frac{1}{2\pi i T}\right)^{3/2} e^{-i\Phi} = 0 \, , \tag{4-56}$$

where the time transformation function is inserted in the form (51). These differential equations are to be solved subject to the initial condition

$$\left(\frac{1}{2\pi i T}\right)^{3/2} e^{-i\Phi} \to \delta(\vec{s}) \quad , \quad \text{for } t \to 0 \tag{4-57}$$

[cf. Eq.(34)], which is satisfied provided that

$$T \to t \quad , \quad \text{for} \quad t \to 0 \quad , \tag{4-58}$$

("tyme → time") and

$$\Phi \to -\frac{s^2}{2t} \quad , \quad \text{for} \quad t \to 0 \quad . \tag{4-59}$$

For the sequel, it is helpful to carry out the differentiations in (55) and (56) formally, and to separate the real and imaginary parts of the resulting equations. This leads us to a system of four partial differential equations determining Φ and T,

$$\{\frac{\partial}{\partial t}\Phi - \frac{3}{4}\nabla_s^2 \log T + \frac{9}{8}(\vec{\nabla}_s \log T)^2 - \frac{1}{2}(\vec{\nabla}_s\Phi)^2\}$$

$$- \frac{1}{8}\{\frac{3}{2}\nabla'^2 \log T - \frac{9}{4}(\vec{\nabla}' \log T)^2 + (\vec{\nabla}'\Phi)^2\} \tag{4-60a}$$

$$= \frac{1}{2}[V(\vec{r}' + \frac{1}{2}\vec{s}) + V(\vec{r}' - \frac{1}{2}\vec{s})] \quad ,$$

$$- \frac{3}{4}\vec{\nabla}'\cdot\vec{\nabla}_s \log T + \frac{9}{8}\vec{\nabla}' \log T \cdot \vec{\nabla}_s \log T - \frac{1}{2}\vec{\nabla}'\Phi\cdot\vec{\nabla}_s\Phi$$

$$= \frac{1}{2}[V(\vec{r}' + \frac{1}{2}\vec{s}) - V(\vec{r}' - \frac{1}{2}\vec{s})] \quad , \tag{4-60b}$$

$$\{3\frac{\partial}{\partial t}\log T + \nabla_s^2\Phi - 3\vec{\nabla}_s\Phi\cdot\vec{\nabla}_s \log T\}$$

$$+ \frac{1}{4}\{\nabla'^2\Phi - 3\vec{\nabla}'\Phi\cdot\vec{\nabla}' \log T\} = 0 \quad , \tag{4-60c}$$

$$\vec{\nabla}'\cdot\vec{\nabla}_s\Phi - \frac{3}{2}\vec{\nabla}'\Phi\cdot\vec{\nabla}_s \log T - \frac{3}{2}\vec{\nabla}' \log T\cdot\vec{\nabla}_s\Phi = 0 \tag{4-60d}$$

Now since

$$\frac{\partial}{\partial t} \sim Z^{4/3} \quad , \quad \vec{\nabla}_s \sim Z^{2/3} \quad , \quad \vec{\nabla}' \sim Z^{1/3} \quad , \tag{4-61}$$

we notice that in (60a) and (60c) the first and second curly-bracket terms are of order $Z^{4/3}$ and $Z^{2/3}$, respectively, whereas the left-hand sides of (60b) and (60d) are of order $Z^{3/3}$ each. The right-hand side of (60a) is equal to

$$\frac{1}{2}\left(e^{\frac{1}{2}\vec{s}\cdot\vec{\nabla}'} + e^{-\frac{1}{2}\vec{s}\cdot\vec{\nabla}'}\right) V(\vec{r}') = \cosh(\frac{1}{2}\vec{s}\cdot\vec{\nabla}')\, V(\vec{r}')$$

$$= V(\vec{r}') + \frac{1}{2}(\frac{1}{2}\vec{s}\cdot\vec{\nabla}')^2\, V(\vec{r}') + \frac{1}{24}(\frac{1}{2}\vec{s}\cdot\vec{\nabla}')^4\, V(\vec{r}') + \ldots \tag{4-62}$$

which are terms of order $Z^{4/3}$, $Z^{2/3}$, $Z^{0/3}$, ... Similarly, we have in (60b)

$$\frac{1}{2}\left(e^{\frac{1}{2}\vec{s}\cdot\vec{\nabla}'} - e^{-\frac{1}{2}\vec{s}\cdot\vec{\nabla}'}\right) V(\vec{r}') = \sinh(\frac{1}{2}\vec{s}\cdot\vec{\nabla}')\, V(\vec{r}')$$

$$= (\frac{1}{2}\vec{s}\cdot\vec{\nabla}')\, V(\vec{r}') + \frac{1}{6}(\frac{1}{2}\vec{s}\cdot\vec{\nabla}')^3\, V(\vec{r}') + \ldots \quad , \tag{4-63}$$

these being terms of order $Z^{3/3}$, $Z^{1/3}$, ... The counting of the powers of $Z^{-1/3}$ is facilitated by introducing a parameter μ that essentially plays the role of $Z^{-1/3}$. It enters Eqs.(60a-d), (61), and (62) via the replacements

$$\frac{\partial}{\partial t} \rightarrow \frac{1}{\mu^4}\frac{\partial}{\partial t} \quad , \quad \vec{\nabla}_s \rightarrow \frac{1}{\mu^2}\vec{\nabla}_s \quad , \quad \vec{\nabla}' \rightarrow \frac{1}{\mu}\vec{\nabla}' \quad ,$$
$$V \rightarrow \frac{1}{\mu^4} V \quad , \quad \vec{s}\cdot\vec{\nabla}' \rightarrow \mu\, \vec{s}\cdot\vec{\nabla}' \quad . \tag{4-64}$$

Then the power of μ multiplying any term indicates its order in powers of $Z^{-1/3}$. All reference to μ can finally be removed by setting μ equal to unity.

After common factors of μ are cancelled, the effect of (64) is to multiply the second curly-bracket terms in (60a) and (60c) by μ^2 and to write, in conjunction with (62) and (63), $\cosh\,(\frac{\mu}{2}\vec{s}\cdot\vec{\nabla}')V(\vec{r}\,')$ and $(1/\mu)\sinh(\frac{\mu}{2}\vec{s}\cdot\vec{\nabla}')V(\vec{r}\,')$ for the right-hand sides of (60a) and (60b). It is then straightforward (and left to the reader) to verify that these equations are solved to order μ^2 by

$$\Phi(\vec{r}\,',\vec{s},t) = \{-\frac{s^2}{2t} + V(\vec{r}\,')t\} + \mu^2\frac{t}{24}\{(\vec{s}\cdot\vec{\nabla}')^2V(\vec{r}\,') + [\vec{\nabla}'V(\vec{r}\,')]^2t^2\} \tag{4-65}$$

and

$$\log T\,(\vec{r}\,',\vec{s},t) = \log t - \mu^2\frac{t^2}{18}\nabla'^2\, V(\vec{r}\,') \quad , \tag{4-66}$$

the latter one being equivalent to

$$T(\vec{r}\,',\vec{s},t) = t[1-\mu^2\frac{t^2}{18}\nabla'^2\, V(\vec{r}\,')] \quad . \tag{4-67}$$

Inasmuch as (65) and (67) are identical with (45), (47), and (50), as soon as μ is put equal to one, we are, indeed, assured that those approximations are correct up to the relative order of $Z^{-2/3}$.

A few comments are in order. The last reasoning, the counting or powers of μ, shows that only even powers of μ emerge; these correspond to corrections of relative orders $Z^{-2/3}$, $Z^{-4/3}$, and so on. Remarkably, there is no $Z^{-1/3}$ term. The Scott correction, however, <u>is</u> of relative size $Z^{-1/3}$. How does this fit in? The answer is both simple and instructive. The strongly bound electrons are exposed to the Coulomb potential without any shielding, so that their energies are measured not in multiples of $Z^{4/3}$ but of Z^2, and their distances are not $\sim Z^{-1/3}$ but $\sim Z^{-1}$, which is also the magnitude of their deBroglie wavelength. In other words: in the Coulomb part of the potential the scale is changed to the effect that, instead of the TF relations (37) and (61), we now have

$$V \sim Z^2 \;,\quad r' \sim s \sim Z^{-1} \;,\quad \vec{\nabla}' \sim \vec{\nabla}_s \sim Z$$
$$t \sim Z^{-2} \;,\quad \frac{\partial}{\partial t} \sim Z^2 \;, \tag{4-68}$$

which has the consequence that all terms in Eqs.(60a-d) are of the same size, namely $\sim Z^4$. This implies that, in contrast to the TF situation, there is no expansion parameter available for a systematic approximate calculation of Φ and T if V is the Coulomb potential. Any scheme based upon disregarding certain terms in Eqs.(60a-d) will inevitably result in a wrong answer.[2] Therefore, the vicinity of the nucleus will not be dealt with correctly if one simply extrapolates the quantum corrections of Eqs.(45), (47), and (50) into this region. There is no way around the special treatment of the strongly bound electrons that we studied in the preceding Chapter.

A second comment is the following. In arriving at the new approximation via the local oscillator potential of Eqs.(26) and (27), terms of order t^3 were kept in Φ and T, those of a higher order in t discarded. Does this imply that one could regard the expansions (41) and (46) as counting the powers of the time t? Of course, they do; the final results [Eqs.(45), (47), (50)], however, are not correct in the sense of displaying all contributions to order t^3. As a matter of fact, they do not even contain all terms of order t, since

$$\Phi(\vec{r}',\vec{s},t) = -\frac{s^2}{2t} + \left[\frac{\sinh(\frac{1}{2}\vec{s}\cdot\vec{\nabla}')}{\frac{1}{2}\vec{s}\cdot\vec{\nabla}'}\right] V(\vec{r}')t + O(t^3) \tag{4-69}$$

is the small-t form of Φ, as can be verified with the aid of Eqs.(60a-d). The extra terms disappear from (69) for $\vec{s}=0$, but the situation is different for the contribution $\sim t^3$, because

$$\Phi(\vec{r}',0,t) = V(\vec{r}')t + \frac{1}{24}\left([\vec{\nabla}'V(\vec{r}')]^2 - \frac{1}{10}(\nabla'^2)^2 V(\vec{r}')\right)t^3 + O(t^5) \tag{4-70}$$

replaces (45) if powers of t instead of $Z^{-1/3}$ are counted. Thus, it is really $Z^{-1/3}$, not t, what is the expansion parameter.[3]

Here is a third comment. If the transition from the dimensional many-particle Hamilton operator (1-1) to the dimensionless one (1-7) is not made, the gradients in Eqs.(60a-d) as well as (45),(47), and (50) come with a factor of $\hbar$ each. Then the $Z^{-2/3}$ terms are all multiplied by $\hbar^2$, so that they can be misunderstood as the beginning of a series in powers of $\hbar$, or $\hbar^2$. It has already been remarked, on the first pages of the Introduction, that $\hbar$ is not a parameter of the theory, so it certainly cannot serve as a measure of the quality of an approximation. What is really meant by the phrase "expanding in powers of $\hbar$" is the process of counting the powers of the $\vec{\nabla}'$ operator. Indeed, the $Z^{-2/3}$ terms are all displaying two $\vec{\nabla}'$s. This is not accidental; there is a simple reason why all terms of order $Z^{-m/3}$ also contain the $\vec{\nabla}'$ differential operator exactly m times. An immediate implication of Eqs.(60a-d) is that both Φ and T are even in $\vec{s}$ and odd in t, so that an arbitrary term in the expansion of either quantity is (symbolically) given by

$$(\nabla'^2)^{\alpha}(\vec{s}\cdot\vec{\nabla}')^{2\beta}\, V^{\gamma}\, t^{1+2\delta} \quad , \tag{4-71}$$

where α, β, γ, and δ are integers. Now, if V is the Coulomb potential, (71) is of the order $\alpha+\gamma-1-2\delta$ in Z^2, and since each such term is of order Z^0, as discussed above, we find

$$\alpha + \gamma = 1 + 2\delta \quad . \tag{4-72}$$

With this restriction, the most general form of (71) is

$$(t\,\nabla'^2)^{\alpha}(\vec{s}\cdot\vec{\nabla}')^{2\beta}(V\,t)^{\gamma} \quad , \tag{4-73}$$

which, for the TF potential, is of order $2(\alpha+\beta)$ in $Z^{-1/3}$. This is, indeed, the number of the $\vec{\nabla}'$ differential operators. We have thus established that, for our application to large atoms, expanding in powers of $Z^{-1/3}$ is equivalent to counting the powers of $\vec{\nabla}'$, or, more colloquially of $\hbar$.[4] It must be emphasized that the situation is likely to be different when the approximation is applied to other physical systems. We have already mentioned the bare Coulomb potential, for which there is no parameter for an expansion in the first place, so that the mechanical counting of the powers of $\vec{\nabla}'$ or $\hbar$ can only be qualified as nonsense.

The final comment answers the question why it is advantageous to write $\langle\vec{r}',t|\vec{r}'',0\rangle$ in the form (51) followed by approximating Φ and T, as compared to the apparently simpler

$$\langle\vec{r}' + \tfrac{1}{2}\vec{s},t|\vec{r}' - \tfrac{1}{2}\vec{s},0\rangle = \left(\frac{1}{2\pi i t}\right)^{3/2} e^{i\frac{s^2}{2t} - iV(\vec{r}')t} \times$$

$$\times\{1 - i\frac{t}{24}(\vec{s}\cdot\vec{\nabla}')^2 V(\vec{r}') + \frac{t^2}{12}\nabla'^2\ V(\vec{r}') - i\frac{t^3}{24}[\vec{\nabla}' V(\vec{r}')]^2 + O(Z^{-4/3})\}\ ,$$
(4-74)

which is known under the name Wigner-Kirkwood expansion.[5] To understand the principal reason it is useful to consider the quantity

$$\begin{aligned} \mathrm{tr}\, e^{-iHt} &= 2\int(d\vec{r}')\ \langle\vec{r}'|e^{-iHt}|\vec{r}'\rangle \\ &= 2\int(d\vec{r}')\ \langle\vec{r}',t|\vec{r}',o\rangle\ . \end{aligned} \qquad (4\text{-}75)$$

The spectral evaluation of this trace,

$$\mathrm{tr}\, e^{-iHt} = \sum_\lambda m(\lambda)\ e^{-iH'(\lambda)t}\ , \qquad (4\text{-}76)$$

identifies the energy eigenvalues $H'(\lambda)$ of H and their multiplicity $m(\lambda)$, both parametrzied by a set of (quantum) numbers, symbolized by λ. It is clear that, since the energy is the fundamental quantity of the system, the spectrum $H'(\lambda)$ is of central interest to us. In the TF approximation (19), λ stands for $\vec{r}'$ and $\vec{p}'$, the semiclassical spectrum being

$$H' \cong \tfrac{1}{2}p'^2 + V(\vec{r}') \qquad (4\text{-}77)$$

and the multiplicity

$$m \cong 2\,\frac{(d\vec{r}')(d\vec{p}')}{(2\pi)^3}\ , \qquad (4\text{-}78)$$

that is: two states per phase space volume of $(2\pi)^3$. The advantage of the approximations (45) and (50) in (51) over the expansion (74) is that the former is easily written in the spectral form (76), whereás the latter is not. It is instructive to evaluate the spectrum and multiplicity corresponding to our new, quantum-corrected, semiclassical approximation. After writing the tyme factor of (51) in analogy to the free-particle momentum integration in Eq.(19),

$$\left(\frac{1}{2\pi iT}\right)^{3/2} = \int\frac{(d\vec{p}')}{(2\pi)^3}\ e^{-i\frac{1}{2}p'^2T}\ , \qquad (4\text{-}79)$$

the trace of (75) is approximated by

$$\mathrm{tr}\, e^{-iHt} \cong 2\int \frac{(d\vec{r}\,')(d\vec{p}\,')}{(2\pi)^3}\, e^{-i\left(\frac{1}{2}p'^2+V(\vec{r}\,')\right)t-\frac{i}{8}\left[\left(\vec{\nabla}'V(\vec{r}\,')\right)^2-\frac{2}{3}p'^2\nabla'^2V(\vec{r}\,')\right]\frac{t^3}{3}} . \tag{4-80}$$

Since we are aiming at an exponent linear in t, as required by (76), it is fitting to represent the exponential function of t^3 with the aid of Airy's function Ai(x),[6] defined by

$$e^{-iy^3/3} = \int_{-\infty}^{\infty} dx\, e^{ixy}\, \mathrm{Ai}(x) \quad . \tag{4-81}$$

The properties of Ai(x) will be of particular interest later, for the moment, however, it suffices that with (81) Eq.(80) reads

$$\mathrm{tr}\, e^{-iHt} \cong 2\int \frac{(d\vec{r}\,')(d\vec{p}\,')}{(2\pi)^3}\, dx\, \mathrm{Ai}(x)\, e^{-iH'(\vec{r}\,',\vec{p}\,',x)t} \quad , \tag{4-82}$$

where

$$H'(\vec{r}\,',\vec{p}\,',x) = \frac{1}{2}p'^2 + V(\vec{r}\,') - \frac{1}{2}x\left[\left(\vec{\nabla}'V(\vec{r}\,')\right)^2 - \frac{2}{3}p'^2\nabla'^2V(\vec{r}\,')\right]^{1/3} \tag{4-83}$$

identifies the effective, quantum-corrected, semiclassical spectrum of H, the multiplicity being

$$m = 2\, \frac{(d\vec{r}\,')(d\vec{p}\,')}{(2\pi)^3}\, dx\, \mathrm{Ai}(x) \quad . \tag{4-84}$$

In contrast to this argument for the form (51), an interpretation of (74) as a natural starting point for a spectral evaluation of tr e^{-iHt} does not seem possible. Therefore, remembering the importance of spectral sums in quantum-mechanical calculations, the form (51) is obviously preferable over the expansion (74).

<u>Quantum corrections II (leading energy correction).</u> Upon setting y=o in Eq.(81), we infer that

$$\int_{-\infty}^{\infty} dx\, \mathrm{Ai}(x) = 1 \quad , \tag{4-85}$$

which permits one to introduce a definition of Airy averaging:

$$\langle f(x)\rangle^\circ \equiv \int_{-\infty}^{\infty} dx\, \mathrm{Ai}(x) f(x) \quad . \tag{4-86}$$

In this notation, Eq.(81) appears as

$$\langle e^{ixy}\rangle^\circ = e^{-iy^3/3} \quad , \tag{4-87}$$

which has the special consequences

$$\langle x\rangle^\circ = 0 \;, \quad \langle x^2\rangle^\circ = 0 \;, \quad \langle x^3\rangle^\circ = 2 \quad . \tag{4-88}$$

In view of the Airy average on the right-hand side of Eq.(82),

$$\mathrm{tr}\, e^{-iHt} \cong 2\int \frac{(d\vec{r}\,')(d\vec{p}\,')}{(2\pi)^3} \langle e^{-iH'(\vec{r}\,',\vec{p}\,',x)t}\rangle^\circ \quad , \tag{4-89}$$

the trace of any function f(H) of the independent-particle Hamilton operator (19) is now approximated by

$$\mathrm{tr}\, f\big(H(\vec{r},\vec{p})\big) \cong 2\int \frac{(d\vec{r}\,')(d\vec{p}\,')}{(2\pi)^3} \langle f\big(H'(\vec{r}\,',\vec{p}\,',x)\big)\rangle^\circ \tag{4-90}$$

which uses the effective spectrum (83). Please note by how little this quantum-corrected version differs from the highly semiclassical phase-space integral (1-43). This original approximation is recovered from (90) by the replacement

$$H'(\vec{r}\,',\vec{p}\,',x) \to H'(\vec{r}\,',\vec{p}\,',o) = H(\vec{r}\,',\vec{p}\,') \quad . \tag{4-91}$$

Concequently, the leading quantum correction to (90), $\Delta_{qu}\mathrm{tr}\, f(H)$, is given by the first non-vanishing term of an expansion in powers of

$$H'(\vec{r}\,',\vec{p}\,',x) - H(\vec{r}\,',\vec{p}\,') = -\tfrac{1}{2}x\big[\big(\vec{\nabla}'V(r')\big)^2 - \tfrac{2}{3}p'^2\nabla'^2V(\vec{r}\,')\big]^{1/3} \quad . \tag{4-92}$$

Because of Eqs.(88) this is the cubic term. Thus

$$\Delta_{qu}\mathrm{tr}\, f(H) = 2\int \frac{(d\vec{r})(d\vec{p})}{(2\pi)^3} f'''\big(\tfrac{1}{2}p^2 + V(\vec{r})\big)\big(-\tfrac{1}{24}\big)\big[(\vec{\nabla}V)^2 - \tfrac{2}{3}p^2\nabla^2 V\big] \quad , \tag{4-93}$$

where the primes denote differentiation with respect to the argument. (The primes on the integration variables have been dropped, since a confusion of numbers with quantum-mechanical operators is no longer likely.) The partial integrations

$$\begin{aligned}\int(d\vec{r})\,f'''(\tfrac{1}{2}p^2+V)\,(\vec{\nabla}V)^2 &= \int(d\vec{r})\,\vec{\nabla}V\cdot\vec{\nabla}f''(\tfrac{1}{2}p^2+V)\\ &= \int(d\vec{r})\,(-\nabla^2V)\,f''(\tfrac{1}{2}p^2+V)\end{aligned} \tag{4-94}$$

and

$$\begin{aligned}\int(d\vec{p})\,f'''(\tfrac{1}{2}p^2+V)\,p^2 &= \int(d\vec{p})\,\vec{p}\cdot\frac{\partial}{\partial\vec{p}}\,f''(\tfrac{1}{2}p^2+V)\\ &= \int(d\vec{p})\,(-3)\,f''(\tfrac{1}{2}p^2+V)\end{aligned} \tag{4-95}$$

simplify the right-hand side of (93), producing

$$\Delta_{qu}\operatorname{tr} f(H) = 2\int\frac{(d\vec{r})\,(d\vec{p})}{(2\pi)^3}\,f''\big(\tfrac{1}{2}p^2+V(\vec{r})\big)\,(-\tfrac{1}{24}\nabla^2V)\ . \tag{4-96}$$

When applied to

$$E_{\zeta\zeta_s} = -\int_{\zeta}^{\zeta_s}d\zeta'\,(\zeta'-\zeta)\operatorname{tr}\delta(-H-\zeta') \tag{4-97}$$

[which combines Eqs. (9) and (13)] this gives

$$\Delta_{qu}E_{\zeta\zeta_s} = \frac{1}{24}\int(d\vec{r})\,\nabla^2V\int_{\zeta}^{\zeta_s}d\zeta'\,(\zeta'-\zeta)\,(\frac{\partial}{\partial\zeta'})^2\frac{1}{\pi^2}[-2(V+\zeta')]^{1/2}\ , \tag{4-98}$$

which uses

$$2\int\frac{(d\vec{p})}{(2\pi)^3}\,\delta(-\tfrac{1}{2}p^2-V-\zeta') = \frac{1}{\pi^2}[-2(V+\zeta')]^{1/2}\ . \tag{4-99}$$

Aiming at a perturbative evaluation of (98), we first dispose of the delta function at the origin in ∇^2V. Of course, $[-2(V+\zeta')]^{1/2}$ is singular at that point. But, the two derivatives with respect to ζ' wipe that term out. Indeed, the whole structure of the second derivative $\sim[-2(V+\zeta')]^{-3/2}\sim r^{3/2}$ is thoroughly zero at the origin. [This is also essential for the partial integration of (94).] So we can, without changing anything, replace ∇^2V in (98) by $\nabla^2(V+Z/r)$. Then we equate this to $-4\pi n$ through the Poisson equation and, with the ζ' integral evaluated, arrive at

$$\Delta_{qu}E_{\zeta\zeta_s} = -\frac{1}{6\pi}\int(d\vec{r})\,n(\vec{r})\{[-2(V+\zeta)]^{1/2} - [-2(V+\zeta_s)]^{1/2} - (\zeta_s-\zeta)[-2(V+\zeta_s)]^{-1/2}\} . \qquad (4\text{-}100)$$

Now, the density n is composed of the TF density (2-51)

$$n_{TF} = \frac{1}{3\pi^2}[-2(V+\zeta)]^{3/2} \qquad (4\text{-}101)$$

and corrections to it referring to strongly bound electrons, quantum effects etc. Likewise the curly brackets are the sum of the TF term and corrections to it. Consequently, the leading quantum correction to E_1 is obtained by disregarding the modifications of the TF part of Eq. (100), implying

$$\Delta_{qu}\, E_1 = -\frac{1}{18\pi^3}\int(d\vec{r})[-2(V+\zeta)]^{4/2} \quad , \qquad (4\text{-}102)$$

where V and ζ are the TF quantities corresponding to the system under consideration. What has been discarded in going from (100) to (102) are corrections to corrections which, if taken seriously, result in energy contributions of a lower order in $Z^{-1/3}$. In Eq.(102), the power 4/2 instead of 2 is a reminder that the domain of integration is the classically allowed region where $V+\zeta<o$.

For neutral atoms, we have $\zeta=o$ and $V=-(Z/r)F(x)$, so that

$$\Delta_{qu}\, E_1 = -\frac{Z^{5/3}}{(4a)^2}\int_o^\infty dx[F(x)]^2 = -\frac{2}{11}\times 0.269900\ Z^{5/3} \quad , \qquad (4\text{-}103)$$

which makes use of the numerical value given for this integral in Problem 2-3. The reason for exhibiting the factor of 2/11 will be clear later, after the leading exchange energy contribution will have been evaluated.

The first convincing derivation of (103) was given by Schwinger;[7] his original argument differs slightly from the one given above, which follows the reasoning of Ref.8. Historically one associates the name von Weizsäcker[9] with the leading quantum correction to the TF model. Let us briefly halt the further development of the theory in order to establish the connection with von Weizsäcker's work.

The von Weizsäcker term. What is known as the von Weizsäcker term is the leading inhomogeneity correction to the TF density functional for the kinetic energy,

$$E_{kin}(n) = \int(d\vec{r})\ \frac{1}{10\pi^2}(3\pi^2 n)^{5/3} \quad , \tag{4-104}$$

derived in Chapter Two [see Eq. (2-95)]. So far, we have worked out the quantum-corrected potential functional $E_1(V+\zeta)$. To find the corresponding density functional we must follow the instructions given after Eq. (2-434): use Eq. (2-20),

$$\delta_V E_1(V+\zeta) = \int(d\vec{r})\delta V(\vec{r})n(\vec{r}) \quad , \tag{4-105}$$

to express V in terms of n, then this V(n) produces $E_{kin}(n)$ when inserted into Eq. (2-428),

$$E_{kin} = E_1(V+\zeta) - \int(d\vec{r})\,(V+\zeta)n \quad . \tag{4-106}$$

Aiming at the von Weizsäcker correction we first forget about the special treatment of the strongly bound electrons, so that ζ_s is chosen arbitrarily large in Eq. (98), implying that

$$\Delta_{qu}\ E_1 = \int(d\vec{r})\ \frac{1}{24\pi^2}[-2(V+\zeta)]^{1/2}\,\nabla^2 V \tag{4-107}$$

supplements the TF result (2-44),

$$(E_1)_{TF} = \int(d\vec{r})\,(-\frac{1}{15\pi^2})\,[-2(V+\zeta)]^{5/2} \quad . \tag{4-108}$$

Now (105) produces

$$n = \frac{1}{3\pi^2}[-2(V+\zeta)]^{3/2} - \frac{1}{24\pi^2}[-2(V+\zeta)]^{-1/2}\ \nabla^2 V$$
$$+ \frac{1}{24\pi^2}\ \nabla^2[-2(V+\zeta)]^{1/2} =$$

$$= \frac{1}{3\pi^2}[-2(V+\zeta)]^{3/2} - \frac{1}{12\pi^2}[-2(V+\zeta)]^{-1/2}\,\nabla^2 V - \frac{1}{24\pi^2}[-2(V+\zeta)]^{-3/2}\,(\vec{\nabla}V)^2 \quad , \tag{4-109}$$

or

$$(3\pi^2 n)^{2/3} = [-2(V+\zeta)] - \frac{1}{24}[-2(V+\zeta)]^{-1}\,\nabla^2 V - \frac{1}{12}[-2(V+\zeta)]^{-2}\,(\vec{\nabla}V)^2 \quad , \tag{4-110}$$

where the corrections to the leading TF expression are consistently treated as small. Equation (110) is solved for V by using the TF connection

$$-2(V+\zeta) = (3\pi^2 n)^{2/3} \tag{4-111}$$

in the (small) last two terms on the right-hand side. This results in

$$-2(V+\zeta) = (3\pi^2 n)^{2/3} + \frac{1}{72}\left[\left(\frac{\vec{\nabla}n}{n}\right)^2 - \frac{\nabla^2 n}{n}\right] . \tag{4-112}$$

In the TF regime the relative size of n and $\vec{\nabla}$ is $n \sim Z^2$, $\vec{\nabla} \sim Z^{1/3}$, so that (112) displays corrections of relative order $Z^{-2/3}$, as it should. This quantum-corrected expression is now inserted into (108) and the second term of (106); in (107), which in itself is a quantum correction, the TF result (111) suffices. The outcome is

$$E_1 = (E_1)_{TF} + \Delta_{qu}E_1$$

$$= \int(d\vec{r})\;\left\{\left(-\frac{1}{15\pi^2}\right)(3\pi^2 n)^{5/3} + \frac{1}{144}\frac{(\vec{\nabla}n)^2}{n} - \frac{5}{144}\nabla^2 n\right\} \tag{4-113}$$

and

$$-\int(d\vec{r})\,(V+\zeta)n = \int(d\vec{r})\;\left\{\frac{1}{6\pi^2}(3\pi^2 n)^{5/3} + \frac{1}{144}\frac{(\vec{\nabla}n)^2}{n} - \frac{1}{144}\nabla^2 n\right\}. \tag{4-114}$$

The terms that are multiples of $\nabla^2 n$ integrate to a null result, so that according to (106)

$$E_{kin}(n) = \int (d\vec{r}) \{ \frac{1}{10\pi^2} (3\pi^2 n)^{5/3} + \frac{1}{72} \frac{(\vec{\nabla} n)^2}{n} \} , \qquad (4\text{-}115)$$

where the second contribution is the von Weizsäcker correction to the TF result (104).[10]

In deriving (115) from the quantum correction (107) the special treatment of the strongly bound electrons has been "forgotten" about. This just means that we extrapolate the correction, which is known to be the correct one in the TF regime, into the vicinity of the nucleus where, according to the discussion around (68), it must fail. And it does. For, although the von Weizsäcker term in (115) is supposed to be a small correction, it cannot be treated as a perturbation. If we would try to do so by inserting the neutral atom TF density

$$n_{TF} = \frac{1}{3\pi^2} [\frac{2Z}{r} F(x)]^{3/2} = \frac{32}{9\pi^3} Z^2 [F(x)/x]^{3/2} , \qquad (4\text{-}116)$$

the resulting integral

$$"\Delta_{qu} E_{kin}" = \frac{Z^{5/3}}{32a^2} \int_0^\infty \frac{dx}{x^{3/2}} [F(x)]^{-1/2} [F(x) - xF'(x)]^2 \qquad (4\text{-}117)$$

would obviously diverge at x=o due to the singularity of n_{TF} at r=o. When being conscious of the necessary corrections for the strongly bound electrons, it is nevertheless possible to arrive at (103) after starting from (115), as is demonstrated by Schwinger in Ref.7.[11]

On the other hand, it is quite clear that there is no chance of being able to express V in terms of n if the special treatment of the innermost electrons is explicitly included into $E_1(V+\zeta)$ as described in the preceding Chapter. The transition from the potential functional to the density functional is no longer feasible now that we have gone beyond the original TF approximation. These observations are, of course, in agreement with the general anticipation discussed after Eq.(2-408), namely that the potential functional is much better suited for improvements over the TF model than the corresponding density functional.

Quantum corrections III (energy). We pick up the story at Eq.(90) where we left it to study the leading correction. In (90) our new approximation is stretched a little bit too far since the intermediate step (79) conceals the circumstance that our knowledge concerning the dependence on

$\nabla^2 V$ does not extend beyond the linear term. Consequently, consistency requires to expand the right-hand side of (90) in powers of $\nabla^2 V$ and to discard the quadratic and higher-order terms. In view of the later application to energy it is useful to do this not for a function of H but of $-H-\zeta$, for which Eq.(90) reads

$$\begin{aligned} \operatorname{tr} f(-H-\zeta) & \\ &\cong 2\int \frac{(d\vec{r})(d\vec{p})}{(2\pi)^3} \left\langle f\left(-\tfrac{1}{2}p^2 - V - \zeta + \tfrac{x}{2}[(\vec{\nabla}V)^2 - \tfrac{2}{3}p^2\nabla^2 V]^{1/3}\right)\right\rangle_{\mathcal{P}} \quad . \end{aligned} \tag{4-118}$$

With the abbreviation

$$z \equiv -\tfrac{1}{2}p^2 - V - \zeta + \tfrac{x}{2}|\vec{\nabla}V|^{2/3} \quad , \tag{4-119}$$

this is to first order in $\nabla^2 V$

$$\begin{aligned} \operatorname{tr} f(-H-\zeta) & \\ &\cong 2\int \frac{(d\vec{r})(d\vec{p})}{(2\pi)^3} \left\langle f(z) - \tfrac{1}{9}\, x\, p^2 |\vec{\nabla}V|^{-4/3}\, \nabla^2 V\, f'(z)\right\rangle_{\mathcal{P}} \quad . \end{aligned} \tag{4-120}$$

Now, the identity

$$p^2 f'(z) = -\vec{p}\cdot\frac{\partial}{\partial\vec{p}} f(z) \tag{4-121}$$

allows a partial $\vec{p}$-integration, so that an equivalent statement is

$$\begin{aligned} \operatorname{tr} f(-H-\zeta) & \\ &\cong 2\int \frac{(d\vec{r})(d\vec{p})}{(2\pi)^3} \left\langle f(z) - \tfrac{1}{3}\, x\, |\vec{\nabla}V|^{-4/3}\, \nabla^2 V\, f(z)\right\rangle_{\mathcal{P}} \quad . \end{aligned} \tag{4-122}$$

A further simplification is achieved after observing that the differential equation obeyed by the Airy function,

$$\frac{d^2}{dx^2}\operatorname{Ai}(x) = x\operatorname{Ai}(x) \tag{4-123}$$

[which can easily be derived from the defining equation (81)], implies the equivalence of the Airy averages

$$\langle x\ g(x)\rangle^o = \langle \frac{d^2}{dx^2}\ g(x)\rangle^o \quad . \tag{4-124}$$

One application is the recurrence relation

$$\langle x^{k+1}\rangle^o = k(k-1)\langle x^{k-2}\rangle^o \quad ; \tag{4-125}$$

it generalizes (88) to

$$\langle x^k\rangle^o = \begin{cases} 3^{-k/3}\dfrac{k!}{(k/3)!} & , \text{ for } k = 0,3,6,\ldots \quad , \\ 0 \quad , & \text{otherwise} \quad . \end{cases} \tag{4-126}$$

Since

$$\frac{d^2}{dx^2}\ f(z) = \frac{1}{4}|\vec{\nabla}V|^{4/3}\ f''(z) \quad , \tag{4-127}$$

employing (124) in (122) produces

$$\text{tr}\ f(-H-\zeta) \cong 2\int\frac{(d\vec{r})\,(d\vec{p})}{(2\pi)^3} \langle f(z) - \frac{1}{12}\ f''(z)\nabla^2 V\rangle^o \quad , \tag{4-128}$$

which is the desired modification of Eq.(90).

This new injunction for evaluating traces of functions of $-H-\zeta$ is now applied to the trace in Eq.(97). The outcome is

$$E_{\zeta\zeta_s} = -2\ \frac{(d\vec{r})\,(d\vec{p})}{(2\pi)^3}\int_{\zeta}^{\zeta_s} d\zeta'(\zeta'-\zeta)\langle\delta(z') - \frac{1}{12}\ \delta''(z')\nabla^2 V\rangle^o \quad , \tag{4-129}$$

where z' is related to ζ' just like z is to ζ [Eq.(119)]. After performing the ζ' integration, we arrive at

$$E_{\zeta\zeta_s} = -2\int\frac{(d\vec{r})\,(d\vec{p})}{(2\pi)^3}\langle z[\eta(z)-\eta(z_s)] - \frac{1}{12}[\delta(z) - \delta(z_s) + (z_s-z)\delta'(z_s)]\nabla^2 V\rangle^o \quad , \tag{4-130}$$

where, of course,

$$z_s = -\frac{1}{2}p^2 - V - \zeta_s + \frac{1}{2}x\,|\vec{\nabla}V|^{2/3} = z - (\zeta_s-\zeta) \quad . \tag{4-131}$$

Equations (129) and (130) are the quantum corrected versions of the TF expressions (3-19) and (3-44), respectively, to which they reduce upon neglecting the $\vec{\nabla}V$ and $\nabla^2 V$ dependences and performing the Airy averaging and the momentum integration. It is a technical detail of enormous significance that one can both Airy average and integrate over momentum in (129) and (130) explicitly, so that, as before, only the spatial integration is left. Because of the Delta functions the $\vec{p}$-integrals are immediate; the Airy averages, however, are nontrivial.[12] The next section shows how to deal with them.

Airy Averages. It is expedient to first consider

$$\begin{aligned} 2\int\frac{(d\vec{p})}{(2\pi)^3}\langle\delta'(z)\rangle^{o} &= -\langle\frac{\partial}{\partial\zeta}\,2\int\frac{(d\vec{p})}{(2\pi)^3}\,\delta(z)\rangle^{o} \\ &= \frac{1}{\pi^2}\langle[-2(V+\zeta-\frac{1}{2}x\,|\vec{\nabla}V|^{2/3})]^{-1/2}\rangle^{o} \quad , \end{aligned} \tag{4-132}$$

where the second equality is based upon the momentum integral (99). Let us reexpress this in terms of the variable

$$y = \frac{2(V+\zeta)}{|2\vec{\nabla}V|^{2/3}} \quad , \tag{4-133}$$

which has the property of being negative (positive) in the classically allowed (forbidden) region, to get

$$2\int\frac{(d\vec{p})}{(2\pi)^3}\langle\delta'(z)\rangle^{o} = \frac{2}{\pi}\,|2\vec{\nabla}V|^{-1/3}\,F_o(y) \quad , \tag{4-134}$$

where

$$F_o(y) = \frac{1}{2\pi}\langle(2^{-2/3}x-y)^{-1/2}\rangle^{o} \quad . \tag{4-135}$$

Since the fundamental Airy average, Eq.(87), concerns an exponential function of x, we employ the identity

$$(2^{-2/3}x-y)^{-1/2} = \int_{-\infty}^{\infty}d\tau\;\delta(\tau^2+y-2^{-2/3}x) \quad , \tag{4-136}$$

in conjunction with the Fourier integral of the Delta function (12), to arrive at

$$F_o(y) = \left(\frac{1}{2\pi}\right)^2 \int_{-\infty}^{\infty} d\tau \int_{-\infty}^{\infty} d\sigma \langle e^{-i(\tau^2+y-2^{-2/3}x)\sigma} \rangle_{\mathcal{P}}$$

$$= \left(\frac{1}{2\pi}\right)^2 \int d\tau\, d\sigma\, e^{-i(\tau^2+y)\sigma - i\sigma^3/12} \quad , \tag{4-137}$$

the last step using (87). After the substitutions

$$\sigma = x+x' \ , \ \tau = \frac{1}{2}(x-x') \ , \ d\tau d\sigma = dx dx' \ , \tag{4-138}$$

the integrand factorizes,

$$F_o(y) = \frac{1}{2\pi} \int_{-\infty}^{\infty} dx\ e^{-iyx - ix^3/3}$$

$$\times \frac{1}{2\pi} \int_{-\infty}^{\infty} dx'\ e^{-iyx' - ix'^3/3} \ , \tag{4-139}$$

so that the Fourier transformed statement of Eq.(81), namely

$$Ai(y) = \frac{1}{2\pi} \int_{-\infty}^{\infty} dx\ e^{-iyx}\ e^{-ix^3/3} \quad , \tag{4-140}$$

can be used twice, with the final outcome

$$F_o(y) = [Ai(y)]^2 \ . \tag{4-141}$$

Before proceeding to the evaluation of the Airy-averaged momentum-integrals of $\delta(z)$, $\eta(z)$,... , let us supply another, more physically oriented derivation of this result. For this purpose recall that the approximations (47) and (50) are exact in the situation of a constant force potential. Therefore, by simply undoing the steps that introduced the momentum integral [Eq.(79)] and the Airy average [Eqs.(81) and (87)], we have

$$2\int \frac{(d\vec{p})}{(2\pi)^3} \langle \delta'(z_1) \rangle_{\mathcal{P}} = 2\langle \vec{r}|\delta'(-H_1-\zeta)|\vec{r}\rangle \quad , \tag{4-142}$$

where z_1 and the Hamilton operator H_1 refer to the constant-force potential V_1 of Eq.(20). The eigenstates of H_1, characterized by their energy $-E$ and their transverse momentum $\vec{k}_\perp$ (which is to say that the two-dimensional vector $\vec{k}_\perp$ is perpendicular to the constant force $-\vec{\nabla}V$) are, properly normalized

$$\langle\vec{r}|E,\vec{k}_\perp\rangle = \frac{e^{i\vec{k}_\perp\cdot\vec{r}}}{(2\pi)^2}\, 2^{1/2}|2\vec{\nabla}V_1|^{-1/6}\, \mathrm{Ai}\left(\frac{2(V_1+E)+k_\perp^2}{|2\vec{\nabla}V_1|^{2/3}}\right) , \tag{4-143}$$

which reminds us that the differential equation obeyed by Ai(x), Eq. (123), is essentially the Schrödinger equation for the one-dimensional linear potential. The matrix element on the right-hand side of (142) is now evaluated by

$$\begin{aligned} 2\int\frac{(d\vec{p})}{(2\pi)^3}\langle\delta'(z_1)\rangle^\circ &= 2\int_{-\infty}^{\infty}dE\int(d\vec{k}_\perp)\,|\langle\vec{r}|E,\vec{k}_\perp\rangle|^2\,\delta'(E-\zeta) \\ &= 2\int(d\vec{k}_\perp)\left(-\frac{\partial}{\partial\zeta}\right)|\langle\vec{r}|\zeta,\vec{k}_\perp\rangle|^2 , \end{aligned} \tag{4-144}$$

This squared wave function does not depend on ζ and $\vec{k}_\perp$ individually but only on the sum $2\zeta+k_\perp^2$, with the consequence

$$\begin{aligned} 2\int\frac{(d\vec{p})}{(2\pi)^3}\langle\delta'(z_1)\rangle^\circ &= 4\pi\int_0^{\infty}dk_\perp^2\left(-\frac{\partial}{\partial k_\perp^2}\right)|\langle\vec{r}|\zeta,\vec{k}_\perp\rangle|^2 \\ &= 4\pi|\langle\vec{r}|\zeta,0\rangle|^2 = \frac{2}{\pi}|2\vec{\nabla}V_1|^{-1/3}[\mathrm{Ai}(y_1)]^2 . \end{aligned} \tag{4-145}$$

Now, dropping the subscript "$_1$" and comparing with (134) reproduces (141).

It is useful to deal with all the Airy-averaged momentum integrals of $z\eta(z)$, $\eta(z)$, $\delta(z)$, ... as a set, exploiting their relationship through differentiation. To convey this compactly, we shall introduce positive and negative powers of derivatives:

$$\delta''(z) = \frac{d}{dz}\delta'(z) \quad , \quad \delta'''(z) = \left(\frac{d}{dz}\right)^2\delta'(z) \quad , \tag{4-146}$$

and

$$\begin{aligned} \delta(z) = \left(\frac{d}{dz}\right)^{-1}\delta'(z) \quad &, \quad \eta(z) = \left(\frac{d}{dz}\right)^{-2}\delta'(z) \quad , \\ z\eta(z) &= \left(\frac{d}{dz}\right)^{-3}\delta'(z) \quad , \end{aligned} \tag{4-147}$$

and so on. Obviously, $(d/dz)^{-1}$ is short hand for integrating according to

$$(\frac{d}{dz})^{-1} f(z) = \int_{-\infty}^{z} dz' \, f(z') \ , \tag{4-148}$$

which exhibits the specific boundary condition at $z=-\infty$ that selects uniquely one of the indefinite integrals of $f(z)$. With this notation, and in view of the linear dependence of z on the constant $-\zeta$, we can write

$$\begin{aligned} &2\int\frac{(d\vec{p})}{(2\pi)^3} \langle(\frac{d}{dz})^{-m} \delta'(z)\rangle^0 \\ &= [\frac{d}{d(-\zeta)}]^{-m} \, 2\int\frac{(d\vec{p})}{(2\pi)^3} \langle\delta'(z)\rangle^0 \\ &= [\frac{d}{d(-\zeta)}]^{-m} \, \frac{2}{\pi}|2\vec{\nabla}V|^{-1/3} \, F_0(y) \ , \end{aligned} \tag{4-149}$$

where the last equality uses Eq. (134). The definition of y in Eq. (133) has the immediate consequence

$$\frac{d}{dz} = 2|2\vec{\nabla}V|^{-2/3} \frac{d}{dy} \ , \tag{4-150}$$

which implies

$$2\int\frac{(d\vec{p})}{(2\pi)^3} \langle(\frac{d}{dz})^{-m} \delta'(z)\rangle^0 = \frac{2}{\pi} \, 2^{-m}|2\vec{\nabla}V|^{(2m-1)/3} \, F_m(y) \ , \tag{4-151}$$

with the functions $F_m(y)$ defined by

$$\begin{aligned} F_m(y) &= (-\frac{d}{dy})^{-m} F_0(y) \\ &= \frac{1}{2\pi}(-\frac{d}{dy})^{-m} \langle(2^{-2/3}x - y)^{-1/2}\rangle^0 \ . \end{aligned} \tag{4-152}$$

The boundary at $z=-\infty$ in (148) clearly corresponds to $\zeta=\infty$ and $y=\infty$ in (149) and (152), respectively, so that $(d/dy)^{-1}$ signifies

$$(\frac{d}{dy})^{-1} f(y) = -\int_{y}^{\infty} dy' \, f(y') \ . \tag{4-153}$$

An immediate recurrence relation for the functions $F_m(y)$ is

$$-\frac{d}{dy} F_m(y) = F_{m-1}(y) \; . \tag{4-154}$$

To produce another relation we first remark that, for m≥o, the second version of (152) has the significance

$$F_m(y) = \frac{(-\frac{1}{2})!}{(m-\frac{1}{2})!} \frac{1}{2\pi} <(2^{-2/3}x - y)^{m-1/2}_{>}^{\circ} , \tag{4-155}$$

which can be checked against (154). Now observe that

$$\begin{aligned} <(2^{-2/3}x - y)^{m+1/2}_{>}^{\circ} &= <(2^{-2/3}x - y)(2^{-2/3}x - y)^{m-1/2}_{>}^{\circ} \\ &= (\frac{1}{4}\frac{d^2}{dy^2} - y) <(2^{-2/3}x - y)^{m-1/2}_{>}^{\circ} , \end{aligned} \tag{4-156}$$

on applying the Airy averaging relation (124). Accordingly,

$$\begin{aligned} (m + \tfrac{1}{2}) F_{m+1}(y) &= (\frac{1}{4}\frac{d^2}{dy^2} - y) F_m(y) \\ &= \frac{1}{4} F_{m-2}(y) - y F_m(y) , \end{aligned} \tag{4-157}$$

which is compatible with (154) and therefore also valid for m<o. Because of the boundary condition in (153) all $F_m(y)$ must tend to zero for $z\to\infty$, a property that is conserved by the recurrence relations (154) and (157).

Now, beginning with our knowledge of $F_o(y)$, Eq.(141), we first use (154) to compute successively

$$\begin{aligned} F_{-1}(y) &= -2Ai(y)\,Ai'(y) , \\ F_{-2}(y) &= 2\{y[Ai(y)]^2 + [Ai'(y)]^2\} , \end{aligned} \tag{4-158}$$

and so on; then we apply (157) for a purely algebraic computation of

$$\begin{aligned} F_1(y) &= -y[Ai(y)]^2 + [Ai'(y)]^2 , \\ F_2(y) &= \frac{2}{3}\{y^2[Ai(y)]^2 - \frac{1}{2}Ai(y)\,Ai'(y) - y[Ai'(y)]^2\} , \end{aligned} \tag{4-159}$$

and so forth. Of course, one makes use of the differential equation (123); and it is clear that all $F_m(y)$ are sums of polynomials in y multiplying the square of the Airy function, or of its derivative, or their product.

If the gradient of V happens to be zero, then we have, for m≥o,

$$2\int\frac{(d\vec{p})}{(2\pi)^3}\,\langle(\frac{d}{dz})^{-m}\,\delta'(z)\rangle^{\circ}$$

$$= [\frac{d}{d(-\zeta)}]^{-m}\;2\int\frac{(d\vec{p})}{(2\pi)^3}\,\langle\delta'(-\tfrac{1}{2}p^2-V-\zeta)\rangle^{\circ} \tag{4-160}$$

$$= [\frac{d}{d(-\zeta)}]^{-m}\;\frac{1}{\pi^2}\;[-2(V+\zeta)]^{-1/2}$$

$$= \frac{1}{\pi^2}\;2^{-m}\;\frac{(-\frac{1}{2})!}{(m-\frac{1}{2})!}\;[-2(V+\zeta)]^{m-1/2}\quad,$$

where the Airy average is immediate and the momentum integral is the ζ derivative of the one of Eq.(99). In Eq.(151), the limit $|\vec{\nabla}V|\to 0$ means $y\to-\infty$ or $y\to+\infty$ depending on the sign of $V+\zeta$. This implies the asymptotic y dependences

$$F_m(y) \cong \frac{1}{2\pi}\,\frac{(-\frac{1}{2})!}{(m-\frac{1}{2})!}\,(-y)^{m-1/2}\quad,\ \text{for } -y\gg 1\ , \tag{4-161}$$

and

$$F_m(y) \cong 0\ ,\ \text{for}\ \ y\gg 1\ , \tag{4-162}$$

of which the second one holds also for m<o, whereas the first one does not. This becomes more apparent when we use the known asymptotic forms of the Airy function [see Problem 4 , or Ref.6],

$$\mathrm{Ai}(y) \cong \frac{1}{\sqrt{\pi}}(-y)^{-1/4}\,\cos(\tfrac{2}{3}(-y)^{3/2}-\tfrac{\pi}{4})\ ,\ \text{for}\ \ -y\gg 1\ , \tag{4-163}$$

and

$$\mathrm{Ai}(y) \cong \frac{1}{2\sqrt{\pi}}\,y^{-1/4}\,\exp(-\tfrac{2}{3}y^{3/2})\ ,\ \text{for}\ \ y\gg 1\ , \tag{4-164}$$

to check Eqs.(161) and (162). For y>>1, we have

$$F_o(y) \cong \frac{1}{4\pi}\frac{1}{\sqrt{y}}\exp(-\tfrac{4}{3}y^{3/2}) \;, \qquad (4\text{-}165)$$

which upon differentiation and integration produces

$$F_m(y) \cong \frac{1}{2\pi}(4y)^{-(m+1)/2}\exp(-\tfrac{4}{3}y^{3/2}) \;, \qquad (4\text{-}166)$$

consistent with both the recurrence relation (157) and the statement (162). For $y<<-1$, matters are not this simple. For instance, (163) gives

$$F_o(y) \cong \frac{1}{\pi}\frac{1}{\sqrt{-y}}[\cos(\tfrac{2}{3}(-y)^{3/2}-\tfrac{\pi}{4})]^2$$

$$= \frac{1}{2\pi}\frac{1}{\sqrt{-y}}[1+\sin(\tfrac{4}{3}(-y)^{3/2})] \;, \qquad (4\text{-}167)$$

which coincides with the highly semi-classical value (161) only after averaging over the oscillations of the sine function. This invites the physical interpretation of producing the TF result by averaging over quantum oscillations, a procedure to which another and more precise meaning will be given in Chapter Five. When integrating (167) with respect to y, the sine function does not contribute to the leading order, so that the statement (161) is reproduced for $m>o$. Let us see, how it would alternatively work for $F_1(y)$ if (163) and the corresponding asymptotic form of the derivative,

$$Ai'(y) \cong \frac{1}{\sqrt{\pi}}(-y)^{1/4}\sin(\tfrac{2}{3}(-y)^{3/2}-\tfrac{\pi}{4}) \;, \qquad (4\text{-}168)$$

were inserted into (159):

$$F_1(y) \cong \frac{1}{\pi}\sqrt{-y}\,\{[\cos(\tfrac{2}{3}(-y)^{3/2}-\tfrac{\pi}{4})]^2+[\sin(\tfrac{2}{3}(-y)^{3/2}-\tfrac{\pi}{4})]^2\}$$

$$= \frac{1}{\pi}\sqrt{-y} \;; \qquad (4\text{-}169)$$

indeed, there are no oscillations in the leading term. However, when differentiating (167) with respect to y, the dominating contribution is supplied by the sine function, with the consequence that for $m<o$ the leading terms of all $F_m(y)$ are oscillatory:

$$F_m(y) \cong \frac{1}{\pi}(-4y)^{-(m+1)/2}\sin\left(\frac{4}{3}(-y)^{3/2}-m\frac{\pi}{2}\right) , \qquad (4\text{-}170)$$

for $m<o$ and $-y>>1$.

Please note that both the smooth and the oscillating term in (167) are necessary in order to ensure the consistency of the asymptotic forms (161), (167), and (170) with the recursion (157).

For the purpose of illustration, Fig.1 shows plots of the functions $F_m(y)$ for $m=-3,\ldots,2$. In agreement with their known asymptotic behavior, these functions are all rapidly decreasing in the classically forbidden region of $y>o$, and either oscillatory or increasing in the classically allowed region of $y<o$. The major achievement

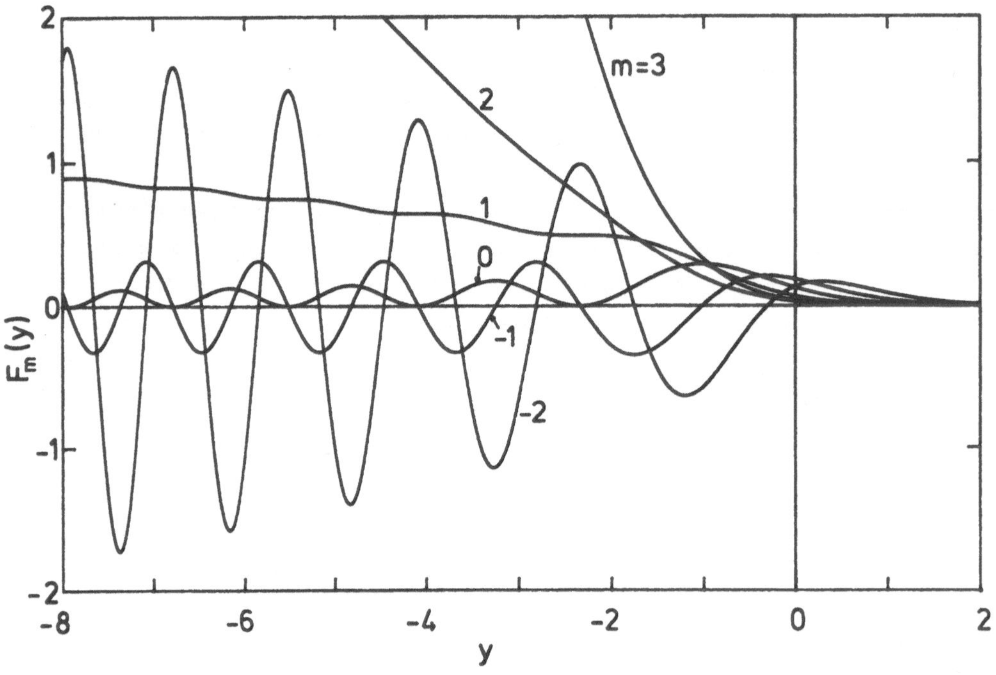

Fig.4-1. Plot of the $F_m(y)$ *for* $m=3,2,1,0,-1,-2$.

is the smooth transition from the classically allowed to the classically forbidden regime, which is to be contrasted to the situation in the TF approximation, where instead of a continuous transition there is typically a sharp boundary associated with the discontinuity of the square root, as is illustrated by the TF density (101).

<u>Validity of the TF approximation.</u> The derivation of Eqs.(161) and (162) from the comparison of (151) with (160) implies that the TF regime is where the functions $F_m(y)$ do not deviate significantly from their asymptotic forms. In order to give a more precise meaning to the requirements

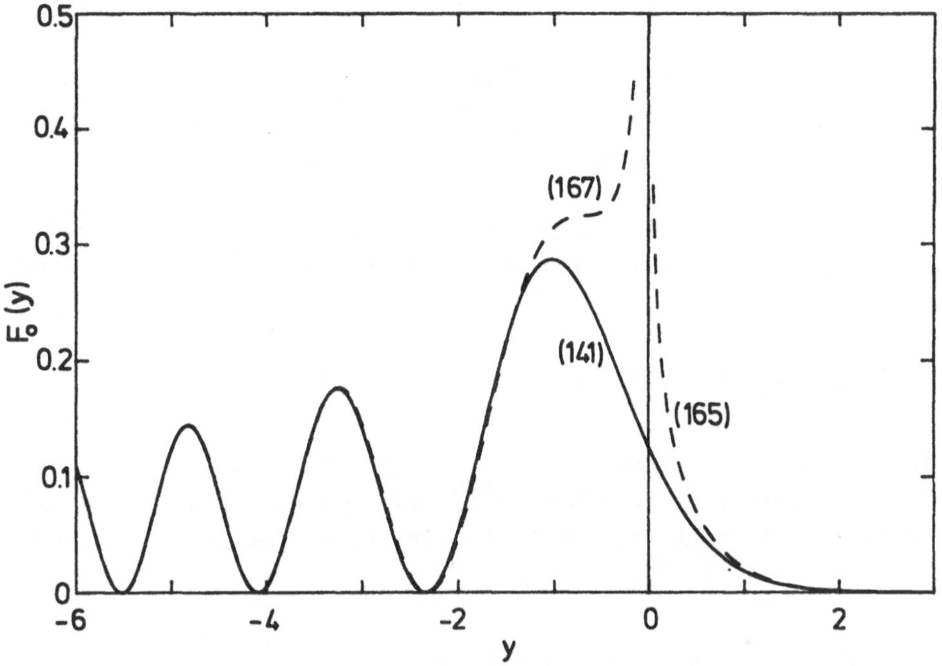

Fig.4-2. Actual $F_o(y)$ *and its asymptotic approximations for* $-6 \leq y \leq 3$.

"y>>1" and "y<<-1" a plot of $F_o(y)$ together with its asymptotic approximations (165) and (167) is presented in Fig.2. We observe that the asymptotic forms differ substantially from the actual function only in the small region $|y| \leq 3/2$. Consequently, the TF model is reliable when

$$|y| = |2(V+\zeta)| \ |2\vec{\nabla}V|^{-2/3} \geq \frac{3}{2} , \tag{4-171}$$

which sharpens the criterion (2-400) used in Chapter Two for the discussion of the range of validity of the TF treatment. [Of course, one is free to pick another number, slightly different from 3/2, on the right-hand side of (171).] If we insert $\zeta = o$ and the neutral-atom TF potential $V = -(Z/r)F(x)$ into (171), we have

$$y_{TF}(x) = -(2a)^{1/3} \ Z^{2/9} \ x^{1/3} F(x) [F(x) - x F'(x)]^{-2/3} =$$

$$= -(2a)^{1/3} Z^{2/9} \left[\frac{x F''(x)}{F(x) - x F'(x)}\right]^{2/3} \tag{4-172}$$

[the latter equality uses the differential equation obeyed by the TF function, Eq. (2-62)]. Upon making use of the small-x and large-x forms of F(x) [$\cong 1$ and $\cong 144/x^3$, respectively], this gives

$$y_{TF}/Z^{2/9} \cong \left\{ \begin{array}{ll} -(2ax)^{1/3} & \text{, for very small } x \\ -(18a/x^2)^{1/3} & \text{, for very large } x \end{array} \right\}, \tag{4-173}$$

showing that y_{TF} tends to -0 both for $x \to o$ and for $x \to \infty$. At x=0.742, the function $x F''(x)/[F(x) - x F'(x)]$ acquires its maximal value of 0.3999, so that

$$y_{TF}/Z^{2/9} \geq -0.657 \quad , \tag{4-174}$$

which implies that the criterion (171) is only met, in a certain range of x, by Z's larger than $(1.5/0.657)^{4.5} = 41.1$. Then, if Z is very large, we learn from (173) that said criterion is obeyed for distances in the region

$$\frac{27}{16a} Z^{-2/3} \lesssim x \lesssim 4\sqrt{a/3}\, Z^{1/3} \quad , \tag{4-175}$$

or

$$\frac{1.7}{Z} \lesssim r \lesssim 1.9 \quad . \tag{4-176}$$

This qualifies our previous statements, extracted from Eq. (2-403), that the range of validity of the TF model is limited by distances of the order of 1/Z and of the order of unity.

Figure 3 shows a plot of $y_{TF}/Z^{9/2}$. The abscissa is chosen linear in $x^{1/3}$, so that the curve is a straight line at small x, as implied by (173). The asymptotic forms (173) are dashed. Further, there are horizontal lines indicating $y_{TF} = -3/2$ for Z = 30, 45, and 90. It is clear, that Z must be much larger than that in order to be able to apply the limits of (175) for the range of validity, where $-y_{TF} > 3/2$. For values of Z corresponding to the Periodic Table, the relevant range of x is substantially smaller than the one of (175). Nevertheless, it is certainly true that this range increases as Z grows, whereby the Z depen-

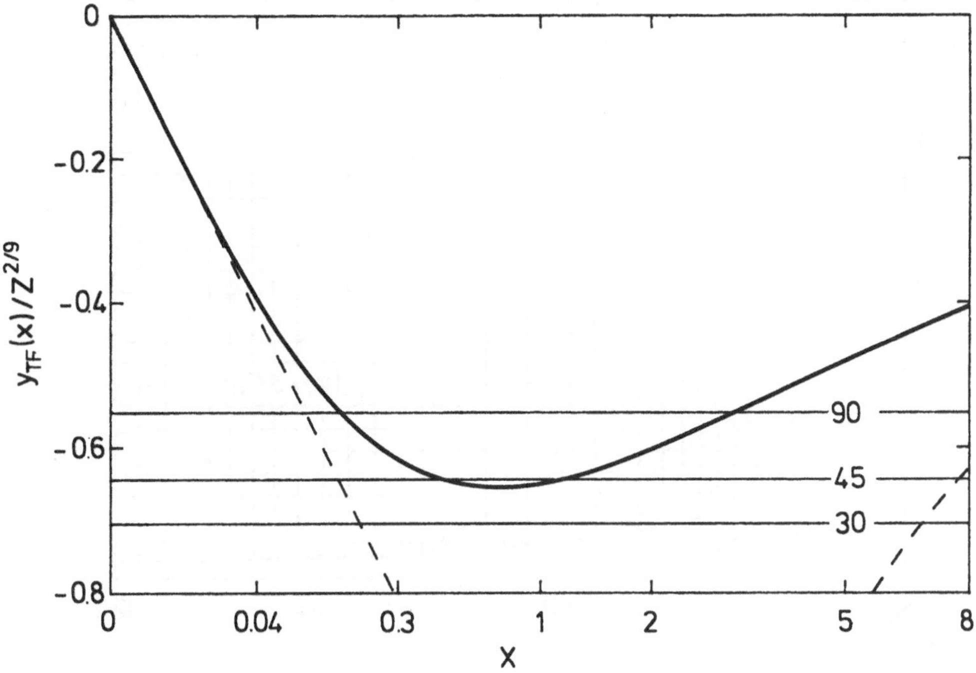

Fig.4-3. $y_{TF}/Z^{9/2}$ *as a function of* x. *The abscissa is linear in the cubic root of* x. *See text.*

dence is basically that of (175). As a further illustration, Fig.4 shows, as a function of $Z^{1/3}$, the range of x in which $-y_{TF}>3/2$. The dashed lines represent the limits in

$$1.91\ Z^{-2/3}(1+9.08\ Z^{-2/3}) \lesssim x \lesssim 2.17\ Z^{1/3}(1-2.24\ Z^{-\gamma/3}), \tag{4-177}$$

which improve (175) by including the next-to-leading terms into the approximations for F(x) at small and large values of x [$F(x) \cong 1 - Bx$ and $F(x) \cong (144/x^3)(1-\beta\, x^{-\gamma})$, respectively].

Of course, all these considerations must not be taken too seriously. Nevertheless, here is the important lesson that the TF approximation can be justified only for rather large values of Z, hardly for $Z \lesssim 60$. One would not expect the TF limit to be particularly accurate for lighter atoms, but as we have observed in Fig.2-2, for instance, the performance of the TF model is not markedly worse for small Z values. It is clear, though, that the refinements of the model, such as the Scott correction or the quantum improvement, have a larger significance

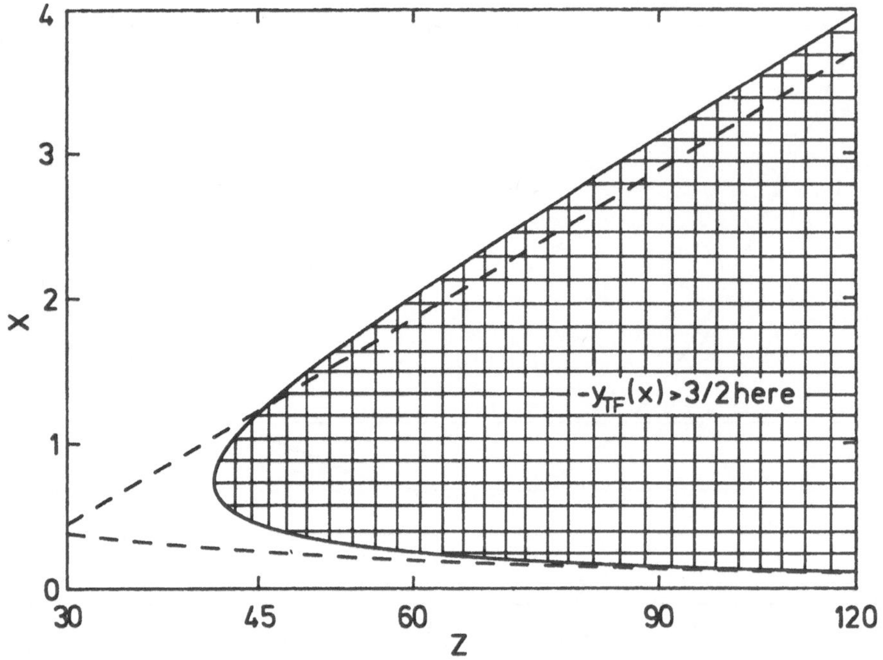

Fig.4-4. The range of x, where $-y_{TF} > 3/2$, is limited by the x values on the solid line. The dashed lines correspond to the limits in Eq.(177). The abscissa is linear in $Z^{1/3}$.

for the small-Z atoms.

Quantum corrected $E_1(V+\zeta)$. Upon employing Eq.(151) in the energy expression (219) [or, equivalently, in (130)], we have

$$E_{\zeta\zeta_s} = -\int(d\vec{r})\,\frac{1}{4\pi}|2\vec{\nabla}V|^{5/3}\int_y^{y_s}dy'\,(y'-y)\,[F_1(y') - \frac{1}{3}|2\vec{\nabla}V|^{-4/3}\nabla^2 V\,F_{-1}(y')] \,, \tag{4-178}$$

where y_s is related to V and ζ_s just like y is to V and ζ:

$$y_s = 2(V+\zeta_s)\,|2\vec{\nabla}V|^{-2/3} \,, \tag{4-179}$$

to be compared with (133). As discussed in Chapter Three, no unique value can be physically assigned to ζ_s, and in order to remove unphysical

Bohr shell oscillations we average over a suitable range of ζ_s, as in Eq.(3-58). This directs our attention to the ζ_s-averaged $F_m(y)$'s, defined by

$$\begin{aligned} \mathcal{F}_m(V,|\vec{\nabla}V|) &\equiv \sum_{j=1}^{J} w_j \int_y^{y_j} dy'(y'-y)F_{m-2}(y') \\ &= \sum_{j=1}^{J} w_j \int_y^{y_j} dy'(y'-y)\left(\frac{d}{dy'}\right)^2 F_m(y') \ , \end{aligned} \tag{4-180}$$

or, after performing the y' integration,

$$\mathcal{F}_m(V,|\vec{\nabla}V|) = \sum_{j=1}^{J} w_j [F_m(y)-F_m(y_j)-(y_j-y)F_{m-1}(y_j)] \ . \tag{4-181}$$

The y_j correspond to the various ζ_j which have the same significance as in Chapter Three, see around Eq.(3-58). Just as in that equation, the potential functional $E_1(V+\zeta)$ is here then given by

$$\begin{aligned} E_1(V+\zeta) &= \mathrm{tr}\left(\tfrac{1}{2}p^2+V+\zeta\right)\eta\left(-\tfrac{1}{2}p^2-V-\zeta\right) \\ &\cong \sum_{j=1}^{J} w_j E_{\zeta\zeta_j} + E_s \end{aligned} \tag{4-182}$$

with

$$\sum_{j=1}^{J} w_j E_{\zeta\zeta_j} = -\int (d\vec{r})\frac{1}{4\pi}\left[|2\vec{\nabla}V|^{5/3}\,\mathcal{F}_3(V,|\vec{\nabla}V|) - \tfrac{1}{3}|2\vec{\nabla}V|^{1/3}\nabla^2 V\,\mathcal{F}_1(V,|\vec{\nabla}V|)\right], \tag{4-183}$$

and [this is Eq.(3-43)]

$$E_s = \int(d\vec{r})V\rho_s + Z^2 n_s + \zeta N_s \ , \tag{4-184}$$

where ρ_s, n_s, and N_s signify what they did in Chapter Three. This is the quantum-corrected E_1.

In Ref.8, a further approximation was introduced aiming at a simplification of E_1, in the sense that it becomes a functional just of V and $|\vec{\nabla}V|$. To this end, the term in (183) with the Laplacian of the potential is integrated by parts whereby the resulting terms containing the gradient of $|\vec{\nabla}V|$ are neglected. All of this amounts to the replacement

$$\frac{1}{4\pi}\,\frac{1}{3}|2\vec{\nabla}V|^{1/3}\nabla^2 V\, F_1(V,|\vec{\nabla}V|) + \frac{1}{4\pi}\,\frac{1}{6}\,|2\vec{\nabla}V|^{5/3}\, F_0(V,|\vec{\nabla}V|) \qquad (4\text{-}185)$$

in (183), so that

$$\sum_{j=1}^{J} w_j E_{\zeta\zeta_j} \cong -\int (d\vec{r})\frac{1}{4\pi}|2\vec{\nabla}V|^{5/3}[F_3 - \frac{1}{6}F_0] \qquad (4\text{-}186)$$

is the expression used in Ref.8. The reason why this was (erroneously) considered to be an approximation more advantageous than (183) is explained below [see the remarks in the paragraph after Eq.(213)].

Before proceeding to construct the new, quantum-corrected density expression, let us briefly remark upon scaling. In Chapter Two it was found that the exact $E_1(V+\zeta)$ responds to transformations of the form (2-472),

$$V(\vec{r}) \to \mu^2 V(\mu\vec{r}) \ , \qquad (4\text{-}187)$$

$$\zeta \to \mu^2 \zeta \ ,$$

by exhibiting a factor of μ^2:

$$E_1(V+\zeta) \to \mu^2 E_1(V+\zeta) \ . \qquad (4\text{-}188)$$

It is reassuring that both terms of (182) possess this scaling property. For E_s this follows from the related discussion in Chapter Three, and for $E_{\zeta\zeta_j}$ it suffices to recall that (187) implies first

$$Z \to \mu Z \ , \qquad (4\text{-}189)$$

[upon using the significance given to Z in (3-111)] and then

$$\zeta_j \to \mu^2 \zeta_j \qquad (4\text{-}190)$$

[in view of both the definition of ζ_j in (3-98) and of the scaling properties of the $|\psi_{n_j}|^2_{av}$ displayed in (3-119)] with the consequences

$$y(\vec{r}) \to y(\mu\vec{r}) \ , \quad y_j(\vec{r}) \to y_j(\mu\vec{r}) \ , \qquad (4\text{-}191)$$

and

$$F_m(V,|\vec{\nabla}V|)(\vec{r}) \to F_m(V,|\vec{\nabla}V|)(\mu\vec{r}) \ , \qquad (4\text{-}192)$$

so that

$$E_{\zeta\zeta_j} \rightarrow -\int (d\vec{r})\frac{1}{4\pi}[\mu^5|2\vec{\nabla}V|^{5/3}(\mu\vec{r})\ F_3(\mu\vec{r})$$

$$-\mu^5\ \frac{1}{3}|2\vec{\nabla}V|^{1/3}(\mu\vec{r})\nabla^2V(\mu\vec{r})\ F_1(\mu\vec{r})] \qquad (4\text{-}193)$$

$$= \mu^2 E_{\zeta\zeta_j}\ ,$$

indeed.

Quantum corrected density. Next, we construct the density corresponding to the quantum corrected E_1 of Eqs.(182) to (184) by employing the fundamental relation (2-14),

$$\delta_V\ E_1 = \int (d\vec{r})\delta V(\vec{r})n(\vec{r})\ . \qquad (4\text{-}194)$$

As in Chapter Three we must not forget about the implicit V dependence, hidden in the ζ_j:

$$\delta_V\ \zeta_j = -\int (d\vec{r})\delta V\ |\psi_{n_j}|^2_{av}\ , \qquad (4\text{-}195)$$

which is Eq.(3-70). The resulting contribution to the density is that of Eq.(3-71),

$$\sum_{j=1}^{J} w_j Q_j |\psi_{n_j}|^2_{av}(r)\ , \qquad (4\text{-}196)$$

where, recalling the definition of Q_j in (3-72),

$$Q_j = -\frac{\partial}{\partial\zeta_j}E_{\zeta\zeta_j} = 2\int\frac{(d\vec{r})\ (d\vec{p})}{(2\pi)^3}(\zeta_j-\zeta)\langle\delta(z_j) - \frac{1}{12}\ \delta''(z_j)\nabla^2V\rangle$$

$$= (\zeta_j-\zeta)\int (d\vec{r})\ \frac{1}{\pi}\ [|2\vec{\nabla}V|^{1/3}\ F_1(y_j) - \frac{1}{3}|2\vec{\nabla}V|^{-1}\ \nabla^2V\ F_{-1}(y_j)]\ , \qquad (4\text{-}197)$$

which uses $E_{\zeta\zeta_j}$ in the form (129) and the Airy integrals (151). This reduces to (3-72) in the TF limit, where the second term vanishes, and in the first one the asymptotic approximation (169) is to be inserted for $F_1(y_j)$.

The variation of the potential in E_s, Eq.(184), exhibits ρ_s, which, combined with (196), gives the density of the inner-most electrons,

$$n_{IME} = \rho_s + \sum_{j=1}^{J} w_j Q_j |\psi_{n_j}|^2_{av} , \tag{4-198}$$

previously seen as Eq.(3-75).

The density of the remaining electrons, denoted by ñ, is obtained by varying V in (183) without taking into account the induced changes of the ζ_j that give rise to (196). Thus, again utilizing the form given for $E_{\zeta\zeta_j}$ in (129), ñ emerges from

$$\begin{aligned} &\int (d\vec{r})\,\delta V(\vec{r})\,\tilde{n}(\vec{r}) \\ &= -\sum_{j=1}^{J} w_j \int_{\zeta}^{\zeta_j} d\zeta'\,(\zeta'-\zeta)\,\delta_V\left[2\int\frac{(d\vec{r})\,(d\vec{p})}{(2\pi)^3} <\delta(z') - \frac{1}{12}\delta''(z')\,\nabla^2 V>^{\circ}\right] , \end{aligned} \tag{4-199}$$

where we must carefully add the three contributions that originate in the explicit change of V and the induced changes of $\vec{\nabla}V$ and $\nabla^2 V$. The corresponding change of z' is [Eq.(119)]

$$\begin{aligned} \delta z' &= -\,\delta V + \frac{x}{2}\,\delta|\vec{\nabla}V|^{2/3} \\ &= -\,\delta V + \frac{x}{6}|\vec{\nabla}V|^{-4/3}\,\delta(\vec{\nabla}V)^2 \\ &= -\,\delta V + \frac{x}{3}|\vec{\nabla}V|^{-4/3}\,\vec{\nabla}V\cdot\vec{\nabla}\delta V , \end{aligned} \tag{4-200}$$

with the consequence

$$\begin{aligned} &\delta_V\left[2\int\frac{(d\vec{r})\,(d\vec{p})}{(2\pi)^3} <\delta(z') - \frac{1}{12}\delta''(z')\,\nabla^2 V>^{\circ}\right] \\ &= \int (d\vec{r})\,(-\delta V)\,2\int\frac{(d\vec{p})}{(2\pi)^3} <\delta'(z') - \frac{1}{12}\delta'''(z')\,\nabla^2 V>^{\circ} \\ &+ \int (d\vec{r})\frac{1}{3}|\vec{\nabla}V|^{-4/3}\,\vec{\nabla}V\cdot\vec{\nabla}\delta V\;\; 2\int\frac{(d\vec{p})}{(2\pi)^3} <x\,\delta'(z') - \frac{x}{12}\delta'''(z')\,\nabla^2 V>^{\circ} \\ &+ \int (d\vec{r})\,(-\frac{1}{12}\nabla^2\delta V)\,2\int\frac{(d\vec{p})}{(2\pi)^3} <\delta''(z')>^{\circ} . \end{aligned} \tag{4-201}$$

The second summand here is then simplified according to Eq. (124), whereafter partial integrations are performed in order to exhibit δV as a factor. The outcome is

$$\tilde{n} = \sum_{j=1}^{J} w_j \int_{\zeta}^{\zeta_j} d\zeta'(\zeta'-\zeta)\, 2\int \frac{(d\vec{p})}{(2\pi)^3} \Big[\langle \delta'(z') - \frac{1}{12}\delta'''(z')\nabla^2 V\rangle^{\circ} + \frac{1}{12}\vec{\nabla}\cdot\big(\langle \delta'''(z') - \frac{1}{12}\delta^{V}(z')\nabla^2 V\rangle^{\circ}\, \vec{\nabla}V\big) + \frac{1}{12}\nabla^2\langle\delta''(z')\rangle^{\circ}\Big] \quad (4\text{-}202)$$

Now, once more employing (124), we have

$$\begin{aligned} \vec{\nabla}\langle\delta''(z')\rangle^{\circ} &= \langle\delta'''(z')\,(-\vec{\nabla}V + \frac{x}{2}\,\vec{\nabla}|\vec{\nabla}V|^{2/3})\rangle^{\circ} \\ &= -\langle\delta'''(z')\rangle^{\circ}\,\vec{\nabla}V + \frac{1}{24}\langle\delta^{V}(z')\rangle^{\circ}\,\vec{\nabla}(\nabla V)^2 , \end{aligned} \quad (4\text{-}203)$$

implying

$$\tilde{n} = \sum_{j=1}^{J} w_j \int_{\zeta}^{\zeta_j} d\zeta'(\zeta'-\zeta)\, 2\int \frac{(d\vec{p})}{(2\pi)^3} \langle \delta'(z') - \frac{1}{12}\delta'''(z')\nabla^2 V - \frac{1}{144}\vec{\nabla}\cdot[\delta^{V}(z')(\nabla^2 V\vec{\nabla}V - \frac{1}{2}\vec{\nabla}(\nabla V)^2)]\rangle^{\circ} . \quad (4\text{-}204)$$

The particular bilinear combination of derivatives of the potential,

$$\nabla^2 V\vec{\nabla}V - \frac{1}{2}\vec{\nabla}(\nabla V)^2 = [\nabla^2 V\overleftrightarrow{1} - \vec{\nabla}\vec{\nabla}V]\cdot\vec{\nabla}V , \quad (4\text{-}205)$$

is such that its divergence,

$$\begin{aligned} &\vec{\nabla}\cdot([\nabla^2 V\overleftrightarrow{1} - \vec{\nabla}\vec{\nabla}V]\cdot\vec{\nabla}V) \\ &= [\nabla^2 V\overleftrightarrow{1} - \vec{\nabla}\vec{\nabla}V]\cdot\cdot\vec{\nabla}\vec{\nabla}V \\ &= (\nabla^2 V)^2 - \vec{\nabla}\vec{\nabla}V\cdot\cdot\vec{\nabla}\vec{\nabla}V , \end{aligned} \quad (4\text{-}206)$$

does not contain derivatives of the potential of higher than second order. The horizontal double dot symbolizes, of course, the double scalar product of the dyadics, generally illustrated by

$$\overleftrightarrow{A}\cdot\cdot\overleftrightarrow{B} = \sum_{j,k=1}^{3} A_{jk}B_{kj} \quad . \tag{4-207}$$

The gradient of $\delta^{V}(z')$ being evaluated in analogy to (203), Eq.(204) then leads to

$$\begin{aligned}\tilde{n} = \sum_{j=1}^{J} w_j \int_{\zeta}^{\zeta_j} d\zeta'(\zeta'-\zeta)\, 2\int\frac{(d\vec{p})}{(2\pi)^3} \langle \delta'(z') - \tfrac{1}{12}\delta'''(z')\nabla^2 V \\ - \tfrac{1}{144}\delta^{V}(z')\,[(\nabla^2 V)^2 - \vec{\nabla}\vec{\nabla}V\cdot\cdot\vec{\nabla}\vec{\nabla}V] \\ + \tfrac{1}{144}\delta^{V\prime}(z')\,\vec{\nabla}V\cdot[\nabla^2 V\overleftrightarrow{1} - \vec{\nabla}\vec{\nabla}V]\cdot\vec{\nabla}V \\ - \frac{1}{(12)^3}\delta^{V'''}(z')\,\vec{\nabla}V\cdot\vec{\nabla}\vec{\nabla}V\cdot[\nabla^2 V\overleftrightarrow{1} - \vec{\nabla}\vec{\nabla}V]\cdot\vec{\nabla}V\rangle^{\circ} ,\end{aligned} \tag{4-208}$$

which, after employing this combination of Eqs.(151) and (180):

$$\begin{aligned}\sum_{j=1}^{J} w_j \int_{\zeta}^{\zeta_j} d\zeta'(\zeta'-\zeta)\, 2\int\frac{(d\vec{p})}{(2\pi)^3} \langle \left(\frac{d}{dz'}\right)^{2-m} \delta'(z')\rangle^{\circ} \\ = \frac{2^{1-m}}{\pi} \,|2\vec{\nabla}V|^{(2m-1)/3}\, F_m(V, |\vec{\nabla}V|) \quad ,\end{aligned} \tag{4-209}$$

reads

$$\begin{aligned}\tilde{n} = \frac{1}{2\pi}\,|2\vec{\nabla}V|\, F_2 - \frac{1}{6\pi}\,|2\vec{\nabla}V|^{-1/3}\,\nabla^2 V\, F_0 \\ - \frac{1}{18\pi}\,|2\vec{\nabla}V|^{-5/3}[(\nabla^2 V)^2 - \vec{\nabla}\vec{\nabla}V\cdot\cdot\vec{\nabla}\vec{\nabla}V]\, F_{-2} \\ + \frac{1}{9\pi}\,|2\vec{\nabla}V|^{-7/3}[\nabla^2 V(\nabla V)^2 - \vec{\nabla}V\cdot\vec{\nabla}\vec{\nabla}V\cdot\vec{\nabla}V]\, F_{-3} \\ - \frac{1}{27\pi}\,|2\vec{\nabla}V|^{-11/3}[\nabla^2 V\vec{\nabla}V\cdot\vec{\nabla}\vec{\nabla}V\cdot\vec{\nabla}V - \vec{\nabla}V\cdot\vec{\nabla}\vec{\nabla}V\cdot\vec{\nabla}\vec{\nabla}V\cdot\vec{\nabla}V]\, F_{-5} \quad .\end{aligned} \tag{4-210}$$

This is the quantum corrected $\tilde{n}$. It reduces to (3-76) in the TF limit, as it should.

The last three terms in (210), the ones involving F_{-2}, F_{-3}, and F_{-5}, together are a total divergence [see Eq.(204)], so that they integrate to zero. Thus,

$$\int(d\vec{r})n = \int(d\vec{r})(n_{IME} + \tilde{n}) \qquad (4\text{-}211)$$

$$= N_s + \sum_{j=1}^{J} w_j Q_j + \int(d\vec{r})\,[\tfrac{1}{2\pi}|2\vec{\nabla}V|\,F_2 - \tfrac{1}{6\pi}|2\vec{\nabla}V|^{-1/3}\,\nabla^2 V\,F_0] \ ,$$

or, after inserting the definition of the Q_j [Eq.(197)] and of the F_m [Eq.(181)],

$$\int(d\vec{r})n = N_s + \sum_{j=1}^{J} w_j \int(d\vec{r})\,\{\tfrac{1}{2\pi}|2\vec{\nabla}V|\,[F_2(y)-F_2(y_j)] \qquad (4\text{-}212)$$

$$- \tfrac{1}{6\pi}|2\nabla V|^{-1/3}\,\nabla^2 V[F_0(y)-F_0(y_j)]\} \ .$$

Consistency requires that this is equal to the count of electrons. Indeed, it is, since Eqs.(182), (184), and (129) imply

$$N = \frac{\partial}{\partial\zeta}E_1 = \frac{\partial}{\partial\zeta}E_s + \sum_{j=1}^{J} w_j \frac{\partial}{\partial\zeta}E_{\zeta\zeta_j} \qquad (4\text{-}213)$$

$$= N_s + \sum_{j=1}^{J} w_j \int(d\vec{r}) \int_{\zeta}^{\zeta_j} d\zeta'\; 2\int\frac{(d\vec{p})}{(2\pi)^3} \langle\delta(z') - \tfrac{1}{12}\delta''(z')\nabla^2 V\rangle^{\circ} \ ,$$

which is identical to the right-hand side of (212) after the Airy-averaged momentum integral is evaluated and integrated over ζ'.

Very remarkably, $\tilde{n}$ depends on V, $\vec{\nabla}V$, and $\vec{\nabla}\vec{\nabla}V$, but not on any higher derivatives of the potential, because the third derivatives cancel in the divergence of Eq.(206). In fact, whenever E_1 is linear in $\nabla^2 V$, such a cancellation will occur. This little observation was somehow missed when the work on Ref.8 was in progress, with the consequence that the approximate functional (186) was considered preferable because, inasmuch as it contains only V and $\vec{\nabla}V$, it ensures the absence of derivatives of the potential of higher than second order. For most applications it should not make a big difference if one employs (183) or (186), but for principal reasons, preference is given to (183).

Another comment on the quantum corrected density harkens back to a remark in Chapter Two, after Eq.(2-24). There is the statement that the densities calculated according to Eqs.(2-14) and (2-20) need not be identical in a certain approximation. Such is the situation here, indeed, as far as $\tilde{n}$ is concerned. In Eq.(202) we have the result that corresponds to (2-14). When employing (2-20), the exact $\tilde{n}$ appears as

$$\tilde{n}(\vec{r}\,') = \sum_j w_j\; 2 < \vec{r}\,' \big| \big(\eta(-H-\zeta) - \eta(-H-\zeta_j) - (\zeta_j-\zeta)\delta(-H-\zeta_j)\big) \big| \vec{r}\,'>$$

$$= \sum_j w_j\; 2\int_{\zeta}^{\zeta_j} d\zeta'(\zeta'-\zeta) <\vec{r}\,'|\delta'(-H-\zeta')|\vec{r}\,'> \quad . \tag{4-214}$$

In order to be able to use the approximations (47), (45), and (50) in (51), this is rewritten with the aid of the Fourier integral for the Delta function:

$$<\vec{r}\,'|\delta'(-H-\zeta')|\vec{r}\,'> = -\frac{\partial}{\partial\zeta'}\int\frac{dt}{2\pi} <\vec{r}\,'|e^{-i(H+\zeta')t}|\vec{r}\,'> \quad , \tag{4-215}$$

where said approximations produce

$$<\vec{r}\,'|\delta'(-H-\zeta')|\vec{r}\,'>$$

$$\cong -\frac{\partial}{\partial\zeta'}\int\frac{dt}{2\pi}\left(\frac{1}{2\pi i t}\right)^{3/2}\left(1+\frac{t^2}{12}\nabla'^2 V(\vec{r}\,')\right)e^{-i\left(V(\vec{r}\,')+\zeta'\right)t}$$
$$\times \exp\left[-i\left(V(\vec{r}\,')+\zeta'\right)t - \frac{i}{24}[\vec{\nabla}'V(\vec{r}\,')]^2 t^3\right] \tag{4-216}$$

to first order in the Laplacian of the effective potential, as always. The factor of t^2 multiplying this Laplacian can equivalently be replaced by $(-\partial/\partial\zeta')^2$ which operation is advantageously performed after the integration. Then the remaining integral is simplified by means of the identities (79) and (87), followed by explicitly integrating over t and differentiating with respect to ζ'. At this stage, we have

$$<\vec{r}\,'|\delta'(-H-\zeta')|\vec{r}\,'> \tag{4-217}$$

$$\cong \left(1-\frac{1}{12}\nabla'^2 V\frac{\partial^2}{\partial\zeta'^2}\right)\int\frac{(d\vec{p}\,')}{(2\pi)^3} <\delta'\left(-\tfrac{1}{2}p'^2 - V(\vec{r}\,') - \zeta' + \frac{x}{2}|\vec{\nabla}'V|^{2/3}\right)>^{\circ}$$

$$= \int\frac{(d\vec{p}\,')}{(2\pi)^3} <\delta'(z') - \frac{1}{12}\,\delta'''(z')\nabla'^2 V(\vec{r}\,')>^{\circ} \quad ;$$

the last step uses the definition of z in (119), some variables being primed now. Consequently, the approximation for $\tilde{n}(\vec{r})$ is here

$$\tilde{n}(\vec{r}) \cong \sum_j w_j \int_{\zeta}^{\zeta_j} d\zeta'(\zeta-\zeta')\,2\int\frac{(d\vec{p})}{(2\pi)^3} <\delta'(z') - \frac{1}{12}\,\delta'''(z')\nabla^2 V>^{\circ} \quad , \tag{4-218}$$

to be compared with our previous result (202). We notice that in (218) those terms of (202) are missing which originate in the induced changes of $\vec{\nabla}V$ and $\nabla^2 V$ when V is infinitesimally varied in (199). As discussed above, these additional contributions to $\tilde{n}$ in (202) integrate to zero and do not affect the count of electrons. They effect a redistribution of the electrons, or, in other words: describe fluctuations around the (expectedly) smooth density of (218), which in terms of the F_m's is

$$\tilde{n} = \frac{1}{2\pi}|2\vec{\nabla}V|F_2 - \frac{1}{6\pi}|2\vec{\nabla}V|^{-1/3}\nabla^2 V\, F_0 \quad , \tag{4-219}$$

simply the first line of Eq.(210).

Of the two approximations for $\tilde{n}$, Eqs.(210) and (219), the first is preferable on principal grounds, because it is the one for which the energy is stationary, whereas the second is more attractive for a practical calculation, because it is simpler. It is obviously consistent to employ (210), but is it equally justifiable to use (219)? Yes, for the following reason. Because of (37) we have $y \sim Z^{2/9}$ in the TF regime, implying

$$F_m \sim Z^{(2m-1)/9} \tag{4-220}$$

as far as the smooth part of the $F_m(y)$ is concerned [see Eq.(161)]. Then the terms of (219) are of the order Z^2 and $Z^{4/3}$, the additional ones of (210) being proportional to $Z^{2/3}$, $Z^{2/3}$, and $Z^{0/3}$, respectively. Therefore, to first order in the quantum correction, (219) is as good as (210). In particular, it suffices for the reproduction of the leading correction, derived earlier in Eq.(109). (For details consult Problem 6.) However, for m<0, the oscillatory behavior of Eq.(170) is dominant, so that

$$F_m \sim Z^{-(m+1)/9} \quad \text{for} \quad m < 0 \quad , \tag{4-221}$$

with the consequence, that the additional terms of (210) are of the order $Z^{4/3}$, $Z^{5/3}$, and $Z^{5/3}$, respectively. The situation is even more complicated if we take into account that in realistic systems y never is large negative (recall Fig.3). Then the counting of powers of Z is confined to the multiplying derivatives of the potential, resulting in $Z^{5/3}$ and $Z^{13/9}$ for the terms of (219), and in $Z^{11/9}$, $Z^{13/9}$, and $Z^{11/9}$ for the additional terms of (210). Summing up: in a general sense, (210) and (219) agree within the accuracy of the model, nevertheless their difference may be substantial for atoms with small Z. Incidentally we remark that in applications to situations more complicated than an isolated atom, such as a molecule where spherical symmetry is not available, practicality is likely to force the use of the simpler approximation that Eq.(219) represents.

To get a feeling for the relative size of the various contributions to $\tilde{n}$, let us consider a simple example. For this purpose it suffices to use a simple, but somewhat realistic potential. For instance, the Tietz potential of a neutral atom (see Problem 2-5),

$$V(r) = -\frac{Z}{r}\left(\frac{1}{1+r/R}\right)^2 , \quad R = \left(\frac{9}{2Z}\right)^{1/3} , \tag{4-222}$$

does the job. Also, just one ζ_s is good enough to produce the strong cancelations at small distances which are the essence of Eq.(181). When correcting for one Bohr shell, we can choose

$$\zeta_s = \frac{Z^2}{2\nu_s^2} - (V + \frac{Z}{r})(r=0) \tag{4-223}$$

[cf. Eq.(3-50)] with $\nu_s = 3/2$, thus

$$\zeta_s = \frac{2}{9}Z^2 - 2Z\left(\frac{2}{9}Z\right)^{1/3} . \tag{4-224}$$

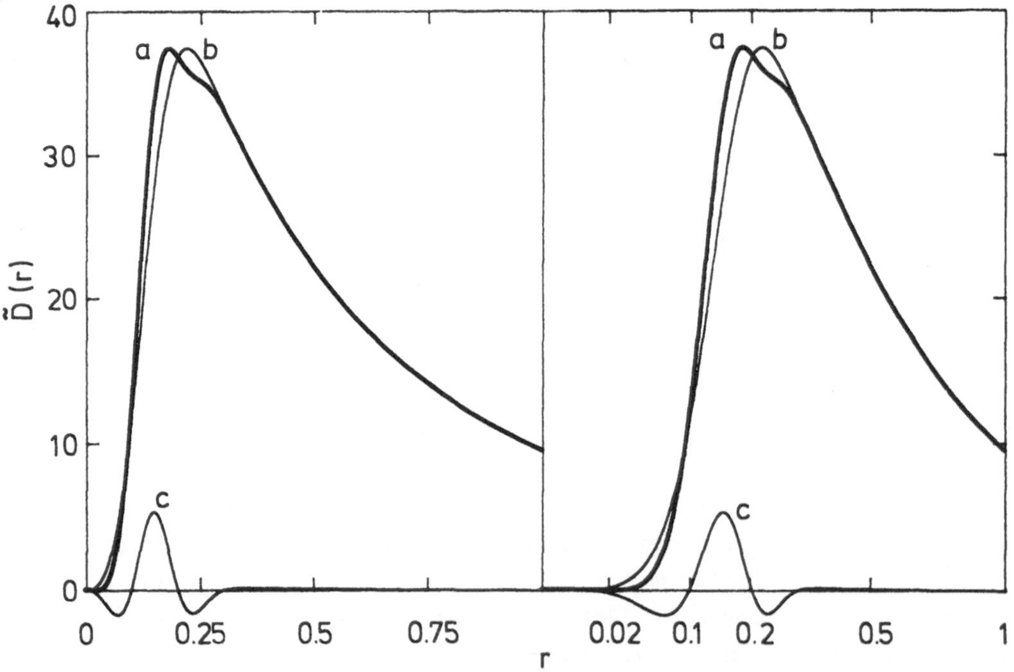

Fig.4-5. Comparison of the radial densities obtained from Eqs.(210) and (219); see text. The range of r is 0≤r≤1, the abscissa being linear in r on the left, linear in the square root of r on the right.

For this choice for V and ζ_s, supplemented by $\zeta=0$, Fig.5 shows the radial density

$$\tilde{D}(r) = 4\pi r^2 \tilde{n}(r) \tag{4-225}$$

corresponding to Z=36 (krypton). The abscissa is linear in r in the left plot, linear in the square root of r in the right one, as in Fig. 3-6. The thick curves (a) represent the $\tilde{D}$ of the full density (210). Curves (b) refer to the leading terms (219), curves (c) to the difference between (a) and (b), given by the additional terms of Eq.(210). We observe that the relative size of this difference is small, except for $r \lesssim 0.1$, where it results in a significant difference between curves (a) and (b). In this region, however, the density is dominated by the contribution n_{IME} from the innermost electrons. Consequently, curves (c) are a small modification of the total density in this range, too. We conclude that using (219) instead of (210) introduces an error which is typically of the order of the physical approximations already present in the model, so that such a procedure is certainly not inconsistent a priori, although one must not forget that under special circumstances the difference between (210) and (219) can, indeed, be large.

Exchange I (general). Early in the development, Eq.(2-36), we split the electron-electron interaction energy E_{ee} into its classical electrostatic part, E_{es}, and the remainder E'_{ee},

$$E_{ee} = E_{es} + E'_{ee} \quad . \tag{4-226}$$

It was sufficient until now to keep only E_{es}, which expressed as a density functional is given by

$$E_{es}(n) = \frac{1}{2} \int (d\vec{r}\,')(d\vec{r}\,'')\ \frac{n(\vec{r}\,')n(\vec{r}\,'')}{|\vec{r}\,'-\vec{r}\,''|} \quad , \tag{4-227}$$

of course. Time has come to concern the remainder E'_{ee}, which is a quantal correction to the classical interaction energy E_{es}.

The interaction energy E_{ee} is the expectation value of the interaction operator H_{ee} in the ground state $|\psi_o\rangle$ of the many particle Hamilton operator (2-409). Thus

$$E_{ee} = \langle\psi_o|\ \frac{1}{2} \sum_{j,k=1}^{N}{}' \frac{1}{r_{jk}}\ |\psi_o\rangle = \tag{4-228}$$

$$= \frac{1}{2} N(N-1) \int (d\vec{r}_1{}')(d\vec{r}_2{}')\ldots(d\vec{r}_N{}') \frac{\psi_o^*(\vec{r}_1{}',\ldots,\vec{r}_N{}')\psi_o(\vec{r}_1{}',\ldots,\vec{r}_N{}')}{|\vec{r}_1{}' - \vec{r}_2{}'|} ,$$

where the tracing over the spin indices is left implicit, and the anti-symmetry of the wave function has been used in equating the sum over j and k to N^2-N times the j=1, k=2 contribution. Upon introducing the two-particle density matrix

$$n^{(2)}(\vec{r}_1{}',\vec{r}_2{}';\vec{r}_1{}'',\vec{r}_2{}'') = N(N-1)\int (d\vec{r}_3{}')\ldots(d\vec{r}_N{}')\,\psi_o^*(\vec{r}_1{}'',\vec{r}_2{}'',\vec{r}_3{}',\ldots,\vec{r}_N{}')\,\psi_o(\vec{r}_1{}',\vec{r}_2{}',\vec{r}_3{}',\ldots,\vec{r}_N{}') , \tag{4-229}$$

Eq.(228) reads

$$E_{ee} = \frac{1}{2}\int (d\vec{r}\,')(d\vec{r}\,'')\,\frac{n^{(2)}(\vec{r}\,';\vec{r}\,'';\vec{r}\,';\vec{r}\,'')}{|\vec{r}\,'-\vec{r}\,''|} \quad . \tag{4-230}$$

We make contact with the one-particle density matrix defined in Eq. (2-422) by stating that

$$n^{(1)}(\vec{r}\,';\vec{r}\,'') = \frac{1}{N-1}\int (d\vec{r}_2{}')\,n^{(2)}(\vec{r}\,';\vec{r}_2{}';\vec{r}\,'';\vec{r}_2{}') \tag{4-231}$$

relates the two density matrices to each other.

The line of thought that led to the Hohenberg-Kohn theorem in Chapter Two implies that $n^{(1)}$ and $n^{(2)}$ are functionals of the density n, which functional dependence is unknown to us. We shall therefore strive for an approximate treatment, one that is consistent with what one does know and exhibits E_{es} as the leading contribution to E_{ee}.

For this purpose we return to Eq.(2-424) where the effective potential V is introduced with the defining property that the matrix elements of the corresponding density operator equal the density matrix $n^{(1)}(\vec{r}\,';\vec{r}\,'')$. As a matter of fact this equality holds only if $\vec{r}\,'$ and $\vec{r}\,''$ are equal or differ by an infinitesimal amount. Nothing more is, indeed, required in order to ensure that the trace of (2-426) produces the kinetic energy correctly. If we thus write $n_V^{(1)}$ for said matrix element

$$\begin{aligned} n_V^{(1)}(\vec{r}\,';\vec{r}\,'') &\equiv 2\,\langle\vec{r}\,'|\eta(-\tfrac{1}{2}p^2 - V(\vec{r})-\zeta)|\vec{r}\,''\rangle \\ &= 2\,\langle\vec{r}\,'|\eta(-H-\zeta)|\vec{r}\,''\rangle \quad , \end{aligned} \tag{4-232}$$

then

$$n^{(1)}(\vec{r}\,';\vec{r}\,'') = n_V^{(1)}(\vec{r}\,';\vec{r}\,'') \tag{4-233}$$

if $\vec{r}'-\vec{r}''$ is infinitesimal. Clearly, the equal sign in (233) cannot be true for arbitrary $\vec{r}'$ and $\vec{r}''$, since $n_V^{(1)}$ originates in a Slater determinant:

$$n_V^{(1)}(\vec{r}';\vec{r}'') = \sum_{j=1}^{N} \psi_j^*(\vec{r}'')\psi_j(\vec{r}')$$
$$= N \int (d\vec{r}_2{}')\ldots(d\vec{r}_N{}')\,\psi_V^*(\vec{r}'',\vec{r}_2{}',\ldots,\vec{r}_N{}')\,\psi_V(\vec{r}',\vec{r}_2{}',\ldots,\vec{r}_N{}') \tag{4-234}$$

with

$$\psi_V(\vec{r}_1{}',\ldots,\vec{r}_N{}') = \frac{1}{\sqrt{N!}} \det_{j,k}\left(\psi_j(\vec{r}_k{}')\right), \tag{4-235}$$

whereas the true ground state wave function $\psi_o(\vec{r}_1{}',\ldots,\vec{r}_N{}')$ is certainly not of this simple structure. The ψ_j's in (234) and (235) are, of course, the N lowest-energy eigenstates of the effective Hamilton operator H, counting different spin states separately. In Eq.(2-509), these were denoted by $\tilde{\psi}_j$.

The fact that $n^{(1)}$ and $n_V^{(1)}$ agree for $\vec{r}'\cong\vec{r}''$ suggests the use of $n_V^{(1)}$ as an approximation to $n^{(1)}$ for arbitrary values of $\vec{r}'$ and $\vec{r}''$. Likewise, $n_V^{(2)}$ should not differ much from $n^{(2)}$ if $\vec{r}_1{}'=\vec{r}_1{}''\cong\vec{r}_2{}'=\vec{r}_2{}''$, that is for the range of arguments which contributes most to E_{ee}. Upon inserting the contruction (235) into (229), we obtain

$$n_V^{(2)}(\vec{r}_1{}',\vec{r}_2{}';\vec{r}_1{}'',\vec{r}_2{}'') = \sum_{j,k=1}^{N} [\psi_j^*(\vec{r}_1{}'')\psi_k^*(\vec{r}_2{}'')\,\psi_j(\vec{r}_1{}')\psi_k(\vec{r}_2{}')$$
$$- \psi_j^*(\vec{r}_1{}'')\,\psi_k^*(\vec{r}_2{}'')\,\psi_k(\vec{r}_1{}')\,\psi_j(\vec{r}_2{}')] \ . \tag{4-236}$$

Before proceeding it is necessary to recall that in both Eq.(234) and (236) tracing over spin indices is implicit. For instance, in (234) the spin matrix is $\delta_{\sigma'\sigma''}$ with σ' and σ'' taking on the values + and - ("up" and "down") each. Thus the trace in question is

$$\sum_{\sigma'} \delta_{\sigma'\sigma'} = 2 \ , \tag{4-237}$$

which is, of course, the factor of two that reflects the spin multiplicity in Eq.(232). Now the spin matrix for the first summand in (236) is

$$\delta_{\sigma_1\sigma_1''}\ \delta_{\sigma_2\sigma_2''} \ , \tag{4-238}$$

the trace of which is

$$\sum_{\sigma_1',\sigma_2'} \delta_{\sigma_1'\sigma_1'} \; \delta_{\sigma_2'\sigma_2'} = 2 \times 2 \tag{4-239}$$

two factors of two. In contrast, the spin matrix of the second summand is

$$\delta_{\sigma_2'\sigma_1''} \; \delta_{\sigma_1'\sigma_2''} \tag{4-240}$$

with the trace

$$\sum_{\sigma_1',\sigma_2'} \delta_{\sigma_2'\sigma_1'} \; \delta_{\sigma_1'\sigma_2'} = 2 = \frac{1}{2}(2\times 2) \quad , \tag{4-241}$$

just one factor of two. This explains why in combining (234) and (236) into

$$\begin{aligned} n_V^{(2)}(\vec{r}_1{}',\vec{r}_2{}';\vec{r}_1{}'',\vec{r}_2{}'') &= n_V^{(1)}(\vec{r}_1{}';\vec{r}_1{}'')\, n_V^{(1)}(\vec{r}_2{}';\vec{r}_2{}'') \\ &\quad -\frac{1}{2}n_V^{(1)}(\vec{r}_1{}';\vec{r}_2{}'')\, n_V^{(1)}(\vec{r}_2{}';\vec{r}_1{}'') \end{aligned} \tag{4-242}$$

the factor of 1/2 appears. Its physical significance is obvious: this factor expresses the fact, that only half of the electron pairs have their spins parallel.

The approximation for $n^{(2)}$ to be used in (230) for the evaluation of E_{ee} is then

$$\begin{aligned} n^{(2)}(\vec{r}', \vec{r}''; \vec{r}', \vec{r}'') &\cong n_V^{(2)}(\vec{r}', \vec{r}''; \vec{r}', \vec{r}'') \\ &= n_V^{(1)}(\vec{r}'; \vec{r}')n_V^{(1)}(\vec{r}'';\vec{r}'') - \frac{1}{2}n_V^{(1)}(\vec{r}';\vec{r}'')n_V^{(1)}(\vec{r}'';\vec{r}') \quad , \end{aligned} \tag{4-243}$$

or with Eqs.(233) and (2-423) combined to

$$n_V^{(1)}(\vec{r}';\vec{r}') = n(\vec{r}') \quad , \tag{4-244}$$

simply

$$n^{(2)}(\vec{r}',\vec{r}'';\vec{r}',\vec{r}'') \cong n(\vec{r}')n(\vec{r}'') - \frac{1}{2}n_V^{(1)}(\vec{r}';\vec{r}'')n_V^{(1)}(\vec{r}'';\vec{r}') . \tag{4-245}$$

This inserted into (230) gives

$$E_{ee} \cong E_{es} + E_{ex} \quad , \tag{4-246}$$

where E_{es} is the electrostatic energy (227) and E_{ex} the exchange energy

$$E_{ex} = -\frac{1}{4}\int (d\vec{r}\,')(d\vec{r}\,'')\,\frac{n_V^{(1)}(\vec{r}\,';\vec{r}\,'')\,n_V^{(1)}(\vec{r}\,'';\vec{r}\,')}{|\vec{r}\,'-\vec{r}\,''|} \quad . \tag{4-247}$$

Note that it is $n_V^{(1)}$, not $n^{(1)}$, appearing in this definition of the exchange energy.

The equality in (246) is clearly approximate. The exchange energy E_{ex} takes into account the antisymmetry of the wave function. It does this in a simple way by approximating the true wave function by a Slater determinant. There is no doubt that the main effect of the antisymmetry is correctly incorporated this way. However, more subtle consequences - such as the influence of a third electron, in the neighborhood of two electrons, upon the interaction of these two - are certainly not contained in E_{ex}. All three, four, five,... particle contributions to the energy are usually called "correlation energy", a term that seems to have as many definitions as there are investigators. For us, it means no more than the difference between the two sides of Eq.(246),

$$E_{corr} \equiv E_{ee} - E_{es} - E_{ex} \; . \tag{4-248}$$

In particular, this E_{corr} is not equal to the difference between the true ground state energy and its HF approximation, because the HF potential differs from the true effective potential V; the deviation between E_{corr} and this difference is, if course, small.

Self energy. In Chapter One there appeared, between Eqs.(1-62) and (1-63), the statement that one should not worry about the electron self-energy, because "as soon as we shall have included the exchange interaction into the picture, the electronic self energy will be exactly canceled by the equally unphysical self-exchange energy." We are now able to justify this remark.

The respective self-energies originate in the j=k terms of Eq.(236), the first summand contributing to E_{es}, the second to E_{ex}. Since these terms cancel exactly in (236) the electrostatic self-energy and the exchange self-energy are identical, but differ in sign. Consequently, their sum vanishes. In other words: the errors introduced by including the self-energies into E_{es} and E_{ex} compensate for each other perfectly. There is absolutely no room for explicit self-energy corrections, once exchange is included into the description.[13]

Exchange II (leading correction). The density matrix $n_V^{(1)}$ is easily related to the time transformation function of Eq.(51) by means of

$$n_V^{(1)}(\vec{r}\,';\vec{r}\,'') = 2\ \langle\vec{r}\,'|\eta(-H-\zeta)|\vec{r}\,''\rangle$$
$$= 2\int_\zeta^\infty d\zeta'\ \langle\vec{r}\,'|\delta(-H-\zeta')|\vec{r}\,''\rangle$$
$$= 2\int_\zeta^\infty d\zeta' \int\frac{dt'}{2\pi}\ \langle\vec{r}\,'|e^{-i(H+\zeta')t'}|\vec{r}\,''\rangle \qquad (4\text{-}249)$$
$$= 2\int_\zeta^\infty d\zeta' \int\frac{dt'}{2\pi}\ e^{-i\zeta't'}\ \langle\vec{r}\,',t|\vec{r}\,'',0\rangle\quad ,$$

or after denoting the difference between $\vec{r}\,'$ and $\vec{r}\,''$ by $\vec{s}$ and half their sum by $\vec{r}\,'$,

$$n_V^{(1)}(\vec{r}\,'+\tfrac{1}{2}\vec{s};\vec{r}\,'-\tfrac{1}{2}\vec{s})$$
$$= 2\int_\zeta^\infty d\zeta' \int\frac{dt'}{2\pi}\left(\frac{1}{2\pi i T}\right)^{3/2} e^{-i\Phi-i\zeta't'} \qquad (4\text{-}250)$$

where $\Phi(\vec{r}\,',\vec{s},t')$ and $T(\vec{r}\,',\vec{s},t')$ are the phase and tyme of Eq.(51). Up to corrections of relative size $Z^{-2/3}$ they are approximately given in (45), (47), and (48). The leading contribution to E_{ex} is thus obtained if we employ the TF approximations to Φ and T, which are

$$\Phi \cong V(\vec{r}\,')t' - \frac{s^2}{2t'}\quad ,\qquad T \cong t'\quad . \qquad (4\text{-}251)$$

Consequently,

$$n_V^{(1)}(\vec{r}\,'+\tfrac{1}{2}\vec{s};\vec{r}\,'-\tfrac{1}{2}\vec{s})$$
$$\cong 2\int_\zeta^\infty d\zeta' \int\frac{dt'}{2\pi}\left(\frac{1}{2\pi i t'}\right)^{3/2} e^{-i(V(\vec{r}\,')+\zeta')t'+is^2/(2t')}\ , \qquad (4\text{-}252)$$

which for s=0 gives the TF density

$$n(\vec{r}\,') = n_V^{(1)}(\vec{r}\,';\vec{r}\,') \cong \frac{1}{3\pi^2}[-2(V(\vec{r}\,')+\zeta)]^{3/2}\ . \qquad (4\text{-}253)$$

After changing the integration variables in (247) to the difference and half the sum of $\vec{r}\,'$ and $\vec{r}\,''$,

$$E_{ex} = -\frac{1}{4}\int(d\vec{r}\,')\ \frac{(d\vec{s})}{s}\ n_V^{(1)}(\vec{r}\,'+\tfrac{1}{2}\vec{s};\vec{r}\,'-\tfrac{1}{2}\vec{s})\ n_V^{(1)}(\vec{r}\,'-\tfrac{1}{2}\vec{s},\vec{r}\,'+\tfrac{1}{2}\vec{s})\ , \qquad (4\text{-}254)$$

the two-fold insertion of (252) draws our attention to the $\vec{s}$ integral

$$\int\frac{(d\vec{s})}{s}\ \exp\left(i\,\frac{s^2}{2}\left(\frac{1}{t'}+\frac{1}{t''}\right)\right) =$$

$$= 2\pi \int_0^\infty ds^2 \exp\left(-\frac{s^2}{2i}\frac{t'+t''}{t't''}\right) \tag{4-255}$$

$$= 4\pi i \frac{t't''}{t'+t''} \quad .$$

At this stage, we have

$$\begin{aligned} E_{ex} \cong -\frac{1}{4} \int (d\vec{r}\,')\, 2\int\frac{dt'}{2\pi}\left(\frac{1}{2\pi i t'}\right)^{3/2} e^{-iV(\vec{r}\,')t'} \\ \times\, 2\int\frac{dt''}{2\pi}\left(\frac{1}{2\pi i t''}\right)^{3/2} e^{-iV(\vec{r}\,')t''} \\ \times \int_\zeta^\infty d\zeta' \int_\zeta^\infty d\zeta''\, e^{-i(\zeta' t'+\zeta'' t'')}\, 4\pi i\, \frac{t't''}{t'+t''} \quad , \end{aligned} \tag{4-256}$$

where the ζ' and ζ'' integrations result in

$$\begin{aligned} 4\pi i \frac{t't''}{t'+t''} \int_\zeta^\infty d\zeta'\, e^{-i\zeta' t'} \int_\zeta^\infty d\zeta''\, e^{-i\zeta'' t''} \\ = \frac{4\pi}{it'+it''} e^{-i\zeta(t'+t'')} = 4\pi \int_\zeta^\infty d\zeta'\, e^{-i\zeta'(t'+t'')} \quad . \end{aligned} \tag{4-257}$$

With this identity, Eq.(256) is simplified considerably:

$$E_{ex} \cong -\pi \int (d\vec{r}\,') \int_\zeta^\infty d\zeta' \left[2\int\frac{dt'}{2\pi}\left(\frac{1}{2\pi i t'}\right)^{3/2} e^{-iV(\vec{r}\,')t'-\zeta' t'}\right]^2 , \tag{4-258}$$

or with (253) and (252),

$$E_{ex} \cong -\pi \int (d\vec{r}\,') \int_\zeta^\infty d\zeta' \left[\frac{\partial}{\partial\zeta'} n(\vec{r}\,')\right]^2 \quad . \tag{4-259}$$

Here the density

$$\begin{aligned} n(\vec{r}\,') &= 2 \langle \vec{r}\,' | \eta(-H-\zeta') | \vec{r}\,' \rangle \\ &\cong \frac{1}{3\pi^2}[-2(V(\vec{r}\,')+\zeta')]^{3/2} \end{aligned} \tag{4-260}$$

is still expressed in terms of the effective potential. The exchange energy should, however, be given as a functional of the density itself. In view of the simple TF relation between n and V this can be achieved easily, as shown by

$$\int_\zeta^\infty d\zeta' \left[\frac{\partial}{\partial\zeta'} n(\vec{r}\,')\right]^2 = \int_\zeta^\infty d\zeta' \left(-\frac{1}{\pi^2}[-2(V(\vec{r}\,')+\zeta')]^{1/2}\right)^2 =$$

$$= \frac{1}{4\pi^4}[-2(V(\vec{r}\,')+\zeta)]^{4/2} = \frac{1}{4\pi^4}[3\pi^2 n(\vec{r}\,')]^{4/3} \quad . \tag{4-261}$$

Thus the leading contribution to E_{ex} is given by

$$E_{ex} \cong - \int (d\vec{r}\,') \frac{1}{4\pi^3}[3\pi^2 n(\vec{r}\,')]^{4/3} \quad , \tag{4-262}$$

which is known as the Dirac approximation to the exchange energy, a choice of name that slightly distorts history (more about this shortly).

The density n is composed of the TF density (260) and corrections to it referring to strongly bound electrons, quantum effects, exchange, and more. Consequently, to first order the exchange energy of an atom is obtained by inserting the uncorrected TF density into (262), with the outcome

$$\Delta_{ex} E = - \frac{1}{4\pi^3} \int (d\vec{r}) [-2(V+\zeta)]^{4/2} \quad , \tag{4-263}$$

where V and ζ are the TF quantities corresponding to the atom in question. This $\Delta_{ex} E$ differs from the $\Delta_{qu} E$ of Eq.(102) only by a numerical factor,

$$\Delta_{ex} E = \frac{9}{2} \Delta_{qu} E \quad , \tag{4-264}$$

so that the total $Z^{5/3}$ contribution to the binding energy of an atom is

$$- \tilde{E} \equiv - (\Delta_{ex}E + \Delta_{qu}E) = \frac{11}{36\pi^3} \int (d\vec{r}) [-2(V+\zeta)]^{4/2} \quad ; \tag{4-265}$$

in particular for a neutral atom, when $\zeta=0$ and $V(r) = -(Z/r)F(x)$,

$$- \tilde{E} = \frac{11}{32} \frac{Z^{5/3}}{a^2} \int_0^\infty dx [F(x)]^2 = 0.269900 \; Z^{5/3} \quad . \tag{4-266}$$

We have seen this numerical factor before, in Eq.(103), and now we understand the reason for exhibiting the factor 2/11 there. Exchange supplies the remaining $9/11 = (9/2)\times(2/11)$.

The extended TF model that includes the corrections for the strongly bound electrons, the quantum correction to E_1, and the exchange energy (to leading order) could be called the Thomas-Fermi-Scott-Weizsäcker-Dirac model. As always, such a christening does not do justice to all those other people who authored significant contributions as well. We shall therefore simply use the name "statistical model", which has the advantage of not distorting history, while at the same time suggesting a higher precision for a larger number of electrons. As we know, this suggestion is right; one must not forget, however, that the approximations employed in developing and improving the description are not

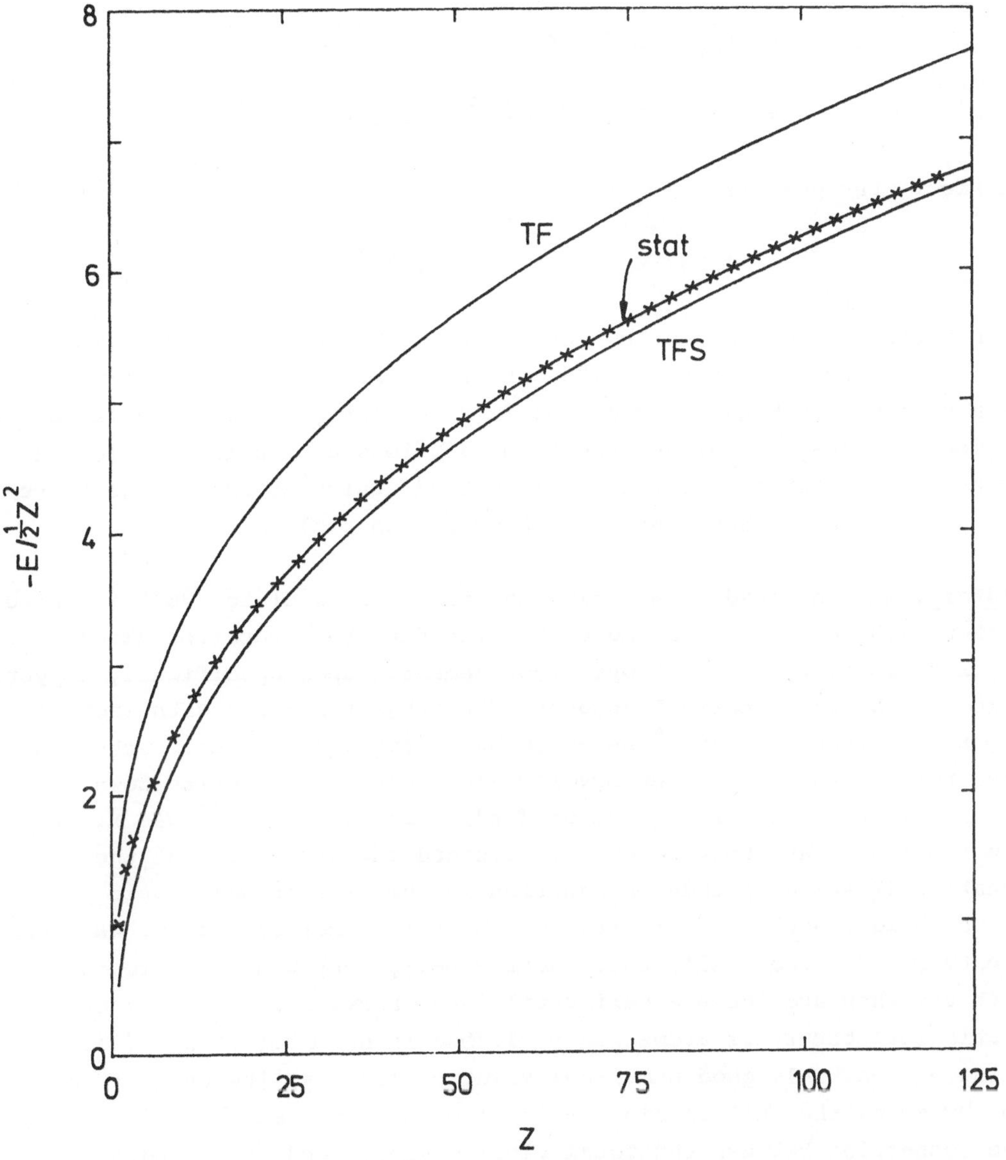

Fig.4-6. Comparison of the predictions for the neutral-atom binding energies made by the TF, the TFS, and the statistical model, as well as by the HF approximation (crosses); see also Figs. 2-2 and 3-3.

at all statistical but semiclassical ones. The label "statistical" is of historical origin; for instance, it occurs in the title of Gombás' textbook of 1949 (see Footnote 1 to Chapter One).

The prediction of the statistical model for the neutral-atom binding energies is obtained by adding the $\tilde{E}$ of (266) to the TFS prediction (3-33), the result being

$$\begin{aligned} -E_{\text{stat}} &= -E_{TF} - \frac{1}{2}Z^2 - \tilde{E} \\ &= 0.768745\ Z^{7/3} - \frac{1}{2}Z^2 + 0.269900\ Z^{5/3} \quad . \end{aligned} \tag{4-267}$$

In Fig.6, the quantity

$$\frac{-E_{\text{stat}}}{\frac{1}{2}Z^2} = 1.537\ Z^{1/3} - 1 + 0.540\ Z^{-1/3} \tag{4-268}$$

is plotted in addition to the corresponding TF and TFS curves. The crosses for $Z = 1,2,3,6,9,\ldots,120$ are the HF predictions, which we have seen earlier, in Figs.2-2 and 3-3. In this plot the remaining deviations between the statistical and the HF predictions are indiscernable - a great triumph for the semiclassical method, which at this stage is recognized to have turned into a high-precision tool.

History. As mentioned repeatedly, the three terms of Eq.(267) are associated with certain names, most of which have been reported already. In order to do justice to the ones, not remakred upon specifically as yet, here is a brief historical account. The subject started with Thomas' paper of November, 1926.[14] He could have, but did not derive the leading term of the binding energy formula. The first to write down Eq. (2-159) [that is the $Z^{7/3}$ term of (267)], in July 1927,[15] was Milne who - being an astrophysicist - recognized the similarity of the TF equation (2-62) with Emden's equation for spheres of polytropic perfect gases, held together by gravitation. Milne's numerical factor was about twenty percent too small, which accidentally improved the agreement with the then available experimental data. Fermi's first paper on the statistical theory of atoms was published in December, 1927.[14] It contains a remarkably good numerical solution for $F(x)$ [he calls it $\varphi(x)$]; for example, the initial slope B is given as 1.58. Fermi also noticed the connection between the total binding energy and this constant, so that he can claim fatherhood of Eq.(2-67). His numerical factor is, of course, much better than Milne's - only half a percent short of the modern value. We are told that Fermi was unaware of Thomas' work until late in 1928, "when it was pointed out to him by one (now unidentified) of the foreign theoreticians visiting Rome."[16] There are two probable candidates for this anonymous person: Bohr and Kramers, whose encouragement is acknowledged by Thomas in his paper.[14]

The credit for the first highly accurate calculation of F(x) belongs to Baker.[17] His work was published in 1930, long before the age of high-speed computers, and contains a value for B which is exact to 0.03%. We honor Baker by assigning his initial to this number. Incidentally, one of the first (if not the first) application of the MIT Differential Analyzer, a mechanical device for solving ordinary differential equations, was the computation of the neutral-atom TF function by Bush and Caldwell in 1931.[18] [A more accurate table of F(x) was only given 24 years later by Kobayashi and co-workers.[19]]

Now to the next term in (267), the correction for the strongly bound electrons. While it has, of course, always been recognized how badly the innermost electrons are represented by the TF model, it would take the surprisingly long time of 25 years until Scott came up with the energy correction of Eq.(3-32), in 1952.[20] In Chapter Three, we already mentioned that his derivation - recall the "boundary effect" argument - has not been widely accepted. Let us quote March once more, who expressed, in 1957,[21] the general feeling concerning Scott's correction, in writing that "it seems difficult to give a completely clear-cut demonstration of the case." As pointed out in Chapter Three, just this was delivered by Schwinger in 1980,[22] another 28 years later. The more sophisticated treatment of the strongly bound electrons presented in that Chapter was published in 1984.[23]

Scott, in the very same paper,[20] was also the first to give a $Z^{5/3}$ term in the energy formula. However, being unaware of the quantum corrections, he considered merely the exchange contribution to $\tilde{E}$, thus accounting for nine eleventh of the last term of (267). Again it took many years before, in 1981, the quantum correction (103) was evaluated by Schwinger.[7] From then on, the statistical energy formula (267) was known. [Strictly speaking, Eq.(267) can already be found in a 1978 paper by Plindov and Dmitrieva; for a comment see Footnote 11.] Of course, there has been important work on extensions of the TF model by other authors. The exchange interaction was first considered by Dirac, as early as 1930,[24] who was possibly reacting to a remark by Fermi at the end of a talk presented at a 1928 conference in Leipzig,[25] which Dirac also attended. But Dirac did not deal with exchange energy, just with the implied modifications of the TF equation. An expression for this energy, namely Eq.(262), was first given by Jensen in 1934,[26] who also on this occasion corrected for an inadvertance of Dirac, whose exchange effect was too large by a factor of two. However, there is no doubt that it was Scott who for the first time evaluated the exchange energy perturbatively, arriving at the neutral-atom version of (263). Maybe both Dirac and Jensen were just thinking that one should not talk about the second correction before the first one is known...

The first attempt at including the nonlocality of quantum mechanics was preformed by von Weizsäcker in 1935.[9] He derived a correction to the kinetic energy [nine times the second term of Eq.(115)], which, as we have observed above, has the serious drawback that it cannot be evaluated in perturbation theory - the outcome would be infinite. The derivation of (102) makes it clear that a consistent treatment requires a simultaneous, correct handling of both the quantum corrections and the corrections for the strongly bound electrons. Why didn't Scott do exactly that? There are two reasons: First, Scott's "boundary effect" theory of the vicinity of the nucleus cannot be directly implemented into the energy functional. And second, the language used by von Weizsäcker, Scott, and others is based on the electron density as the fundamental quantity, whereas these problems are most conveniently discussed by giving the fundamental role to the effective potential, as we have emphasized repeatedly.

Energy correction for ions. For an ion with a degree of ionization $q = 1 - N/Z$, the energy correction of order $Z^{5/3}$ is, according to Eq. (265), given by

$$-\tilde{E} = \frac{11}{32}\frac{Z^{5/3}}{a^2}\int_0^\infty dx\,[f_q(x)]^{4/2}$$
$$\equiv \frac{11}{32}\frac{Z^{5/3}}{a^2}\,\tilde{e}(q)\,, \tag{4-269}$$

where f_q is the corresponding TF function. Since f_q turns negative at $x = x_o(q)$, we have

$$\tilde{e}(q) = \int_0^{x_o(q)} dx\,[f_q(x)]^2\,. \tag{4-270}$$

In the two situations of high ionization, $N \ll Z$, and weak ionization, $q \ll 1$, the analytic dependence of $\tilde{e}$ on the degree of ionization can be studied with the aid of the expansions (2-275) and (2-361), respectively.

When switching from $f_q(x)$ to $\phi_\lambda(t)$, related to each other as in (2-261) and (2-264),

$$f_q(x) = q\,\phi_\lambda(x/x_o)\ , \quad \phi_\lambda(t) = \frac{1}{q}\,f_q(tx_o)\ ,$$
$$\lambda(q) = q^{1/2}[x_o(q)]^{3/2}\ , \tag{4-271}$$

Eq.(270) reads

$$\tilde{e} = q^2\,x_o\int_o^1 dt[\phi_\lambda(t)]^2\ . \tag{4-272}$$

Here we can insert the expansion (2-275) to produce

$$\tilde{e} = q^2\,x_o\int_o^1 dt\Big[(1-t)^2 + 2(1-t)\sum_{k=1}^{\infty}\lambda^k\phi_k(t) + \sum_{j,k=1}^{\infty}\lambda^{k+j}\phi_j(t)\phi_k(t)\Big] \tag{4-273}$$

All $\phi_k(t)$ and their first derivatives vanish at $t = 1$ [this is an implication of the recurrence relation (2-278)], so that two partial integrations establish the identity

$$\int_o^1 dt\ 2(1-t)\phi_k(t) = \int_o^1 dt\ t^2(1-\frac{t}{3})\phi_k''(t)\ . \tag{4-274}$$

In particular, differentiation of Eq.(2-277) gives

$$\phi_1''(t) = t^{-1/2}(1-t)^{3/2}\ , \tag{4-275}$$

with the consequence

$$\int_o^1 dt\ 2(1-t)\phi_1(t) = \int_o^1 dt(1-\frac{t}{3})t^{3/2}(1-t)^{3/2} = \frac{5\pi}{256}\ . \tag{4-276}$$

[In terms of Euler's Beta function this is $B(\frac{5}{2},\frac{5}{2}) - \frac{1}{3}B(\frac{7}{2},\frac{5}{2})$.] Thus

$$\tilde{e} = q\,x_o^2\Big[\frac{1}{3} + \frac{5\pi}{256}\,\lambda + O(\lambda^2)\Big]\ , \tag{4-277}$$

or with [Eqs.(2-285) and (2-287)],

$$\lambda(q) = \frac{16}{\pi}\,\frac{N}{Z}[1 + O(N/Z)]\ ,$$
$$x_o(q) = (\frac{16}{\pi}\,\frac{N}{Z})^{2/3}[1 + O(N/Z)]\ , \tag{4-278}$$

finally

$$\tilde{e} = \frac{1}{3}\left(\frac{16}{\pi}\,\frac{N}{Z}\right)^{2/3}\left[1 - \left(\frac{1024}{45\pi^2} - \frac{31}{16}\right)\frac{N}{Z} + O\left(\left(\frac{N}{Z}\right)^2\right)\right] . \qquad (4\text{-}279)$$

Naturally, one can determine the subsequent coefficients in this expansion in powers of N/Z as well; we quote from Ref.27:

$$O\left(\left(\frac{N}{Z}\right)^2\right) = -\left(\frac{2\,883\,584}{2025\pi^4} - \frac{105\,421}{630\pi^2} + \frac{7}{3}\right)\left(\frac{N}{Z}\right)^2 + O\left(\left(\frac{N}{Z}\right)^3\right) . \qquad (4\text{-}280)$$

The corresponding numerical statement about $\tilde{E}$ is then

$$-\tilde{E} = 0.4327\ Z\ N^{2/3}\left[1 - 0.3681\ \frac{N}{Z} - 0.0026\left(\frac{N}{Z}\right)^2 + O\left(\left(\frac{N}{Z}\right)^3\right)\right] . \qquad (4\text{-}281)$$

This series converges as well as the one for E_{TF} does [Eq.(2-289) and Problem 2-7]. In particular, Fig.2-5 shows that neglecting the $O\big((N/Z)^3\big)$ terms does not result in a significant error for $N = Z/2$. For this degree of ionization the analog of the neutral-atom binding energy (267) is, therefore,

$$-E_{stat}(Z,N{=}Z/2) = 0.7368\ Z^{7/3} - \frac{1}{2}\,Z^2 + 0.2223\ Z^{5/3} . \qquad (4\text{-}282)$$

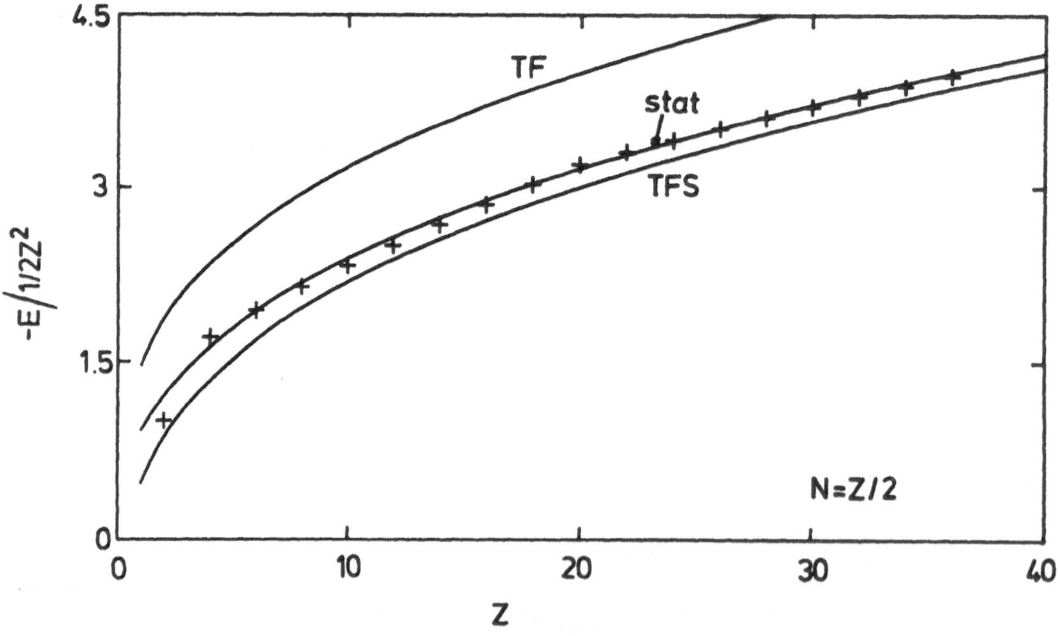

Fig. 4-7. Comparison of the predictions for the binding energy of ions with N/Z = 1/2 made by the TF, the TFS, and the statistical model, as well as by the HF approximation (crosses).

In Fig.7, the successive approximations of

$$\frac{-E_{stat}(Z,N=Z/2)}{\frac{1}{2}Z^2} = 1.474\ Z^{7/3} - 1 + 0.4445\ Z^{5/3} \ , \tag{4-283}$$

which are the TF, the TFS, and the statistical-model predictions, are compared to the respective HF results,[28] for Z=2N=2,3,6,...,36. The agreement is just as impressive as for the neutral atoms in Fig.6.

We turn to weakly ionized systems now. Here it is useful to look at the difference between $\tilde{e}(q)$ and its neutral atom value

$$\tilde{e}(0) = \int_0^\infty dx[F(x)]^2 = 0.615435 \ , \tag{4-284}$$

the numerical value of which was obtained in Problem 2-3. This difference

$$\tilde{e}(0) - \tilde{e}(q) = \int_{x_o(q)}^\infty dx[F(x)]^2 + \int_0^{x_o(q)} dx\left([F(x)]^2 - [f_q(x)]^2\right) \tag{4-285}$$

can be expanded in powers of $\beta[x_o(q)]^{-\gamma}$, after the expansions (2-349) and (2-361) are used to rewrite the integrands. The outcome is

$$\begin{aligned}
&\tilde{e}(0) - \tilde{e}(q) \\
&= \frac{(12)^4}{[x_o(q)]^5}\left\{\left[\frac{1}{5} + h_q^5\int_0^{h_q}\frac{dt}{t^6}\left(1 - (\tfrac{\Lambda}{12})^4 t^6[\Phi(t)]^2\right)\right]\right. \\
&+ 2\sum_{k=1}^\infty c_k(\beta x_o^{-\gamma})^k\left[\frac{1}{5+k\gamma} + h_q^{5+k\gamma}\int_0^{h_q}\frac{dt}{t^{6+k\gamma}}\left(1-(\tfrac{\Lambda}{12})^4 t^6\Phi(t)\psi_k(t)\right)\right] \\
&+ \sum_{j,k=1}^\infty c_j c_k(\beta x_o^{-\gamma})^{j+k}\left[\frac{1}{5+(j+k)\gamma}\right. \\
&\left.\left. + h_q^{5+(j+k)\gamma}\int_0^{h_q}\frac{dt}{t^{6+(j+k)\gamma}}\left(1-(\tfrac{\Lambda}{12})^4 t^6\psi_j(t)\psi_k(t)\right)\right]\right\} \ .
\end{aligned} \tag{4-286}$$

The known expansions of $x_o(q)$ and h_q in powers of $q^{\gamma/3}$, Eqs. (2-339) and (2-372), are now employed in identifying powers of $q^{\gamma/3}$ in (286). The result is

$$\tilde{e}(0) - \tilde{e}(q) = (\tfrac{12}{\Lambda})^4\ \Lambda^{2/3}\ q^{5/3}\Big[J_o +$$

$$+\left(\frac{5}{3}\gamma\frac{e_2}{e_1}J_o - 2\beta\Lambda^{-2\gamma/3}J_1\right)q^{\gamma/3} + O(q^{2\gamma/3})\Big] , \tag{4-287}$$

where J_o and J_1 denote the integrals

$$J_o = \frac{1}{5} + \int_o^1 \frac{dt}{t^6}\left(1-(\tfrac{\Lambda}{12})^4 t^6[\Phi(t)]^2\right) = 0.844849,$$

$$J_1 = \frac{1}{5+\gamma} + \int_o^1 \frac{dt}{t^{6+\gamma}}\left(1-(\tfrac{\Lambda}{12})^4 t^6\Phi(t)\psi_1(t)\right) = 0.078777 . \tag{4-288}$$

Their numerical values are most precisely calculated by the technique that produced Λ and α in Eq.(2-312) and (2-313); we recall that it is based upon expanding the integrands in powers of αt^σ and $(1-t)^{1/2}$ and matching the results of term-by-term integration around $t = 0.9$ where both expansions converge.

The numerical version of (287) is

$$\tilde{e}(0) - \tilde{e}(q) = 0.156210\, q^{5/3}[1+0.628951\, q^{\gamma/3} +0.414174\, q^{2\gamma/3}+O(q^{3\gamma/3})] , \tag{4-289}$$

where we also report the coefficient of the second power of $q^{\gamma/3}$.[29] When inserted into (269) this supplies

$$\tilde{E}(Z,N) - \tilde{E}(Z,Z) = 0.06851(Z-N)^{5/3}(1+0.6290\, q^{\gamma/3}+0.4142\, q^{2\gamma/3}+\ldots) , \tag{4-290}$$

which is the analog of Eq.(2-395),

$$E_{TF}(Z,N) - E_{TF}(Z,Z) = 0.04731(Z-N)^{7/3}(1+0.8259\, q^{\gamma/3}+0.6676\, q^{2\gamma/3}+\ldots) . \tag{4-291}$$

The main application of these results lies in predicting the ionization energy of neutral atoms, when $N=Z-1$ and $q=1/Z$.

Ionization energies. Since the Scott correction $\frac{1}{2}Z^2$ does not depend on the number of electrons, N, the entire difference between the statistical-model energy of an ion and the corresponding neutral atom is given by the sum of the differences in (290) and (291),

$$E_{stat}(Z,N) - E_{stat}(Z,Z)$$

$$= [E_{TF}(Z,N) - E_{TF}(Z,Z)] \qquad (4\text{-}292)$$

$$+ [\tilde{E}(Z,N) - \tilde{E}(Z,Z)] \quad .$$

In particular, when asking for the energy needed to remove just one electron from the neutral atom - this is the ionization energy I(Z) - we find

$$I_{stat}(Z) \equiv E_{stat}(Z,Z-1) - E_{stat}(Z,Z)$$

$$= 0.04731(1+0.8259\ Z^{-\gamma/3} + 0.6676\ Z^{-2\gamma/3} + \ldots)$$

$$+ 0.06851(1+0.6290\ Z^{-\gamma/3} + 0.4142\ Z^{-2\gamma/3} + \ldots), \qquad (4\text{-}293)$$

or,

$$I_{stat}(Z) = 0.1158(1+0.7094\ Z^{-\gamma/3} + 0.5177\ Z^{-2\gamma/3} + \ldots) \ . \qquad (4\text{-}294)$$

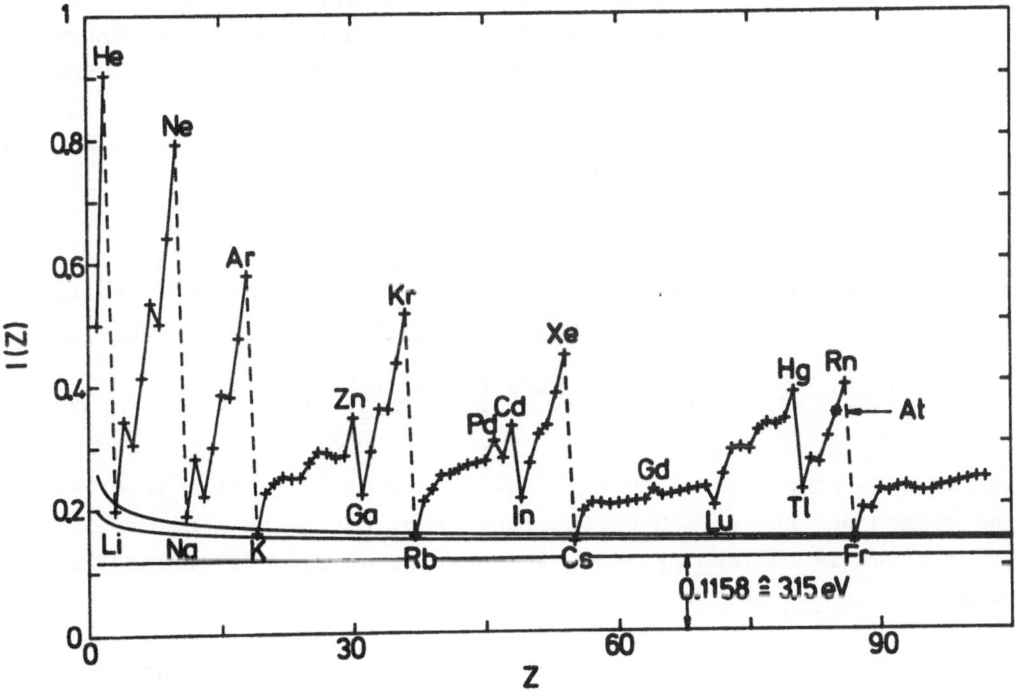

Fig.4-8. Comparison of experimental ionization energies with the successive approximations to $I_{stat}(Z)$ in Eq.(294).

This statistical-model ionization energy makes no reference to the shell structure of real atoms. The same is, of course, true for the total energy E_{stat}, where shell effects are insignificant because the few most weakly bound electrons do not contribute a large amount to the total energy. In contrast, when considering the ionization energy, it is the one most weakly bound electron that we are interested in. The experimental ionization energies must reveal the atomic shells as a consequence. The statistical-model contribution I_{stat} is, therefore, expected to be a numerically reliable prediction only for those atoms in which the one electron to be removed does not have partners in its shell. These are the alkaline metals Li(Z=3), Na(11), K(19), Rb(37), Cs(55), and Fr(87). Indeed, the successive approximations represented by Eq.(294) agree quite well with the experimental data[30] for the five lighter ones, as illustrated in Fig.8. For francium the statistical model prediciton of about 3.9 eV [using two or three terms of (294) yields 3.86 eV and 4.02 eV, respectively.]

An important remark ist the following. The contributions to the ionization energy from E_{TF} and $\tilde{E}$ are of roughly the same magnitude; for instance, $\tilde{E}$ supplies 59% to the large-Z ionization energy of 0.1158 ≙ 3.15 eV, whereas E_{TF} supplies 41%. How does this fit into the general picture of $\tilde{E}$ being the second correction to E_{TF}, smaller by a factor of $Z^{-2/3}$? Indeed, Eqs.(2-240) and (269) tell us that

$$\frac{\tilde{E}(Z,N)}{E_{TF}(Z,N)} = \frac{11}{32a} Z^{-2/3} \tilde{e}(q)/e(q) , \tag{4-295}$$

which is proportional to $Z^{-2/3}$ for a given degree of ionization. If we, however, compare the respective contributions not to the total energy but to the energy required to remove a given number of electrons,

$$\frac{\tilde{E}(Z,N) - \tilde{E}(Z,Z)}{E_{TF}(Z,N) - E_{TF}(Z,Z)} = \frac{11}{32a} Z^{-2/3} \frac{\tilde{e}(0) - \tilde{e}(q)}{e(0) - e(q)} , \tag{4-296}$$

this ratio is proportional to $Z^{-2/3} q^{-2/3} = (Z-N)^{-2/3}$, which involves the fixed common net charge Z-N of the ions under consideration. More precisely, for weakly ionized systems, the numbers of Eqs. (290) and (291) produce

$$\frac{\tilde{E}(Z,N) - \tilde{E}(Z,Z)}{E_{TF}(Z,N) - E_{TF}(Z,Z)} = 1.448(Z-N)^{-2/3}(1-0.197q^{\gamma/3}-0.091q^{2\gamma/3}+\ldots) \tag{4-297}$$

In view of

$$q^{\gamma/3} = (Z-N)^{\gamma/3} Z^{-\gamma/3} , \tag{4-298}$$

this is an expansion in powers of $Z^{-\gamma/3}$ for a <u>given net charge Z-N</u>. Such is the situation when asking about the ionization energy, the net charge being one then.

Since our derivation and evaluation of $\tilde{E}$ made extensive use of the fact that it is a small correction to E_{TF}, its application to circumstances in which the contributions of E_{TF} and $\tilde{E}$ are of comparable magnitude need not always produce reliable results. Fortunately, nothing went wrong when the ionization energies (294) were calculated, as we are reassured by Fig.8.

Here is a historical remark. Attempts at calculating ionization energies in extended Thomas-Fermi theories have been made already in the 1930's. Since the analytic treatment of the energy of weakly ionized atoms was not available then, one had to rely upon rather crude numerical solutions of the TF equation for various degrees of ionization. The results thus obtained did not agree well with the experimental data, unless some ad-hoc modifications of the description were introduced. The discussion presented above shows that the essential ingredients are the coefficients in the expansions (290) and (291). These are hard to get by if one depends upon evaluating expressions like the integral (269) by purely numerical means.[31] From looking at the results reported by Gombás[32] I get the definite impression that said ad-hoc modifications were aimed at producing a statistical-model prediction that does not primarily agree with the experimental ionization energies of the inert gases, but would go through the oscillations of Fig.8 in a symmetrical way, instead. Certainly, there is good intuitive reason to expect the statistical-model prediction to average over the oscillations due to the atomic shell structure. However, as pointed out above, the anticipation of good agreement for the inert gases is supported by physical intuition quite as well, if not more so. Anyway, now fifty years later, we do not have to resort to guesswork any longer, because numerical solutions of the TF equation ceased to be the main tool for the study of weakly ionized systems.

<u>Minimal binding energies (chemical potentials).</u> In the discussion of ionization energies we were interested in the energy change due to the removal of one atom, that is due to changing the number of electrons N by unity from N=Z to N=Z-1. A related question concerns the energy change caused by an infinitesimal variation of the number of electrons, which exhibits the quantity ζ, the minimal binding energy of the electrons:

$$E(Z,N+\delta N) - E(Z,N) = \delta N \frac{\partial}{\partial N} E(Z,N) = -\delta N\, \zeta(Z,N) . \qquad (4\text{-}299)$$

In thermodynamics, the derivative of the energy with respect to the particle number is called the chemical potential, usually denoted by μ. Thus we have the simple relation

$$\mu(Z,N) = -\zeta(Z,N) , \qquad (4\text{-}300)$$

stating that the minimal binding energy and the chemical potential differ only by their sign, whereby $\zeta(Z,N)$ is a non-negative quantity (because the effective potential tends to zero at large distances).

In the statistical model, we have

$$\zeta_{\text{stat}} = \frac{\partial}{\partial N}(E_{TF} + \tilde{E}) , \qquad (4\text{-}301)$$

since the Scott correction does not depend on N. For N<<Z, combining Eqs.(2-289) and (281) with Problem 2-7 produces

$$\zeta_{\text{stat}} = 0.3816\, Z^2 N^{-2/3}(1-1.6944\,\frac{N}{Z} + 0.6360(\frac{N}{Z})^2 + \ldots)$$
$$+ 0.2885\, Z\, N^{-1/3}(1 - 0.9203\,\frac{N}{Z} - 0.0102(\frac{N}{Z})^2 + \ldots) . \qquad (4\text{-}302)$$

In the circumstance of weak ionization, $q = 1-N/Z<<1$, the respective result is obtained from Eqs.(291) and (290); it is[33]

$$\zeta_{\text{stat}} = 0.1104(Z-N)^{4/3}(1+0.9170q^{\gamma/3}+0.8148q^{2\gamma/3}+\ldots)$$
$$+ 0.1142(Z-N)^{2/3}(1+0.7261q^{\gamma/3}+0.5421q^{2\gamma/3}+\ldots) . \qquad (4\text{-}303)$$

In the previous section, it was emphasized that our results obtained by treating $\tilde{E}$ as a small correction to E_{TF} are only applicable in situations where the dominant contribution is that of the TF part, indeed. For Eqs.(302) and (303) this means that Z^3/N or Z-N must (roughly) exceed unity, respectively. As a consequence, Eq.(303) does not imply $\zeta_{\text{stat}} = 0$ for neutral atoms. Knowledge of the chemical potential for neutral atoms can only come from solving the new differential equation for the effective potential. What we can gain here is some insight about the relative size of $\zeta(Z,Z)$. Since the neutral-atom value of ζ is zero in the TF model, we have

$$\zeta_{\text{stat}}(Z,Z) = \tilde{\zeta}(Z,Z) , \qquad (4\text{-}304)$$

where $\tilde{\zeta}$ denotes the contribution to ζ due to the exchange energy and the quantum corrections. Now

$$\tilde{\zeta}(Z,Z) < \tilde{\zeta}(Z,N) \quad \text{for} \quad N < Z \quad , \tag{4-305}$$

where according to (303) the right-hand side is proportional to $Z^{2/3}$ for large Z and a given degree of ionizations. We can therefore infer, that the neutral-atom value of $\tilde{\zeta}$, and thus of ζ_{stat}, is of a smaller order in powers of $Z^{1/3}$, presumably

$$\zeta_{stat}(Z,Z) \sim Z^{1/3} \quad , \quad \text{or} \quad Z^{0/3} \quad . \tag{4-306}$$

Note that this statement is nothing more than an educated guess. There is some numerical support for the notion that $\zeta_{stat}(Z,Z)$ is constant in Z, this constant being about 0.009; details will be supplied below, when numerical solutions of the new differential equation are discussed.

The minimal binding energy ζ (or the chemical potential μ) is a purely theoretical quantity. It cannot be measured experimentally, and making contact with the concept of electronegativity[34] does not change this situation, it's only a different name. Furthermore, in view of the discrete nature of the bound-state spectrum of the independent-particle Hamilton operator, ζ is not a uniquely defined quantity; in an exact treatment any value out of a (small) range is equally good. In the models discussed so far, the energy depends continuously on the (large) number of electrons, and since the formalism allows for non-integer values of N, there is no difficulty in considering infinitesimal variations δN, as in (299). The physical situation that justifies such a procedure is, of course, that of so many electrons that the change of electron number by one can be regarded as a tiny change of N.

<u>Shielding of the nuclear magnetic moment.</u> In contrast to the N-derivative of the energy E(Z,N), the Z-derivative

$$\frac{\partial}{\partial Z} E(Z,N) = -\left\langle\frac{1}{r}\right\rangle \quad , \tag{4-307}$$

[see Eq.(1-96)] has an experimental manifestation. It determines the magnetic shielding of the nuclear magnetic moment by the electrons. To make this point, we repeat the argument given by Lamb.[35]

An external magnetic field $\vec{B}_0$, described by a vector potential $\vec{A}_0$,

$$\vec{B}_0 = \vec{\nabla} \times \vec{A}_0 \quad , \tag{4-308}$$

induces an electric current in the atomic electrons, given by

$$\vec{j}(\vec{r}) = -\alpha \vec{A}_o(\vec{r}) n(\vec{r}) , \tag{4-309}$$

where the minus sign reflects the negative charge of the electrons, and the fine structure constant $\alpha = 1/137.036..$ shows up because the speed of light equals $1/\alpha$ in our atomic units. The corresponding induced magnetic field is

$$\begin{aligned} \vec{B}_{ind}(\vec{r}) &= \vec{\nabla} \times \alpha \int (d\vec{r}') \frac{\vec{j}(\vec{r}')}{|\vec{r}-\vec{r}'|} \\ &= \alpha \int (d\vec{r}') \vec{j}(\vec{r}') \times \vec{\nabla}' \frac{1}{|\vec{r}-\vec{r}'|} . \end{aligned} \tag{4-310}$$

In particular, at the site of the nucleus at $\vec{r} = 0$, we have

$$\vec{B}_{ind}(0) = \alpha \int (d\vec{r}) \vec{j}(\vec{r}) \times \vec{\nabla} \frac{1}{r} . \tag{4-311}$$

Now, if $\vec{B}_o$ is constant (over the volume of the atom), we can use

$$\vec{A}_o(\vec{r}) = \frac{1}{2} \vec{B}_o \times \vec{r} , \tag{4-312}$$

implying

$$\vec{B}_{ind}(0) = -\frac{1}{2} \alpha^2 \vec{B}_o \cdot \int (d\vec{r}) \frac{n(\vec{r})}{r} [\overleftrightarrow{1} - \frac{\vec{r}}{r} \frac{\vec{r}}{r}] . \tag{4-313}$$

The electron density in an isolated atom is spherically symmetric, $n(\vec{r}) = n(r)$, and therefore the replacement

$$\overleftrightarrow{1} - \frac{\vec{r}}{r} \frac{\vec{r}}{r} \rightarrow \frac{2}{3} \overleftrightarrow{1} \tag{4-314}$$

can be performed in the integrand. Thus

$$\vec{B}_{ind}(0) = -\sigma_m \vec{B}_o , \tag{4-315}$$

with the shielding coefficient σ_m defined by[36]

$$\sigma_m \equiv \frac{1}{3} \alpha^2 \int (d\vec{r}) n(r)/r = \frac{1}{3} \alpha^2 \langle \frac{1}{r} \rangle . \tag{4-316}$$

The numerical value of the factor in front is 1.7750×10^{-5}. The total magnetic field at the site of the nucleus being $\vec{B}_o + \vec{B}_{ind}(0)$, the interaction energy of the nuclear magnetic moment $\vec{\mu}_N$ with this magnetic field is

$$- \vec{\mu}_N \cdot [\vec{B}_o + \vec{B}_{ind}(0)] = - \vec{\mu}_N \cdot [(1-\sigma_m)\vec{B}_o]$$
$$= - [(1-\sigma_m)\vec{\mu}_N] \cdot \vec{B}_o \quad , \qquad (4\text{-}317)$$

showing that due to the shielding by the electrons, the nuclear magnetic moment appears to be smaller by a factor $1-\sigma_m$ than its actual value.

For a neutral atom, the Z-derivative of the energy is to be computed by

$$\frac{\partial}{\partial Z} E(Z,N)\Big|_{N=Z} = \frac{d}{dZ} E(Z,Z) - \frac{\partial}{\partial N} E(Z,N)\Big|_{N=Z}$$
$$= \frac{d}{dZ} E(Z,Z) + \zeta(Z,Z) \quad . \qquad (4\text{-}318)$$

In the statistical model, we know $E(Z,Z)$ in Eq.(267), and the reasoning of the preceding section taught us that $\zeta(Z,Z)$ is of higher order in $Z^{-1/3}$ than the terms of dE/dZ. Consequently,

$$(\sigma_m)_{stat} = (3.1840\, Z^{4/3} - 1.7750\, Z + 0.7985\, Z^{2/3}) \times 10^{-5} \quad , (4\text{-}319)$$

for neutral atoms. Since highly precise measurements of nuclear magnetic moments are performed on atomic beams, the neutral-atom predictions for σ_m are the only ones of immediate interest. Of course, the corresponding expressions for ions can also be produced, by differentiating the respective energy formulae.

In these experiments one always deals with neutral atoms, so that the magnetic shielding is present all the time. The magnetic moment of the bare nucleus cannot be measured independently. As a consequence, the shielding factor $1-\sigma_m$ itself is not available experimentally. One must entirely rely upon theoretical predictions of its value. In other words: the test of Eq.(319) is, once more, performed by comparing with another theoretical calculation, for which the HF method is the natural choice. This is done in Table 1.[37] The relative deviation is noticed to be less than 1% for Z>10, less than $\frac{1}{2}$% for Z>25, less than $\frac{1}{4}$% for Z>62. Concerning the shielding factor $1-\sigma_m$ the two theoretical values differ at most by 0.00003; a small amount, indeed. This happens for mercury (Z=80), for which the HF calculation yields $1-\sigma_m=0.99027$, whereas the statistical model gives 0.99030. Again the semiclassical treatment has proven to not only supply an understanding of the analytical dependence on the atomic number, as expressed in Eq.(319), but

Table 4-1. Comparison of <u>HF</u> and <u>S</u>tatistical-<u>M</u>odel predictions for $10^5\sigma_m$.

Z	HF	SM	Z	HF	SM
1	1.8	2.2	30	252.2	251.3
2	5.99	5.74	35	312.1	310.9
3	10.15	10.11	40	374.2	373.9
4	14.93	15.13	45	440.0	439.9
5	20.20	20.68	50	508.6	508.6
6	26.07	26.70	56	591.9	594.4
7	32.55	33.13	62	681.7	683.8
8	39.51	39.94	68	776.4	776.3
9	47.07	47.09	74	873.6	871.9
10	55.23	54.55	80	973.0	970.4
15	96.12	96.02	86	1072.8	1071.6
20	142.3	143.2	94	1210.	1210.5
25	194.2	195.2	102	1355.	1353.8

also to produce highly accurate numerical predictions, not inferior to results obtained by the much more involved HF method.

<u>Simplified new differential equation (ES model).</u> The evaluation of the energy correction $\tilde{E}$ as a small correction to the TFS energy has proven quite successful. Let us now try to incorporate the exchange energy and the quantum correction to E_1 into the energy functional itself. At the present stage, we shall be satisfied by keeping only first order terms, so that we get a simplified picture without the detail supplied by the plethora of Airy functions that one meets, for instance, in the density expression (210).

We are aiming at a description that enables us to study the outer reaches of the atom. Consequently, one simplification will consist of paying little attention to the details of the special treatment of the innermost electrons. This is illustrated by combining the TF version of $E_1(V+\zeta)$ and the leading correction (97) into

$$E_1(V+\zeta) \cong \int (d\vec{r}) \left\{-\frac{1}{15\pi^2}[-2(V+\zeta)]^{5/3} + \frac{1}{24\pi^2}[-2(V+\zeta)]^{1/2}\nabla^2 V\right\} + \text{CSBE} \quad , \tag{4-320}$$

where the initials CSBE are to remind us of the necessity of making corrections for the strongly bound electrons eventually. The electron-electron interaction energy $E_{ee}(n)$ includes the exchange energy in the form of the Dirac approximation (262) (do not forget that it was Jensen who actually found this expression); thus

$$E_{ee}(n) = \frac{1}{2}\int(d\vec{r})(d\vec{r}\,')\frac{n(\vec{r})n(\vec{r}\,')}{|\vec{r}-\vec{r}\,'|} - \int(d\vec{r})\frac{1}{4\pi^3}[3\pi^2 n(\vec{r})]^{4/3} . \tag{4-321}$$

The stationary property of the energy functional (2-434),

$$E(V,n,\zeta) = E_1(V+\zeta) - \int(d\vec{r})(V+\frac{Z}{r})n + E_{ee}(n) - \zeta N , \tag{4-322}$$

under infinitesimal variations of both V and n implies

$$\begin{aligned} -\frac{1}{4\pi}\nabla^2\left(V+\frac{1}{\pi}(3\pi^2 n)^{1/3}+\frac{Z}{r}\right) \\ = n = \frac{1}{3\pi^2}[-2(V+\zeta)]^{3/2} - \frac{1}{24\pi^2}[-2(V+\zeta)]^{-1/2}\nabla^2 V \\ + \frac{1}{24\pi^2}\nabla^2[-2(V+\zeta)]^{1/2} + \text{CSBE} . \end{aligned} \tag{4-323}$$

What is subtracted from V on the left-hand side is, of course, the exchange potential V_{ex}, defined by

$$\delta_n E_{ex} = \int(d\vec{r})\delta n\, V_{ex} , \tag{4-324}$$

in the Dirac approximation, that is

$$V_{ex} \cong -\frac{1}{\pi}(3\pi^2 n)^{1/3} , \tag{4-325}$$

as it results from the exchange energy term of (321).

Consistent with the strategy of keeping the corrections only to first order, we bring the Laplacian on the right-hand side of (323) to the left-hand one, where it supplements the exchange potential by one sixth. At this stage, we have

$$\begin{aligned} -\frac{1}{4\pi}\nabla^2\left(V+\frac{7}{6\pi}(3\pi^2 n)^{1/3}+\frac{Z}{r}\right) \\ = \frac{1}{3\pi^2}[-2(V+\zeta)]^{3/2}\left\{1+\frac{-\frac{1}{4\pi}\nabla^2(V+\frac{Z}{r})}{\frac{1}{3\pi^2}[-2(V+\zeta)]^{3/2}}\frac{1}{6\pi}[-2(V+\zeta)]^{-1/2}\right\} \\ + \text{CSBE} . \end{aligned} \tag{4-326}$$

The inclusion of Z/r into the Laplacian on the right-hand side does not change anything, because the resulting Delta function is multiplied by an expression that vanishes at the origin. (Incidentally, one might remark that this would refer to the domain of the strongly bound electrons anyway.) Now, the second term in the curly brackets is already a correction to the unity to which it is added, so that the TF evaluation

$$-\frac{1}{4\pi}\nabla^2\left(V+\frac{Z}{r}\right)=\frac{1}{3\pi^2}[-2(V+\zeta)]^{3/2} \tag{4-327}$$

is justified. Then we obtain

$$\begin{aligned}-\frac{1}{4\pi}\nabla^2&\left(V+\frac{7}{6\pi}(3\pi^2n)^{1/3}+\frac{Z}{r}\right)\\&=\frac{1}{3\pi^2}[-2(V+\zeta)]^{3/2}\left(1+\frac{1}{6\pi}[-2(V+\zeta)]^{-1/2}\right)\\&\cong\frac{1}{3\pi^2}[-2(V+\zeta)]^{3/2}\left(1+\frac{1}{9\pi}[-2(V+\zeta)]^{-1/2}\right)^{3/2},\end{aligned} \tag{4-328}$$

the last step once more being a consistent first order approximation, or finally,

$$\begin{aligned}-\frac{1}{4\pi}\nabla^2&\left(V+\frac{7}{6\pi}(3\pi^2n)^{1/3}+\frac{Z}{r}\right)\\&=\frac{1}{3\pi^2}\left([-2(V+\zeta)]+\frac{1}{9\pi}[-2(V+\zeta)]^{1/2}\right)^{3/2}+\text{CSBE}.\end{aligned} \tag{4-329}$$

This equation invites the introduction of a pseudo-density ρ,

$$\rho\equiv\frac{1}{3\pi^2}\left([-2(V+\zeta)]+\frac{1}{9\pi}[-2(V+\zeta)]^{1/2}\right)^{3/2} \tag{4-330}$$

in terms of which the physical density appears as

$$n=\rho+\frac{1}{24\pi^2}\nabla^2[-2(V+\zeta)]^{1/2}+\text{CSBE} \tag{4-331a}$$

$$=\rho+\frac{1}{24\pi^2}\nabla^2(3\pi^2\rho)^{1/3}+\text{CSBE}\quad. \tag{4-331b}$$

When this is inserted into Eq.(321), E_{ee} reads

$$\begin{aligned}E_{ee}&=\frac{1}{2}\int(d\vec{r})(d\vec{r}\,')\frac{\rho(\vec{r})\rho(\vec{r}\,')}{|\vec{r}-\vec{r}\,'|}\\&+\frac{1}{24\pi^2}\int(d\vec{r})\frac{\rho(\vec{r})}{|\vec{r}-\vec{r}\,'|}\nabla'^2[3\pi^2\rho(\vec{r}\,')]^{1/3}\\&-\int(d\vec{r})\frac{1}{4\pi^3}[3\pi^2\rho(\vec{r})]^{4/3}+\text{CSBE}\quad=\end{aligned}$$

$$= \frac{1}{2} \int (d\vec{r})\,(d\vec{r}\,')\ \frac{\rho(\vec{r})\,\rho(\vec{r}\,')}{|\vec{r} - \vec{r}\,'|}$$

$$- \frac{11}{9} \int (d\vec{r})\ \frac{1}{4\pi^3}\ [3\pi^2 \rho(\vec{r})]^{4/3} + \text{CSBE} \quad , \tag{4-332}$$

where a two-fold partial integration has been performed, and only first order modifications are taken into account. Please note the factor of 11/9 that multiplies the pseudo exchange-energy term. It is, of course, the same 11/9 factor that we have seen in Eq.(265),

$$\tilde{E} = (\Delta_{ex}E + \Delta_{qu}E) = \left(1 + \frac{2}{9}\right) \Delta_{ex}E$$

$$= \frac{11}{9}\ \Delta_{ex}E \quad . \tag{4-333}$$

Infinitesimal variations of ρ in (332) exhibit a pseudo exchange-potential

$$U_{ex} = - \frac{11}{9\pi}\ (3\pi^2 \rho)^{1/3} \quad , \tag{4-334}$$

to be contrasted with the exchange potential V_{ex} in Eq.(325). It is thus fitting to supplement the pseudo density ρ with a pseudo potential U such that

$$- \frac{1}{4\pi}\ \nabla^2 (U - U_{ex} + \frac{Z}{r}) = \rho \quad . \tag{4-335}$$

Comparison with (329) implies

$$U - U_{ex} = V + \frac{7}{6\pi}\ (3\pi^2 n)^{1/3} \quad , \tag{4-336}$$

or (to first order, again)

$$U = V - \frac{1}{18\pi}\ [-2(V+\zeta)]^{1/2} \quad . \tag{4-337}$$

We use this U and the ρ of (330) to rewrite the second term on the right-hand side of (322) consistently,

$$- \int (d\vec{r})\ (V + \frac{Z}{r})\, n$$

$$= - \int (d\vec{r})\ (V + \frac{Z}{r})\, \rho - \int (d\vec{r})\ (3\pi^2 \rho)^{1/3}\ \frac{1}{24\pi^2}\ \nabla^2 (V + \frac{Z}{r}) + \text{CSBE} =$$

$$= -\int(d\vec{r})\,(U+\frac{Z}{r})\rho - \int(d\vec{r})\,\frac{1}{18\pi}\,[-2(V+\zeta)]^{1/2}\rho$$
$$- \int(d\vec{r})\,(3\pi^2\rho)^{1/3}\,\frac{1}{24\pi^2}\,\nabla^2 V \tag{4-338}$$
$$+ Z\,\frac{1}{6\pi}\,[3\pi^2\rho(0)]^{1/3} + \text{CSBE}\ .$$

The term referring to r=0 is part of the CSBE, where we incorporate it. Further, the second and third terms are already corrections to the first, so we can rewrite them by employing the TF relations (for $r \neq 0$)

$$\rho \cong \frac{1}{3\pi^2}\,[-2(V+\zeta)]^{3/2} = -\frac{1}{4\pi}\,\nabla^2 V\ . \tag{4-339}$$

Then

$$-\int(d\vec{r})\,(V+\frac{Z}{r})n = -\int(d\vec{r})\,(U+\frac{Z}{r})\rho$$
$$- \int(d\vec{r})\,\frac{1}{36\pi^2}\,[-2(V+\zeta)]^{1/2}\nabla^2 V \tag{4-340}$$
$$+ \text{CSBE}\ .$$

Likewise, $E_1(V+\zeta)$ of Eq.(320) can be expressed in terms of $U+\zeta$, the outcome being

$$E_1 = \int(d\vec{r})\,(-\frac{1}{15\pi^2})\,[-2(U+\zeta)]^{5/2} \tag{4-341}$$
$$+ \int(d\vec{r})\,\frac{1}{36\pi^2}\,[-2(V+\zeta)]^{1/2}\nabla^2 V + \text{CSBE}\ .$$

After adding the contributions (332), (340), and (341) the new stationary energy functional is (tentatively)

$$E_{ES}(U,\rho,\zeta) = \frac{1}{2}Z^2 + \int(d\vec{r})\,(-\frac{1}{15\pi^2})\,[-2(U+\zeta)]^{5/2}$$
$$- \int(d\vec{r})\,(U+\frac{Z}{r})\rho + \frac{1}{2}\int(d\vec{r})\,(d\vec{r}\,')\,\frac{\rho(\vec{r})\rho(\vec{r}\,')}{|\vec{r}-\vec{r}\,'|} \tag{4-342}$$
$$- \frac{11}{9}\int(d\vec{r})\,\frac{1}{4\pi^3}\,(3\pi^2\rho)^{4/3} - \zeta N\ .$$

The CSBE have here been made explicit by exhibiting the Scott term $1/2\ Z^2$. It is, therefore, understood that (342) should not be employed for calculating atomic properties which require an accurate density at small distances.

The functional (342) plus the relations (331b) and (337) define an extension of the TFS model which has aquired the name ES model, reflecting the initials of the authors of Ref.38, where a variant of (342) is derived.[39] This model incorporates the quantum corrections to E_1 and the exchange energy in the simplest, though accurate fashion. It is particularly well fitted to studying such properties of atoms that are sensitive to the outer reaches of the atom, inasmuch as in the vicinity of the nucleus the pseudo density ρ has the characteristics of the TF density. With the necessary changes,[40] this is the first model in the development that has a fair chance of predicting realistic electronic structures in molecules; to my knowledge, no one has worked on such an application as yet.

Before continuing the discussion of the ES model, I must offer a remark on the Thomas-Fermi-Dirac (TFD) model. It is obtained by extending the TF picture by accounting for the exchange energy only, without making CSBE or including the quantum corrections to E_1. The stationary TFD energy functional is therefore

$$\begin{aligned} E_{TFD}(V,n,\zeta) &= \int(d\vec{r})\left(-\frac{1}{15\pi^2}\right)[-2(V+\zeta)]^{5/2} \\ &\quad - \int(d\vec{r})\left(V+\frac{Z}{r}\right)n + \frac{1}{2}\int(d\vec{r})(d\vec{r}\,')\frac{n(\vec{r})n(\vec{r}\,')}{|\vec{r}-\vec{r}\,'|} \\ &\quad - \int(d\vec{r})\frac{1}{4\pi^3}(3\pi^2 n)^{4/3} - \zeta N \quad . \end{aligned} \tag{4-343}$$

It looks quite similar to the ES functional (342). It is essential to appreciate the enormous differences between the two models. It is not only the factor of 11/9 and the Scott term $\frac{1}{2}Z^2$ that distinguish (343) from (342); there are the additional relations (331b) and (337) which state that U and ρ are not the effective potential and the physical density themselves, in contrast to the variables V and n in the TFD functional. As the whole development shows, the TFD model is somewhat physically inconsistent because the leading correction is left out and only part of the second correction is accounted for. Keeping the promise given at the beginning of this Chapter, we shall not spend time on investigating the implications of the rather irrelevant TFD model.

The stationary property of the ES functional (342) is stated above, but we still have to demonstrate it. The change in E_{ES} due to infinitesimal variations of U, ρ, and ζ is

$$\delta E_{ES} = \int(d\vec{r})\delta U\left\{\frac{1}{3\pi^2}[-2(U+\zeta)]^{3/2} - \rho\right\} +$$

$$+ \int (d\vec{r})\delta\rho\{-(U+\frac{Z}{r})+\int (d\vec{r}')\frac{\rho(r')}{|\vec{r}-\vec{r}'|} - \frac{11}{9\pi}(3\pi^2\rho)^{1/3}\}$$
$$- \delta\zeta\{N - \int (d\vec{r})\frac{1}{3\pi^2}[-2(U+\zeta)]^{3/2}\} \quad . \tag{4-344}$$

Now, Eqs.(330) and (337) can be combined into

$$\rho = \frac{1}{3\pi^2}[-2(U+\zeta)]^{3/2} \quad , \tag{4-345}$$

so that the curly brackets multiplying δU vanish in (344). Further, Eqs. (334) and (335) imply that the contents of the curly brackets multiplying $\delta\rho$ are a constant; and in view of the boundary condition obeyed by V, and therefore also by U, at infinity, this constant equals zero. Finally, the Laplacian term in (331) integrates to a null result, with the consequence that

$$N = \int (d\vec{r})\,\rho = \int (d\vec{r})\frac{1}{3\pi^2}[-2(U+\zeta)]^{3/2} \tag{4-346}$$

Therefore, E_{ES} is indeed stationary under infinitesimal variations of U, ρ, and ζ.

Upon introducing an electrostatic pseudo-potential

$$U_{es} \equiv U - U_{ex} = U + \frac{11}{9\pi}(3\pi^2\rho)^{1/3}$$
$$= U + \frac{11}{9\pi}[-2(U+\zeta)]^{1/2} \quad , \tag{4-347}$$

the differential equation (335) reads

$$-\frac{1}{4\pi}\nabla^2(U_{es}+\frac{Z}{r}) = \rho \ . \tag{4-348}$$

This will be a useful equation determining U_{es} only if ρ can be expressed in terms of U_{es}. In the first place, both U_{es} and ρ are given as algebraic functions of U and ζ, Eqs.(345) and (347). Since Eq.(345) already presents ρ as a function of $-2(U+\zeta)$, we need to invert the relation

$$-2(U_{es}+\zeta) = -2(U+\zeta) - \frac{22}{9\pi}[-2(U+\zeta)]^{1/2} \tag{4-349}$$

to exhibit $-2(U+\zeta)$ in terms of $-2(U_{es}+\zeta)$. (Recall that, as always, square roots of negative numbers are zero.) For $\underline{-2(U+\zeta) \leq 0}$, we have simply

$$-2(U+\zeta) = -2(U_{es}+\zeta) \quad , \tag{4-350}$$

whereas in the situation $\underline{-2(U+\zeta) > 0}$ we obtain, after completing the square,

$$- 2(U_{es}+\zeta) = \left([-2(U+\zeta)]^{1/2} - \frac{11}{9\pi}\right)^2 - \left(\frac{11}{9\pi}\right)^2 \qquad (4\text{-}351)$$

$$\geq - \left(\frac{11}{9\pi}\right)^2 ,$$

or

$$[-2(U+\zeta)]^{1/2} = \frac{11}{9\pi} \pm \sqrt{\left(\frac{11}{9\pi}\right)^2 + [-2(U_{es}+\zeta)]} . \qquad (4\text{-}352)$$

The left-hand side being positive, the lower sign is an option only for $-(\frac{11}{9\pi})^2 < -2(U_{es}+\zeta) < 0$. For $-2(U_{es}+\zeta) < -(\frac{11}{9\pi})^2$, the unique relation is (350); for $-2(U_{es}+\zeta) > 0$ we have equally uniquely

$$- 2(U+\zeta) = \left[\frac{11}{9\pi} + \sqrt{\left(\frac{11}{9\pi}\right)^2 + [-2(U_{es}+\zeta)]}\right]^2 . \qquad (4\text{-}353)$$

However, in the range $-(\frac{11}{9\pi})^2 < -2(U_{es}+\zeta) < 0$ there is the choice between (350), (353), and

$$- 2(U+\zeta) = \left[\frac{11}{9\pi} - \sqrt{\left(\frac{11}{9\pi}\right)^2 + [-2(U_{es}+\zeta)]}\right]^2 . \qquad (4\text{-}354)$$

Thus, there is no unique relation between ρ and $-2(U_{es}+\zeta)$ in the first place. This is troublesome. (The same difficulty occurs, of course, in the TFD model.)We have noticed here that Eqs.(342), (331b), and (337) do not suffice to define the ES model. It has to be supplemented by an injunction for relating ρ to U_{es} in a unique way. The natural and usual procedure is to pick a certain U_o in the range

$$0 \leq U_o \leq \frac{1}{2}\left(\frac{11}{9\pi}\right)^2 = 0.0757 \qquad (4\text{-}355)$$

and to use (350) for $U_{es}+\zeta > U_o$ and (353) for $U_{es}+\zeta < U_o$, whereas (354) is never employed. The new differential equation, obeyed by U_{es}, is then

$$- \frac{1}{4\pi} \nabla^2 \left(U_{es} + \frac{Z}{r}\right) = \rho$$

$$= \begin{cases} \frac{1}{3\pi^2}\left[\frac{11}{9\pi} + \sqrt{\left(\frac{11}{9\pi}\right)^2 + [-2(U_{es}+\zeta)]}\right]^3 , & \text{for } U_{es}+\zeta < U_o , \\ 0 , & \text{for } U_{es}+\zeta > U_o , \end{cases} \qquad (4\text{-}356)$$

the two ranges being the interior and the exterior of the atom.

Just inside the edge of the atom, the value of ρ is

$$\rho_o = \frac{1}{3\pi^2}\left[\frac{11}{9\pi} + \sqrt{\left(\frac{11}{9\pi}\right)^2 - 2U_o}\right]^3 , \qquad (4\text{-}357)$$

just outside it is zero. Obviously, this model does not provide a realistic description at the edge of the atom. Nevertheless, it represents an enormous improvement over the TF model, as far as the outer regions of the atom are concerned. (We shall report the results of a specific calculation shortly.)

For the two limiting values for U_o in (355) the corresponding ρ_o's are the limits in the relation

$$\frac{1}{3\pi^2}\left(\frac{22}{9\pi}\right)^3 \geq \rho_o \geq \frac{1}{3\pi^2}\left(\frac{11}{9\pi}\right)^3 \quad , \tag{4-358}$$

their decimal versions being 0.0159 and 0.0020, respectively. Thus ρ_o is a small number that does not depend on Z, in contrast to the density of the bulk of the electrons which is proportional to Z^2. This observation is reassuring, because it means that no prediction of the model will be very sensitive to the particular choice made for ρ_o. Or, turning the argument around: any result sensitive to the value of ρ_o must not be trusted.

The appearance of an additional parameter U_o (or, equivalently, ρ_o) is, of course, annoying. The more so, since the model does not select an optimal value for U_o - there is none. The resolution of this problem must come from an improvement of the description which removes this deficiency. Just this was the motivation for developing the treatment of the quantum corrections and of the exchange energy as presented in the earlier sections of this Chapter, and as initially reported in Ref.8. Readers familiar with the usual derivation of the TFD model will recall that there the question of the value of the density at the boundary (that is just inside of the edge of the atom) is positively answered by requiring

$$n_o = \frac{1}{3\pi^2}\left(\frac{5}{4\pi}\right)^3 \quad . \tag{4-359}$$

The corresponding value of the pseudo density in the ES model is obtained by supplying the necessary factors of 11/9,

$$\rho_o = \frac{1}{3\pi^2}\left(\frac{11}{9}\times\frac{5}{4\pi}\right)^3 = 0.0039 \quad . \tag{4-360}$$

This needs some clarification in view of our insistence that there is no best value for ρ_o. For this purpose, we review the argument, that is usually put forward in favor of the "optimal" n_o in the TFD model, in the context of the ES model. We begin with incorporating the parameter U_o into the energy functional, where it is now expedient to use a functional of U_{es} and ζ. It is given by

$$E_{ES}(U_{es},\zeta) = \int (d\vec{r})\,[E(U_{es}+\zeta)-E(U_o)]\,\eta\big(U_o-(U_{es}+\zeta)\big)$$
$$- \frac{1}{8\pi}\int (d\vec{r})\,[\vec{\nabla}(U_{es}+\frac{Z}{r})]^2 - \zeta N + \frac{1}{2}Z^2 \quad , \qquad (4\text{-}361)$$

where the energy density $E(U_{es}+\zeta)$ is

$$E(U_{es}+\zeta) = -\frac{1}{15\pi^2}\Big(\sqrt{(\tfrac{11}{9\pi})^2+[-2(U_{es}+\zeta)]} - \frac{11}{36\pi}\Big)$$
$$\times \left[\frac{11}{9\pi}+\sqrt{(\tfrac{11}{9\pi})^2+[-2(U_{es}+\zeta)]}\right]^4 \quad . \qquad (4\text{-}362)$$

Since [cf. Eq.(356)]

$$\frac{\partial}{\partial U_{es}}\left\{[E(U_{es}+\zeta)-E(U_o)]\,\eta\big(U_o-(U_{es}+\zeta)\big)\right\}$$
$$= \frac{1}{3\pi^2}\left[\frac{11}{9\pi}+\sqrt{(\tfrac{11}{9\pi})^2+[-2(U_{es}+\zeta)]}\right]^3 \eta\big(U_o-(U_{es}+\zeta)\big) \qquad (4\text{-}363)$$
$$= \rho \quad ,$$

the functional (361) is indeed the correct one. Note, in particular, that the term $E(U_o)$ <u>must</u> be subtracted from $E(U_{es}+\zeta)$ to ensure the continuity of the energy density at the boundary. The usual argument[41] is now the requirement that this term is absent in the energy functional, under which circumstances the energy density will be continuous only if $E(U_o) = 0$, implying

$$U_o = \frac{15}{32}\left(\frac{11}{9\pi}\right)^2 \quad , \qquad (4\text{-}364)$$

which produces (360) when inserted into (357). Fine, but there is no physical reason behind this requirement of vanishing $E(U_o)$; the uniqueness of U_o is simply an illusion. Certainly, different values of U_o will lead to different values of the energy, and these energy differences must be irrelevant if we want to take the ES model seriously. Indeed, they are. To make this point, we consider the change of E_{ES} resulting from an infinitesimal variation of U_o by δU_o. Since E_{ES} is stationary under variations of both U_{es} and ζ, their induced changes do not contribute. Consequently,

$$\delta_{U_o} E_{ES} = -\,\delta U_o \int (d\vec{r})\,\frac{\partial E(U_o)}{\partial U_o}\,\eta\big(U_o-(U_{es}+\zeta)\big)$$
$$= -\,\delta U_o \int (d\vec{r})\,\rho_o\,\eta\big(U_o-(U_{es}+\zeta)\big) \quad . \qquad (4\text{-}365)$$

Under the circumstance of spherical symmetry, as is the situation in an isolated atom, the step function limits the domain of integration to a sphere of some radius r_o. For $r > r_o$, the pseudo density ρ vanishes so that the Poisson equation (356) implies

$$U_{es} = -\frac{Z-N}{r} \qquad \text{for} \qquad r > r_o \tag{4-366}$$

which in conjunction with the continuity of U_{es}, at $r = r_o$, produces the statement

$$\zeta = \frac{Z-N}{r_o} + U_o \quad . \tag{4-367}$$

Incidentally, this identifies U_o as the minimal binding energy of neutral ES atoms. [Note that a Z-independent $\zeta(Z,Z)$ is consistent with (306).] In terms of ρ_o and r_o, Eq.(365) gives for an atom

$$\frac{\partial}{\partial U_o} E_{ES} = -\frac{4\pi}{3} r_o^3 \rho_o \quad . \tag{4-368}$$

From the numerical solutions of Eq.(356), reported in Ref. 38 and discussed to some extent in the next section, one can infer a slow Z dependence of r_o^3, as expressed by

$$r_o^3 \cong \begin{cases} 11.5 \times Z^{0.3} & \text{for} \quad U_o = 0 \quad , \\ 30. \quad \times Z^{0.25} & \text{for} \quad U_o = \frac{1}{2}\left(\frac{11}{9\pi}\right)^2 \quad , \end{cases} \tag{4-369}$$

for neutral atoms. With the corresponding values for ρ_o in (358), this says

$$\frac{\partial}{\partial U_o} E_{ES} \cong \begin{cases} -0.77 \times Z^{0.3} & \text{for} \quad U_o = 0 \quad , \\ -0.25 \times Z^{0.25} & \text{for} \quad U_o = \frac{1}{2}\left(\frac{11}{9\pi}\right)^2 \quad , \end{cases} \tag{4-370}$$

implying

$$\begin{aligned} E_{ES}(U_o = 0) &- E_{ES}\left(U_o = \frac{1}{2}\left(\frac{11}{9\pi}\right)^2\right) \\ &\lesssim 0.04 \ Z^{1/3} \quad . \end{aligned} \tag{4-371}$$

This amount is utterly irrelevant at the present level of approximation, where the significant contributions to the energy are proportional to

$Z^{7/3}$, $Z^{6/3}$, and $Z^{5/3}$. The argument supporting (364) and (360) is thus recognized to be quite artificial, indeed.

We have discussed the mathematical aspects of the troublesome boundary problem in the ES (and the TFD) model to some extent. Its resolution requires the recognition of where we have stressed the physical approximation too far. We postpone the necessary remarks on this point for a short while to the benefit of reporting an application of the model first.

An application of the ES model. Diamagnetic susceptibilities. The differential equation for U_{es}, Eq. (356), acquires a simpler appearance upon adopting new scales for r and U_{es}, defined by

$$r = \frac{3\pi}{4}\left(\frac{2}{\pi}\right)^{1/2} y = 1.00468\ y \qquad (4-372)$$

and

$$U_{es}(r) = -\zeta + \frac{1}{2}\left(\frac{11}{9\pi}\right)^2 - \frac{1}{2}\left(\frac{11}{3\pi}\right)^2 \frac{\Psi(y)}{y} \quad , \qquad (4-373)$$

where we have already expressed the spherical symmetry associated with an isolated atom. The radial function $\Psi(y)$ obeys (see also Problem 9)

$$\left(\frac{d}{dy}\right)^2 \Psi(y) = y\left[\left(\Psi(y)/y\right)^{1/2} + \frac{1}{3}\right]^3 \qquad (4-374)$$

for $y \leq y_o \equiv 0.99534\ r_o$, and is subject to the boundary condition

$$\Psi(0) = 6\pi\left(\frac{2}{\pi}\right)^{3/2} Z = 1.43136\ Z \quad , \qquad (4-375)$$

which states $U_{es} \rightarrow -Z/r$ for $r \rightarrow 0$. The known form of U_{es} in the exterior of the atom, Eq. (366), appears as

$$\frac{\Psi(y)}{y} = \frac{1}{9} - 2\left(\frac{3\pi}{11}\right)^2 U_o + q\,\Psi(0)\left(\frac{1}{y} - \frac{1}{y_o}\right) \quad ,$$
$$\text{for } y \geq y_o \quad , \qquad (4-376)$$

where $q = 1-N/Z$ is, as always, the degree of ionization, and Eqs. (367) and (375) have been used. In terms of boundary conditions on $\Psi(y)$ at $y = y_o$, this reads

$$\Psi(y_o)/y_o = \frac{1}{9} - 2\left(\frac{3\pi}{11}\right)^2 U_o \qquad (4-377)$$

and

$$\Psi(y_o) - y_o \frac{d\Psi}{dy}(y_o) = q\,\Psi(0) \quad . \tag{4-378}$$

The range of U_o in (355) gives a corresponding range for $\Psi(y_o)/y_o$ in (377),

$$\frac{1}{9} \geq \Psi(y_o)/y_o \geq 0 \quad , \tag{4-379}$$

where we, incidentally, remark that the value of U_o in (364) implies $\Psi(y_o)/y_o = 1/144$.

The differential equation (374) can easily be solved numerically, whereby integrating inwards from $y = y_o$ to $y = 0$ using $\sqrt{y}$ as the basic variable is the recommended procedure. The values of y_o, reported in Table 2, have been calculated this way.[38] The numbers refer to the purely diamagnetic closed-shell neutral atoms Ne, Ar, Kr, and Xe, along with the singly and doubly charged positive ions having those electronic configurations. The neutral-atom results have been employed in establishing Eq.(369).

Table 4-2. Values of y_o for inert-gas atoms and related ions. The left- and right-hand columns refer to the respective extreme values in (379).

N	Z = N		Z = N + 1		Z = N + 2	
10	2.805	3.737	2.296	2.569	1.928	2.050
18	3.010	3.957	2.551	2.865	2.206	2.359
36	3.231	4.191	2.821	3.179	2.504	2.692
54	3.349	4.316	2.964	3.343	2.662	2.868

The merit of the ES model is the improved description of the outer reaches of the atom. A quantity sensitive to this part of the atom is the expectation value of the squared distance from the nucleus,

$$\overline{r^2} = \frac{1}{N}\int (d\vec{r})\, r^2 n \quad . \tag{4-380}$$

Experimental data are obtained from measurements of the molar diamagnetic susceptibility

$$-\chi_m = \chi_o N \overline{r^2} \quad , \tag{4-381}$$

whose unit χ_o is composed of the fine structure constant α, the Bohr radius a_o, and Avogadro's number N_A according to

$$\chi_o = \frac{1}{6}\alpha^2 a_o^3 N_A = 0.7920\times 10^{-6}\ \mathrm{cm}^3 \quad . \tag{4-382}$$

Table 3 displays the experimental values for $\overline{r^2}$, derived this way. It should be appreciated that the entries for the neutral atoms are well established,[42] those for the ions[43] are unavoidably uncertain owing to the necessity of measuring them in ionic crystals.

In order to express the ES prediction for $\overline{r^2}$ in terms of $\Psi(y)$, we first recall the relation (331b) which, combined with the differential equation (356), reads

$$n = \nabla^2\left[-\frac{1}{4\pi}\left(U_{es} + \frac{Z-N}{r}\right) + \frac{1}{24\pi^2}(3\pi^2\rho)^{1/3}\right] - \frac{1}{4\pi}\nabla^2\left(\frac{N}{r}\right) \quad . \tag{4-383}$$

In writing this structure we exploit the known form of U_{es} outside the atom: the contents of the square brackets equal zero for $r > r_o$. A twofold partial integration, therefore, produces ($\nabla^2 r^2 = 6$)

$$N\,\overline{r^2} = 6\int_0^{r_o} dr\, r^2\left[-U_{es} - \frac{Z-N}{r} + \frac{1}{6\pi}(3\pi^2\rho)^{1/3}\right] \quad , \tag{4-384}$$

where we have made use of the observation that the second Laplacian in (383) is $N\delta(\vec{r})$ and does not contribute to $\overline{r^2}$. Now we employ such relations like

$$\begin{aligned} -U_{es}(r) + U_{es}(r_o) &= \frac{1}{2}\left(\frac{11}{3\pi}\right)^2\left[\frac{\Psi(y)}{y} - \frac{\Psi(y_o)}{y_o}\right] \quad , \\ -U_{es}(r_o) - \frac{Z-N}{r} &= \frac{Z-N}{r_o} - \frac{Z-N}{r} = -\frac{1}{2}\left(\frac{11}{3\pi}\right)^2 q\,\Psi(0)\left(\frac{1}{y} - \frac{1}{y_o}\right) \quad , \\ \frac{1}{6\pi}(3\pi^2\rho)^{1/3} &= \frac{1}{2}\left(\frac{11}{3\pi}\right)^2 \frac{1}{11}\left[\sqrt{\Psi(y)/y} + \frac{1}{3}\right] \end{aligned} \tag{4-385}$$

and

$$6\,dr\,r^2 = \frac{9\pi}{16}\left(\frac{11}{2}\right)^{1/2} 2\left(\frac{3\pi}{11}\right)^2 dy\, y^2 \tag{4-386}$$

to arrive at the final form for numerical integration:

$$\begin{aligned} \overline{r^2} = \frac{1}{N}\frac{9\pi}{16}\left(\frac{11}{2}\right)^{1/2}\Big[\frac{1}{3}y_o^2\left(\frac{y_o}{33} - \Psi(y_o) - \frac{q}{2}\Psi(0)\right) \\ + \int_0^{y_o} dy\, y\left(\Psi(y) + \frac{1}{11}\sqrt{y\Psi(y)}\right)\Big] \quad . \end{aligned} \tag{4-387}$$

Table 4-3. Experimental data for $\overline{r^2}$.

N	Z = N	Z=N+1	Z=N+2
10	0.852	0.768	0.546
18	1.373	1.023	0.750
36	1.010	0.772	0.632
54	1.027	0.821	0.678

Table 4-4. Predictions for $\overline{r^2}$ by the ES model, for $\Psi(y_o)/y_o = 1/9$. In parentheses, the deviations from the experimental values, in percent.

N	Z = N	Z = N + 1	Z = N + 2
10	1.626 (91.)	1.050 (37.)	0.732 (34.)
18	1.413 (2.9)	1.013 (-1.0)	0.767 (2.3)
36	1.152 (14.)	0.903 (17.)	0.737 (17.)
54	1.001 (-2.5)	0.819 (-0.2)	0.691 (1.9)

The results are shown in Tables 4 and 5 for the two extreme boundary values $\Psi(y_o)/y_o = 1/9$ and $\Psi(y_o)/y_o = 0$, respectively. One gets the distinct impression that the first boundary condition, corresponding to $U_o = 0$, outperforms the other one. The agreement with experimental values is within 3% for Z = 18 and 54 in Table 4. A larger error for Z = 10 is understandable; Z = 36 exhibits a quantum oscillation. Indeed, in each column of Table 3 we witness a succession of increase, decrease, and increase with growing N. This oscillatory behavior is clearly a manifestation of the atomic shell structure. As such it cannot be reproduced by the ES model. The numbers reported in Tables 4 and 5 are, nevertheless, an enormous improvement over the predictions of the TF model, in which we have, for neutral atoms,

Table 4-5. Like Table 4, for $\Psi(y_o)/y_o = 0$.

N	Z = N	Z = N + 1	Z = N + 2
10	1.992 (134)	1.110 (45.)	0.744 (36.)
18	1.667 (21.)	1.063 (3.9)	0.783 (4.4)
36	1.305 (29.)	0.943 (22.)	0.752 (19.)
54	1.116 (8.7)	0.851 (3.7)	0.703 (3.7)

$$\overline{(r^2)}_{TF} = \frac{1}{Z}\int(d\vec{r})\; r^2 \frac{1}{3\pi^2}[-2V_{TF}(r)]^{3/2}$$
$$= \frac{6}{Z}\int_0^\infty dr\; r^2\left(-V_{TF}(r)\right) \qquad (4\text{-}388)$$
$$= \frac{6a^2}{Z^{2/3}}\int_0^\infty dx\; x\,F(x) \quad ,$$

or with the numerical value of the integral of 9.194,[44]

$$\overline{(r^2)}_{TF} = 43.2/Z^{2/3} \quad . \qquad (4\text{-}389)$$

This gives 9.3, 6.3, 4.0, and 3.0 for Ne, Ar, Kr, and Xe, respectively - too large by factors of 11., 4.6, 4.0, and 2.9 when compared with the experimental numbers of Table 3. No doubt, the ES model improves matters considerably. Incidentally, we remark that a slightly modified version of the TF formula (389), namely,

$$\overline{(r^2)}_{ES} \cong \frac{43.2}{\left(Z^{1/3}+Z_o^{1/3}\right)^2} \quad , \qquad (4\text{-}390)$$

reproduces the ES numbers for neutral atoms, if one choses $Z_o \cong 20$.

So we found satisfactory agreement between the ES predictions for $\overline{r^2}$ and the experimental data. It is true, that a further refinement of the theory is required in order to correctly reproduce the measurements within the experimental uncertainties (of, typically, a few percent) for all the atoms in Tables 3, 4, and 5. But already at the present level of accuracy the ES model does not perform worse than HF calculations,[45] which yield the numbers listed in Table 6. The rather large deviations from the experimental values, even for the neutral atoms, are somewhat unexpected. Could it be that this is an artifact of the spurious nodes that always occur in the numerical HF wave functions at large distances, where, as a consequence, the HF densities are too large ?[46] Please observe further that the N = 10 numbers do not fit into the gene-

Table 4-6. HF predictions for $\overline{r^2}$ and, in parentheses, their deviations from the experimental values, in percent.

N	Z = N	Z = N+1	Z = N+2
10	0.937 (10.)	0.641 (-17.)	0.472 (-14.)
18	1.446 (5.3)	1.086 (6.1)	0.857 (14.)
36	1.098 (8.7)	0.884 (15.)	0.741 (17.)
54	1.160 (13.)	0.973 (19.)	0.843 (24.)

ral pattern in Table 6. This casts some doubt upon the reliability of the corresponding experimental data.

Since we are at it, let us also report the predictions for $\overline{r^2}$ obtained from the description which keeps all the structure offered by the Airy functions in Eq.(210), for instance, along with the necessary corrections for the innermost electrons. The corresponding new differential equation will be given below [see Eq.(504)]. Here we just take a look at Table 8 which displays the numbers obtained for $\overline{r^2}$ in Ref.47. With the sole exception of argon (N=Z=18), these agree perfectly with the experimental data, given that their uncertainties are larger for the ions. Incidentally, owing to some numerical difficulties with the new differential equation for neutral atoms (see below), the Z=N predictions in Table 8 are less accurate than the ones for ions with Z=N+1 and Z=N+2. Please note, in particular, that the oscillatory functions F_m in Eq.(210) supply just the right amount of structure in the densities to reproduce the quantum oscillation at N=36: there is a decrease - increase phenomenon in Table 8 just like in Table 3 (experiment) and Table 6 (HF).

We leave the ES model here, being content with demonstrating its usefulness in just one simple application;[48] but not yet quite, inasmuch as we still have to deliver the discussion concerning the troublesome boundary condition, as promised at the end of the preceding section. We shall thereby be led to models that differ only slightly from the ES model and offer a more realistic description of the edge of the atom without employing a differential equation much more involved than Eqs. (356), or (374).

Table 4-7. Predicitons for $\overline{r^2}$ by the new theory (not simplified to the ES model). The deviations from the experimental data are given in the parentheses, in percent.

N	Z = N	Z = N+1	Z = N+2
18	1.46 (6.3)	1.036 (1.3)	0.786 (4.8)
36	1.03 (2.0)	0.813 (5.3)	0.664 (5.1)
54	1.01 (-1.7)	0.831 (1.2)	0.704 (3.8)

Improved (?) ES model. Electric polarizabilities. We now return to the discussion of the troublesome boundary problem in the ES model. What is the origin of the unrealistic behavior of the pseudo density ρ, which decreases continuously until it reaches the value of ρ_o at which point it instantly drops to zero ? [By the way, the physical density $n(\vec{r})$ is even less realistic, because the Laplacian in (331b) produces Delta functions at the location of the discontinuity of $\rho(\vec{r})$.] Clearly we are paying a price for the simplicity of the model whose main feature is the inclusion of the exchange energy in form of the Dirac approximation (262),

$$E_{ex}(n) \cong - \int (d\vec{r}) \frac{1}{4\pi^3}[3\pi^2 n(\vec{r})]^{4/3} \quad . \tag{4-391}$$

This is a good approximation in the dense interior of the atom. In employing it without modification for calculating the contribution from the outer regions of the atom, where the density is small, we have stressed (391) too far. In other words: where the density is sufficiently large to ensure that the Dirac approximation of the exchange energy density ($\sim n^{4/3}$) is small compared to, for instance, the kinetic energy density ($\sim n^{5/3}$), that approximation is reliable; in low density regions it is not good enough. Since (391) comes from inserting the TF density into the more generally valid expression (259),

$$E_{ex}(n) \cong -\pi \int (d\vec{r}\,') \int_{\zeta}^{\infty} d\zeta' [\frac{\partial}{\partial\zeta'} n(\vec{r}\,')]^2 \tag{4-392}$$

where the relation

$$n(\vec{r}\,') = 2 < \vec{r}\,' |\eta(-H-\zeta')| \vec{r}\,' > \tag{4-393}$$

is to be used both to evaluate the integrals of (392) and to express E_{ex} as a functional of the actual density (for which $\zeta'=\zeta$) after these integrations. Going through this procedure with the TF approximation for the right-hand side of (393) gives (391). This TF density is unrealistic at the edge of the atom where $V+\zeta \cong 0$. There the corrections for the strongly bound electrons are irrelevant and the gradient of the potential is practically constant, so that higher derivatives do not matter. Consequently, at the edge the density is well approximated by [cf. Eq.(219)][49]

$$n(\vec{r}) = \frac{1}{2\pi} |2\vec{\nabla}V(\vec{r})| F_2(y(\vec{r})) \tag{4-394}$$

with

$$y(\vec{r}) = \frac{2(V(\vec{r})+\zeta)}{|2\vec{\nabla}V(\vec{r})|^{2/3}} \quad . \tag{4-395}$$

Now observe that replacing the gradient here by its value at the edge

$$|2\vec{\nabla}V(\vec{r})|^{2/3} \rightarrow V_o = \text{const.} \quad , \tag{4-396}$$

results in an expression for the density

$$n(\vec{r}) \cong \frac{1}{2\pi} V_o^{3/2} F_2\big(2(V+\zeta)/V_o\big) \quad , \tag{4-397}$$

which is equivalent to (394) at the edge and reduces to the TF approximation when applied to the dense interior of the atom, where $V+\zeta$ is a large negative number and the asymptotic form of $F_2(y)$, namely [Eq.(161)]

$$F_2(y) \cong \frac{2}{3\pi}(-y)^{3/2} \quad \text{for} \quad -y \gg 1 \quad , \tag{4-398}$$

is available.

We are thus invited to insert (397) into (392), thereby thinking of V_o as a parameter somewhat related to the gradient at the edge of the atom - not, of course, meaning that (396) becomes an identity at this edge, but merely concerning the order of magnitude of V_o. Just as for the U_o of the ES model, there is no best value for V_o. To some extent it can be regarded as an adjustable parameter. There is, again, a price to be paid for the simplicity gained in the transition from Eq. (210) [or (219)] to (397).

The differentiation of $F_2(y)$ produces $F_1(y)$, see Eq.(154), so that the new approximation for the exchange energy is (see also Problem 10)

$$E_{ex}(n) \cong -\int (d\vec{r}) \frac{1}{2\pi} V_o^2 \int_{y(\vec{r})} dy' [F_1(y')]^2 \tag{4-399}$$

where $y(\vec{r})$ is given in terms of the density $n(\vec{r})$ by means of

$$n(\vec{r}) = \frac{1}{2\pi} V_o^{3/2} F_2\big(y(\vec{r})\big) \quad . \tag{4-400}$$

This reduces to the Dirac expression (391) where (398) is applicable, which is the situation for $y \lesssim -3/2$ according to (171). Since $F_2(-3/2) = 0.371$ this requires

$$n \gtrsim (0.15\, V_o)^{3/2} \quad . \tag{4-401}$$

In particular, for $V_o \to 0$, Eq.(391) is regained for all values of $n(\vec{r})$.

The exchange potential that corresponds to (399) is obtained by considering infinitesimal variations of the density. They cause a change of y by δy, given by

$$\delta n = -\frac{1}{2\pi} V_o^{3/2} F_1(y)\ \delta y \quad , \tag{4-402}$$

with the consequence

$$\delta_n E_{ex}(n) = \int (d\vec{r})\ \delta y\ \frac{1}{2\pi} V_o^2 [F_1(y)]^2$$
$$= \int (d\vec{r})\ \delta n [-V_o^{1/2} F_1(y)] \quad . \tag{4-403}$$

This identifies the new exchange potential [cf. Eq.(324)]

$$V_{ex}(n) = - V_o^{1/2} F_1\big(y(n)\big)$$
$$= - V_o^{1/2} F_1\big(F_2^{-1}(2\pi V_o^{-3/2} n)\big) \quad , \tag{4-404}$$

where we have, for once, inverted Eq.(400) formally. For densities that are large in the sense of (401) this reduces to the Dirac expression (325), as it must do.

So much about an improved local treatment of the exchange correction. Before discussing the corresponding improvement of the independent-particle energy $E_1 - \zeta N$, it is necessary to establish a criterion for our judgement whether a modified, local relation between the (pseudo) density and the electrostatic (pseudo) potential is more realistic at the edge of the atom. A quantity that is very sensitive to the dependence of the density upon the potential is the electric polarizability α_p of the atom.[50] It measures the effectiveness of a weak external electric field $\vec{E}$ in inducing a dipole moment

$$\vec{d} = \int (d\vec{r})\ \vec{r}\ n(\vec{r}) \tag{4-405}$$

of the charge distribution inside the atom. We shall confine the present discussion to the circumstance of no permanent electric dipole moment in the absence of external electric field. In isolated atoms, this is the actual situation, since the density is spherically symmetric as long as there are no external fields.

The induced dipole moment is proportional to the applied electric field, if this field is sufficiently weak, and the factor expressing this linear relation is the polarizability α_p,

$$\vec{d} = \alpha_p \vec{E} \quad . \tag{4-406}$$

It is available experimentally from measurements of the static dielectric constant ϵ of (not too dense) gases. With n_{gas} being the density of atoms in the gas, the Clausius-Mosotti formula connects ϵ to α_p according to

$$\frac{\epsilon-1}{\epsilon+2} = \frac{4\pi}{3} n_{gas} \alpha_p \quad . \tag{4-407}$$

The polarizability of a conducting sphere is simply the radius of the sphere cubed, so that we can interpret

$$r_p \equiv \alpha_p^{1/3} \tag{4-408}$$

as an "effective polarization radius" of the atom. Experimental values of r_p (in atomic units) are listed in Table 8 for those neutral atoms, for which α_p is known somewhat accurately.[51] Also reported are HF predictions,[45] which occasionally agree very well with the experimental data, but deviate substantially for many Z values in the Table.

Let us now discuss α_p in the context of ES-type models, in which the physical density is calculated from a pseudo density as in Eq. (331), which more generally reads

$$n = \rho - \frac{9/11}{24\pi} \nabla^2 U_{ex}(\rho) + \text{CSBE} \quad , \tag{4-409}$$

and ρ is an (algebraic) function of $U_{es} + \zeta$, as in Eq.(356),

Table 4-8. Experimental values for polarization radii r_p(EXP), compared with HF predictions.

Z	EXP	HF	Z	EXP	HF
2	1.114	1.14	16	2.6	2.85
3	5.45	4.76	17	2.4	2.60
4	3.36	3.74	18	2.23	2.37
5	2.71	2.85	19	6.62	6.33
6	2.28	2.27	20	5.54	6.11
7	1.96	1.89	21	5.37	5.65
8	1.75	1.70	36	2.56	2.76
9	1.56	1.53	37	6.84	6.77
10	1.39	1.38	38	5.7	6.72
11	5.45	5.01	54	3.00	3.31
12	4.2	4.57	55	7.4	7.61
13	3.9	4.20	56	6.5	7.70
14	3.3	3.58	80	3.24	4.34
15	2.9	3.10	82	3.66	4.29

$$\rho(\vec{r}) = \rho\left(U_{es}(\vec{r})+\zeta\right) \quad , \tag{4-410}$$

where U_{es} and ζ are determined by Poisson's equation

$$-\frac{1}{4\pi}\nabla^2(U_{es}-V_{ext}) = \rho \quad , \tag{4-411}$$

subject to the usual boundary conditions, and by the normalization

$$\int(d\vec{r})\rho = N \quad . \tag{4-412}$$

Now, in performing the CSBE in (409), the density is modified by terms that are spherically symmetric and do not contribute in (405). Further, the Laplacian of (409), when inserted into (405), integrates to a null result. Thus, n can be equivalently replaced by ρ in Eq.(405),

$$\vec{d} = \int(d\vec{r})\vec{r}\rho(\vec{r}) \quad . \tag{4-413}$$

Considering a weak constant electric field $\vec{E}$ in addition to the Coulomb field of the nucleus, we have

$$V_{ext} = -\frac{Z}{r} - \vec{E}\cdot\vec{r} \tag{4-414}$$

as the external potential in (411). Consequently, U_{es} consists of the $\vec{E}=0$ term $U_{es}^{(0)}$ and contributions $U_{es}^{(1)}$, $U_{es}^{(2)}$, ... which are linear, quadratic, ... in $\vec{E}$:

$$U_{es} = U_{es}^{(0)} + U_{es}^{(1)} + U_{es}^{(2)} + \ldots \quad , \tag{4-415}$$

Quite analogously, one has

$$\rho = \rho^{(0)} + \rho^{(1)} + \rho^{(2)} + \ldots \tag{4-416}$$

and

$$\zeta = \zeta^{(0)} + \zeta^{(1)} + \zeta^{(2)} + \ldots \tag{4-417}$$

In view of Eq.(410), we have

$$\rho^{(0)} = \rho(U_{es}^{(0)} + \zeta^{(0)}) \tag{4-418}$$

and

$$\rho^{(1)} = \frac{\partial\rho}{\partial U_{es}}\left(U_{es}^{(0)} + \zeta^{(0)}\right)\times\left(U_{es}^{(1)} + \zeta^{(1)}\right) \equiv$$

$$\equiv \rho'(r)\left(U_{es}^{(1)}(\vec{r}) + \zeta^{(1)}\right) \quad . \tag{4-419}$$

Since $\rho^{(0)}$ already integrates to N, the spatial integral of $\rho^{(1)}$ must vanish. This implies $\zeta^{(1)} = 0$, so that

$$\rho^{(1)}(\vec{r}) = \rho'(r)\ U_{es}^{(1)}(\vec{r}) \quad . \tag{4-420}$$

If we measure $U_{es}^{(1)}$ in multiples of $\vec{E}\cdot\vec{r}$,

$$U_{es}^{(1)}(\vec{r}) = -\ \vec{E}\cdot\vec{r}\ v(r) \quad , \tag{4-421}$$

where, of course,

$$v(r) \to 1 \quad \text{for} \quad r \to \infty \ , \tag{4-422}$$

then the induced dipole moment is, to first order in the applied field, given by

$$\vec{d} = \int (d\vec{r})\vec{r}\ \rho^{(1)}(\vec{r}) = \vec{E}\cdot\int (d\vec{r})\ \vec{r}\,\vec{r}[-\rho'(r)]v(r) \quad , \tag{4-423}$$

where $\vec{r}\,\vec{r}$ can be equivalently replaced by $\frac{1}{3}\,r^2\,\overleftrightarrow{1}$, so that we find

$$\alpha_p = \frac{4\pi}{3}\int_0^\infty dr\ r^4[-\rho'(r)]v(r) \quad . \tag{4-424}$$

Because of the large - r behavior of $v(r)$, displayed in (422), $\rho'(r)$ must tend toward zero faster than $1/r^5$ as $r \to \infty$. Further observe that $\rho'(r)$ is negative, provided that the function of (410) is reasonable and gives a smaller density for larger values of $U_{es}(\vec{r}) + \zeta$. Then α_p is ensured to be positive, as it must be.

The radial function $v(r)$ is determined by the first order terms in the Poisson equation (411), which are

$$-\frac{1}{4\pi}\nabla^2(U_{es}^{(1)} + \vec{E}\cdot\vec{r}) = \rho^{(1)} \quad , \tag{4-425}$$

and produce, after inserting Eqs.(420) and (421), the differential equation

$$\left[\frac{d^2}{dr^2} + \frac{4}{r}\frac{d}{dr} + 4\pi\rho'(r)\right]v(r) = 0 \quad . \tag{4-426}$$

The corresponding integral equation

$$v(r) = 1 - \frac{1}{r}\int_0^\infty dr'\ r'^3\ \frac{r_<}{r_>^2}[-\frac{4\pi}{3}\rho'(r')]v(r') \qquad (4\text{-}427)$$

incorporates the boundary condition (422). Here $r_<$ and $r_>$ stand for the smaller and the larger one of r and r', respectively. Upon using the identity

$$\frac{r_<}{r_>^2} = \frac{r'}{r^2} + \left(\frac{r}{r'^2} - \frac{r'}{r^2}\right)\eta(r'-r) \qquad (4\text{-}428)$$

in conjunction with Eq.(424), this integral equation appears as

$$v(r) = 1 - \frac{\alpha_p}{r^3} + \int_r^\infty dr'\ r'\left(\left(\frac{r'}{r}\right)^3 - 1\right)[-\frac{4\pi}{3}\rho'(r')]v(r') \qquad (4\text{-}429)$$

Since $\rho'(r)$ approaches zero faster than $1/r^5$, the remaining integral tends to zero faster than $1/r^3$, with the consequence

$$v(r) \cong 1 - \frac{\alpha_p}{r^3} \quad \text{for large } r \quad , \qquad (4\text{-}430)$$

or, with (421) and (406),

$$U_{es}^{(1)} \cong -\vec{E}\cdot\vec{r} + \vec{d}\cdot\frac{\vec{r}}{r^3} \quad , \qquad (4\text{-}431)$$

which correctly exhibits the dipole potential. At short distances, it is fitting to use the identity

$$\frac{r_<}{r_>^2} = \frac{r}{r'^2} - \left(\frac{r}{r'^2} - \frac{r'}{r^2}\right)\eta(r-r') \quad , \qquad (4\text{-}432)$$

instead of (428), in (427), which gives

$$v(r) = v_o + \int_0^r dr'\ r'\left(1-\left(\frac{r'}{r}\right)^3\right)[-\frac{4\pi}{3}\rho'(r')]v(r') \quad , \qquad (4\text{-}433)$$

where

$$v_o = v(0) = 1 - \int_0^\infty dr\ r[-\frac{4\pi}{3}\rho'(r)]v(r) \quad . \qquad (4\text{-}434)$$

In particular, if the connection (410) between the potential U_{es} and the density ρ has the TF form for $r\to 0$, then

$$-\frac{4\pi}{3}\rho'(r) \cong \frac{4}{3\pi}\left(\frac{2Z}{r}\right)^{1/2} \quad \text{for} \quad r\to 0 \quad , \qquad (4\text{-}435)$$

which, inserted into (433), implies the small-r form

$$v(r) = v_o[1 + \frac{16}{27\pi}(2Zr^3)^{1/2} + \dots] \quad . \tag{4-436}$$

After picking an arbitrary (positive) value $\tilde{v}_o$ as a guess for v_o, one can then use this initial behavior of $v(r)$ to start the numerical integration of the differential equation (426). Since this equation is linear in $v(r)$, the solution $\tilde{v}(r)$ thus obtained will be a multiple of the actual $v(r)$, which has the definite asymptotic form (430), approaching unity for $r \to \infty$. Inasmuch as

$$\left.\begin{aligned} \tilde{v}(r) &= \frac{\tilde{v}_o}{v_o}(1 - \frac{\alpha_p}{r^3}) \\ r\frac{d}{dr}\,\tilde{v}(r) &= \frac{\tilde{v}_o}{v_o}\,\frac{3\alpha_p}{r^3} \end{aligned}\right\} \quad \text{for } r \to \infty \quad , \tag{4-437}$$

we employ the scale invariant expression

$$\alpha_p = \lim_{r\to\infty} \left\{ r^3\,[1 + \frac{3\tilde{v}(r)/r}{d\tilde{v}(r)/dr}]^{-1} \right\} \tag{4-438}$$

to extract the polarizability. In practice, this limiting process simply means that we have to pick a distance r so large that $\rho'(r)$ is essentially zero.

Now, after this preparatory general discussion, let us see what requirements emerge on the $(U_{es}+\zeta)$-dependence of ρ. First observe that $U_{es}+\zeta$ tends to ζ as $r \to \infty$,

$$U_{es} + \zeta \to \zeta \geq \zeta(Z,Z) \equiv \zeta_o \quad , \tag{4-439}$$

while ρ tends to zero. This implies

$$\rho\big(U_{es}(\vec{r})+\zeta\big) = 0 \quad \text{for} \quad U_{es}(\vec{r})+\zeta \geq \zeta_o \quad , \tag{4-440}$$

which condition is satisfied both by the TF and by the ES relation, where $\zeta_o=0$ and $\zeta_o=U_o$, respectively. As a consequence of (440), atomic ions have an edge at $r=r_o$ with

$$U_{es}(r_o) + \zeta(Z,N) = -\frac{Z-N}{r_o} + \zeta(Z,N) = \zeta_o \quad , \tag{4-441}$$

beyond which the density equals zero, whereas neutral atom may extend to infinity, as is the situation in the TF model, or be limited to a finite volume as well, as realized in the ES model. Then consider, as a generalization of the TF relation, power laws

$$\rho(U_{es}+\zeta) \cong \mu[\zeta_o-(U_{es}+\zeta)]^{\nu+1} \qquad \text{for} \quad U_{es}+\zeta \leq \zeta_o \ , \tag{4-442}$$

with constants μ and ν. This produces, in conjunction with the Poisson equation (411),

$$\frac{d^2}{dr^2}(-r\,U_{es}) = 4\pi\mu\, r^{-\nu}(-r\,U_{es})^{\nu+1} \tag{4-443}$$

as the differential equation governing the large-r asymptotic form of the neutral-atom pseudo-potential $U_{es}(r)$. For $0<\nu<2$, this asymptotic form is algebraic,

$$U_{es}(r) \longrightarrow -\left(\frac{1-\nu/2}{\pi\mu\nu}\right)^{1/\nu} r^{-2/\nu} \ , \tag{4-444}$$

and for $\nu = 0$ it is exponential,

$$U_{es}(r) \longrightarrow -\frac{U_\infty}{r}\exp(-\sqrt{4\pi\mu}\ r) \tag{4-445}$$

with an undetermined constant U_∞. The corresponding asymptotic forms of $\rho'(r)$ emerge from

$$-\rho'(r) \cong \mu(\nu+1)[-U_{es}(r)]^{\nu} \tag{4-446}$$

for these neutral atoms, with the outcome

$$-\rho'(r) \longrightarrow \begin{cases} \dfrac{(\nu+1)(1-\nu/2)}{\pi\ \nu^2\ r^2} \sim \dfrac{1}{r^2}\ , & \text{for } 0 < \nu < 2\ , \\[2ex] \mu = \text{const.}\ , & \text{for } \nu = 0\ . \end{cases} \tag{4-447}$$

Such $\rho'(r)$'s result in infinite polarizabilities for neutral atoms, because $\rho'(r)$ must tend to zero faster than $1/r^5$ in order to produce a finite α_p. This we observed at Eq. (424); a different argument, based upon (426) and (438), is the subject of Problem 11.

We thus conclude, that the power-law form (442) cannot be the correct potential-dependence of the density, at least for $0 \leq \nu < 2$. The range $\nu < 0$ is immediately disposed of, because there Eq. (446) implies a growth of $-\rho'(r)$ at large distances. On the other hand, for $\nu \geq 2$ the potential U_{es} itself decreases slower than $1/r$, certainly an unrealistic behavior. The inference is therefore, that ρ must approach zero faster than any power of $\zeta_o - (U_{es}+\zeta)$ as this quantity tends to zero (from positive values, of course). In the ES model, the step function in (356) [or (363)] ensures this. But here the transition through the atomic edge is too rapid. In $\rho'(r)$ we meet a term

$$-\rho'(r) = \ldots + \rho_o \, \delta\big(U_o - (U_{es}+\zeta)\big) , \tag{4-448}$$

which in the differential equation (426) gives rise to a discontinuity of $dv(r)/dr$ where $U_{es} + \zeta = U_o$. For ions, this is at the edge at $r = r_o$ only, for neutral atoms, however, $U_{es} + \zeta = U_o$ in the entire exterior of the atom, that is for all $r \geq r_o$. No sensible interpretation can be given to such a $\rho'(r)$. The only way out is to insist that $\rho'(r) = 0$ for $r > r_o$ also for neutral ES atoms. Then, in view of $U_{es} \sim (r_o - r)^2$ just inside of the edge of a neutral atom, the Delta function in (448) implies $v(r=r_o) = 0$, so that $\alpha_p = r_o^3$, or with (408) and (369):

$$2.2 \times Z^{0.1} \lesssim r_p = r_o \lesssim 3.1 \times Z^{0.08} , \tag{4-449}$$

which - surprisingly enough - roughly reproduces the numbers of Table 8 order-of-magnitude wise. Of course, we are not going to take (449) seriously.

In search for a density-potential relation (410) that is decreasing, at the edge, more rapidly than the power law (characteristic, for instance, for the TF model) but not quite as sudden as the step function present in the ES model, one naturally recalls the smooth transition associated with the Airy functions in Eqs. (210) or (219). With the emphasis on the vicinity of the atomic edge, an improved treatment of the exchange energy was achieved above by replacing the TF relation by (397). In striving for an improved version of the ES model, we shall use this insight about exchange and perform the replacement

$$-\frac{11}{9}\int(d\vec{r})\,\frac{1}{4\pi^3}\,(3\pi^2\rho)^{4/3} \rightarrow -\frac{11}{9}\int(d\vec{r})\,\frac{V_o^2}{2\pi}\int_{y(\vec{r})}^{\infty} dy'\,[F_1(y')]^2 \tag{4-450}$$

in the energy functional (342), where $y(\vec{r})$ is determined by

$$\rho(\vec{r}) = \frac{1}{2\pi}\,V_o^{3/2}\,F_2\big(y(\vec{r})\big) \tag{4-451}$$

This is, of course, simply the exchange energy of (399), supplied with the - by now familiar - factor of 11/9 and treated as a functional of the pseudo density $\rho(\vec{r})$.

Concerning the corresponding modification of the first integral on the right-hand side of (342), the observation of

$$- \frac{1}{4\pi} V_o^{5/2} F_3\big(2(U+\zeta)/V_o\big) \to -\frac{1}{15\pi^2}[-2(U+\zeta)]^{5/2} \ , \qquad (4\text{-}452)$$

$$\text{for} \quad V_o \to 0 \ ,$$

invites the replacement

$$\int (d\vec{r}) \left(-\frac{1}{15\pi^2}\right)[-2(U+\zeta)]^{5/2} \to \int (d\vec{r}) \left(-\frac{1}{4\pi} V_o^{5/2}\right) F_3\big(2(U+\zeta)/V_o\big) \qquad (4\text{-}453)$$

with the consequence

$$\rho = \frac{1}{2\pi} V_o^{3/2} F_2\big(2(U+\zeta)/V_o\big) \ . \qquad (4\text{-}454)$$

Although this looks quite natural, it is simply not good, because $\rho(\vec{r})$ does not approach zero at large distances, where $U \to 0$ and $F_2(\ldots) \to F_2(2\zeta/V_o) > 0$. Instead of (453), I would therefore like to propose

$$\int (d\vec{r}) \left(-\frac{1}{15\pi^2}\right)[-2(U+\zeta)]^{5/2} \to \int (d\vec{r}) \frac{3}{8\pi} V_o^{5/2} [F_3(\bar{y}) - F_2^2(\bar{y})/F_1(\bar{y})]$$

$$(4\text{-}455)$$

with the understanding that $\bar{y}$ is determined by

$$-(U+\zeta) = \frac{3}{4} V_o F_2(\bar{y})/F_1(\bar{y}) \qquad (4\text{-}456)$$

in the classically allowed regime where $-(U+\zeta)>0$, whereas the integrand in (455) is set equal to zero for $-(U+\zeta)<0$. Note that this is also a property of the original integrand with its TF structure. In view of this consequence of (456):

$$\delta U = \frac{3}{4} V_o \left[1 - \frac{F_2(\bar{y}) F_o(\bar{y})}{F_1^2(\bar{y})}\right] \delta\bar{y} \ , \qquad (4\text{-}457)$$

the response of (455) to infinitesimal variations of U is

$$\int (d\vec{r}) \frac{3}{8\pi} V_o^{5/2} [-F_2(\bar{y}) + 2F_2(\bar{y}) - F_2^2(\bar{y}) F_o(\bar{y})/F_1^2(\bar{y})] \ \delta\bar{y}$$

$$= \int (d\vec{r}) \frac{1}{2\pi} V_o^{3/2} F_2(\bar{y}) \ \delta U \ , \qquad (4\text{-}458)$$

which identifies ρ as a function of $U+\zeta$,

$$\rho = \begin{cases} \frac{1}{2\pi} V_o^{3/2} F_2(\bar{y}) \ , & \text{for} \ -(U+\zeta)>0 \ , \\ 0 \ , & \text{for} \ -(U+\zeta)<0 \ , \end{cases} \qquad (4\text{-}459)$$

with $\bar{y}$ from (456), of course.

Before continuing the study of the implications of Eqs.(455) and (456), it seems necessary to offer some motivation for these relations. Consider the elimination of the pseudo potential U from the ES energy functional. It amounts to

$$\int(d\vec{r})\left\{-\frac{1}{15\pi^2}[-2(U+\zeta)]^{5/2}-(U+\zeta)\rho\right\}\rightarrow\int(d\vec{r})\frac{1}{10\pi^2}(3\pi^2\rho)^{5/3} \qquad (4\text{-}460)$$

in (342), quite analogous to the transition from $E_{TF}(V,n,\zeta)$ of Eq. (2-435) to the TF density functional (2-95). The same procedure performed after the replacement (455) gives

$$\begin{aligned}&\int(d\vec{r})\left\{\frac{3}{8\pi}V_o^{5/2}[F_3(\bar{y})-F_2^2(\bar{y})/F_1(\bar{y})]-(U+\zeta)\rho\right\}\\ &\qquad\rightarrow\int(d\vec{r})\,\frac{3}{8\pi}V_o^{5/2}\,F_3(\bar{y})\quad,\end{aligned} \qquad (4\text{-}461)$$

where now $\bar{y}$ is determined by the pseudo density through (459). For large ρ (on the scale set by $V_o^{3/2}$), $\bar{y}$ is large negative, so that with (161)

$$\rho\cong\frac{1}{3\pi^2}V_o^{3/2}(-\bar{y})^{3/2} \qquad (4\text{-}462)$$

and

$$\begin{aligned}\frac{3}{8\pi}V_o^{5/2}\,F_3(\bar{y})&\cong\frac{1}{10\pi^2}V_o^{5/2}(-\bar{y})^{5/2}\\ &\cong\frac{1}{10\pi^2}(3\pi^2\rho)^{5/3}\quad.\end{aligned} \qquad (4\text{-}463)$$

Thus we recognize (461) plus (459) as the natural modification of the right-hand side of (460). Translated into the potential language this produces (455) plus (456), where (459) becomes an implied statement.

We shall now put things together. The modifications (450) and (455) turn the ES energy functional (342) into that of the Modified ES model,

$$\begin{aligned}E_{MES}(U,\rho,\zeta)&=\frac{1}{2}Z^2+\int(d\vec{r})\frac{3}{8\pi}V_o^{5/2}[F_3(\bar{y})-F_2^2(\bar{y})/F_1(\bar{y})]\\ &\quad-\int(d\vec{r})\left(U+\frac{Z}{r}\right)\rho+\frac{1}{2}\int(d\vec{r})(d\vec{r}\,')\frac{\rho(\vec{r})\rho(\vec{r}\,')}{|\vec{r}-\vec{r}\,'|}\\ &\quad-\frac{11}{9}\int(d\vec{r})\frac{V_o^2}{2\pi}\int_y^\infty dy'\,[F_1(y')]^2-\zeta N\quad,\end{aligned} \qquad (4\text{-}464)$$

where $\bar{y}$ and y are related to $U+\zeta$ and ρ by Eqs. (451) and (456). In the limit $V_o\rightarrow o$, the ES functional (342) is reproduced. The stationary property of this new functional with respect to infinitesimal variations of U implies (459), and the consideration of $\delta\rho$ produces

$$-\left(U+\frac{Z}{r}\right)+\int(d\vec{r}\,')\frac{\rho(\vec{r}\,')}{|\vec{r}-\vec{r}\,'|}+U_{ex}=0\quad, \qquad (4\text{-}465)$$

or in differential form

$$-\frac{1}{4\pi}\nabla^2(U-U_{ex}+\frac{Z}{r}) = \rho(\vec{r}) \quad , \tag{4-466}$$

where the new pseudo exchange-potential

$$U_{ex} = -\frac{11}{9}v_o^{1/2}F_1(y) \tag{4-467}$$

reflects the induced modification of (334). In order to establish the potential-density relation (410), first note that Eqs.(451) and (459) yield $y=\bar{y}$, then conclude that for $U+\zeta>0$ the vanishing of U_{ex} is implied, so that the electrostatic pseudo-potential U_{es}, defined as in the ES model, by

$$U_{es} = U - U_{ex} \tag{4-468}$$

agrees with U in the classically forbidden region. For $U+\zeta<0$, we have, after combining (467), (456), and (468),

$$-(U_{es}+\zeta) = \frac{3}{4}v_o\frac{F_2(y)}{F_1(y)} - \frac{11}{9}v_o^{1/2}F_1(y) \quad . \tag{4-469}$$

The trouble of the ES model, namely the not unique relation between $(U+\zeta)$ and $(U_{es}+\zeta)$, which prompted the present discussion, is only avoided if the right-hand side in this equation is (i) positive for all values of y, and (ii) monotically decreasing as y increases. The first requirement is satisfied if

$$\frac{27}{44}v_o^{1/2} > F_1^2(y)/F_2(y) \quad , \quad \text{for all } y \quad , \tag{4-470}$$

the second one if

$$\frac{27}{44}v_o^{1/2} > \frac{F_o(y)F_1^2(y)}{F_1^2(y)-F_2(y)F_o(y)} \quad , \quad \text{for all } y \quad . \tag{4-471}$$

These two ratios of F_m functions acquire their maximal values of 0.4749 and 1.0929 at $y=-1.42$ and $y=-0.50$, respectively, so that both (470) and (471) are obeyed if the latter one is, this being the situation for

$$v_o^{1/2} > \frac{44}{27}\times 1.0929 = 1.781 \quad . \tag{4-472}$$

Then (469) can be uniquely solved for y, which inserted into (451) produces the value of ρ corresponding to the given $U_{es}+\zeta$, to be used in Poisson's equation (411). Let us further recall that the physical density n is then to be computed with the aid of (409), where (467) is needed.

It remains to be demonstrated that this new model is, indeed, more realistic than the previous unmodified ES model, in the sense that the description of the atomic edge is improved. Let us begin with considering the dense interior of the atom, where ρ is large [according to (401) this means nothing more than $\rho \gtrsim (0.15\, V_o)^{3/2}$] and both y and $U_{es}+\zeta$ are large negative numbers. If we employ the asymptotic forms (161) of the $F_n(y)$'s, then Eqs. (469) and (451) read

$$-(U_{es}+\zeta) = \frac{1}{2} V_o(-y)^{2/2} - \frac{11}{9\pi} V_o^{1/2}(-y)^{1/2} \quad , \tag{4-473}$$

or

$$(-V_o y)^{1/2} = \frac{11}{9\pi} + \sqrt{(\frac{11}{9\pi})^2 + [-2(U_{es}+\zeta)]} \quad , \tag{4-474}$$

and

$$\rho = \frac{1}{3\pi^2}(-V_o y)^{3/2} \quad , \tag{4-475}$$

which combine to reproduce the ES relation (356). Thus, the MES density agrees with the ES one as long as one stays away from the edge of the atom. Near the edge, y is large positive, and the asymptotic forms (165) apply. In (469) and (451) they yield

$$-(U_{es}+\zeta) = \frac{3}{8} V_o\, y^{-1/2} \tag{4-476}$$

and

$$\rho = \frac{1}{32\pi^2} V_o^{3/2}\, y^{-3/2} \exp(-\frac{4}{3} y^{3/2}) \quad , \tag{4-477}$$

so that we have

$$\rho = \frac{16}{27\pi^2}[-(U_{es}+\zeta)/V_o^{1/2}]^3 \exp\{-\frac{9}{128}[\frac{V_o}{-(U_{es}+\zeta)}]^3\} \quad . \tag{4-478}$$

Here it is: "a density-potential relation (410) that is decreasing, at the edge, more rapidly than the power law ... but not quite as sudden as the step function ...", which we have been looking for. There is a price for the intrinsic simplicity of the MES model: it contains a parameter, V_o, for which we do not know a reasonable value beforehand, except for the restriction (472), where one should expect sensible values to be considerably larger than this absolute bound. Such an additional parameter, U_o, was already present in the ES model. That U_o could be identified as the minimal binding energy of neutral atoms, about which we have no independent accurate knowledge [notwithstanding the vague statement (306)]. The new parameter V_o can be given a hand-waving interpretation in terms of the gradient of the potential near the atomic edge, as suggested by (396). Therefore, the details of the shape of the density, for instance, will depend sensitively on V_o, and so will the polarizabilities. This offers the possibility of adjusting V_o by comparison with some experimentally, or for this matter independently theoretically, known quantity.

The MES model just proposed has not yet been tested numerically, except for some preliminary results that look encouraging. An extensive study is certainly necessary, and possibly very rewarding. In doing so, one should not forget the possibility of further modifications, such as choosing two different values for V_o in (450), (451), and (455), (456), with the consequence that y and $\bar{y}$ are no longer identical. For the sake of simplicity, we opted for just one V_o, and this should suffice.

<u>Exchange III.</u> (Exchange potential). In order to go finally beyond the simplified description of the ES model, we need a more realistic exchange potential in the Poisson equation

$$-\frac{1}{4\pi}\nabla^2(V-V_{ex}+\frac{Z}{r}) = n \quad . \tag{4-479}$$

where we recall that the task of expressing the density n in terms of the effective potential V has already been performed. The result is the sum of the density of the innermost electrons, n_{IME}, and the remaining ones, $\tilde{n}$,

$$n = n_{IME} + \tilde{n} \quad , \tag{4-480}$$

which are reported in Eqs.(198) and (210) [or (219)].

It is true that the derivation of the exchange energy (259) employed the TF approximations (251), so that upon inserting the quantum corrected density we shall not obtain the correct quantum corrections to the exchange energy. These, however, are corrections of a correction; at the level of accuracy presently considered they are irrelevant. We feel, therefore, justified in using (259) to find the exchange potential needed in (279). With this in mind, let us change ζ infinitesimally, which leads to

$$\delta E_{ex} = \pi\int(d\vec{r})\delta\zeta\,(\frac{\partial n}{\partial\zeta})^2 \quad , \tag{4-481}$$

the corresponding change of the density being

$$\delta n = \delta\zeta\frac{\partial n}{\partial\zeta} \quad . \tag{4-482}$$

Since variations of E_{ex} identify the exchange potential,

$$\delta E_{ex} = \int (d\vec{r})\, \delta n\, V_{ex} \quad , \tag{4-483}$$

the combination of Eqs.(481) and (482) tells us

$$V_{ex} = \pi \frac{\partial n}{\partial \zeta} \quad , \tag{4-484}$$

where n is thought to be expressed in terms of $V+\zeta$, so that the insertion of (484) into (479) supplies the desired differential equation for V.

Before proceeding, let us briefly illustrate Eq.(484) in the context of Dirac's approximation to E_{ex},

$$E_{ex} = -\int (d\vec{r})\, \frac{1}{4\pi^3}\, [n(\vec{r})]^{4/3} \quad , \tag{4-485}$$

for which V_{ex} is given in (325),

$$V_{ex} = -\frac{1}{\pi}\, (3\pi^2 n)^{1/3} \quad . \tag{4-486}$$

Since (485) is valid in the TF regime, we have to use the TF expression for the density in (484), with the result

$$\begin{aligned} V_{ex} &= \pi \frac{\partial}{\partial \zeta}\left(\frac{1}{3\pi^2}[-2(V+\zeta)]^{3/2}\right) \\ &= -\frac{1}{\pi}[-2(V+\zeta)]^{1/2} = -\frac{1}{\pi}(3\pi^2 n)^{1/3} \quad , \end{aligned} \tag{4-487}$$

indeed.

As a preparation for applying (484) to (480), let us consider the dependence of the functions F_m, as given in Eq.(181),

$$F_m(V,|\vec{\nabla}V|) = \sum_{j=1}^{J} w_j [F_m(y) - F_m(y_j) - (y_j - y) F_{m-1}(y_j)] \quad , \tag{4-488}$$

upon their arguments V and $|\vec{\nabla}V|$, which enter via

$$y = 2(V+\zeta)\ |2\vec{\nabla}V|^{-2/3} \quad , \quad y_j = 2(V+\zeta_j)\ |2\vec{\nabla}V|^{-2/3} \quad . \tag{4-489}$$

Changing the potential and its gradient (locally) induces

$$\delta y = 2|2\vec{\nabla}V|^{-2/3}\, \delta V - \frac{4}{3}|2\vec{\nabla}V|^{-2}\ y\ \delta(\vec{\nabla}V)^2 \tag{4-490}$$

and the corresponding variation of y_j. Consequently,

$$\delta F_m = \sum_{j=1}^{J} w_j \{-[F_{m-1}(y)+F_{m-1}(y_j)]\delta y + (y_j-y)F_{m-2}(y_j)\delta y_j\}$$

$$= -2|2\vec{\nabla}V|^{-2/3} F_{m-1}\,\delta V \tag{4-491}$$

$$+ \frac{4}{3}|2\vec{\nabla}V|^{-2}\,\delta(\vec{\nabla}V)^2 \sum_{j=1}^{J} w_j\{yF_{m-1}(y) - yF_{m-1}(y_j) - (y_j-y)y_jF_{m-2}(y_j)\}$$

or, after utilizing the recurrence relation (157),

$$\delta F_m = -2|2\vec{\nabla}V|^{-2/3} F_{m-1}\,\delta V + \frac{1}{3}|2\vec{\nabla}V|^{-2}[F_{m-3} - (4m-2)F_m]\,\delta(\nabla V)^2 \quad , \tag{4-492}$$

where, as the derivation implies, it is understood that ζ and the ζ_j's are meant to be unchanged. With the standard factor of $|2\vec{\nabla}V|^{(2m-1)/3}$, Eq.(492) reads, more compactly,

$$\delta(|2\vec{\nabla}V|^{(2m-1)/3} F_m) = -2|2\vec{\nabla}V|^{2m/3-1} F_{m-1}\delta V + \frac{1}{3}|2\vec{\nabla}V|^{(2m-7)/3} F_{m-3}\delta(\nabla V)^2 \ . \tag{4-493}$$

Another preparatory remark concerns Eq.(484) itself. There n is to be inserted as a (local) function of $V+\zeta$, $\vec{\nabla}V$, and possibly higher derivatives. Note in particular the fundamental dependence on the sum $V+\zeta$, which has the consequence that the derivative with respect to ζ can be equivalently replaced by $\partial/\partial V$ with the implicit understanding that the derivatives of the potential are kept constant,

$$V_{ex} = \pi \frac{\partial}{\partial V} n(V+\zeta, \vec{\nabla}V, \ldots) \quad . \tag{4-494}$$

In applying this statement to the Scott- and quantum-corrected density (480) we shall disregard the contributions from the innermost electrons, so that we do not take into account the Scott correction to exchange; further we shall be content with the first term of (219). Thus we get

$$V_{ex} \cong \pi \frac{\partial}{\partial V}\left(\frac{1}{2\pi}|2\vec{\nabla}V|F_2\right) = -|2\vec{\nabla}V|^{1/3} F_1(V, |\vec{\nabla}V|) \quad , \tag{4-495}$$

where (493) for m=2 has been used for variations of V only. Inasmuch as F_1 is constructed according to Eq.(488), it contains the typical strong cancellations in the vicinity of the nucleus, and the compensating term referring to the innermost electrons is missing. Consequently, this exchange potential is partly corrected for the strongly bound electrons, but not in a fully consistent way. Here applies the same argument that is valid for the quantum corrections [partly manifest in (495) because of the dependence on the gradient of V], namely that we need not be concerned with corrections to exchange. In other words: for the main body of the atom, where exchange effects make themselves feel, the approximation (495) suffices.

In a certain sense, even the simple expression (495) is still too complicated for our purposes. When inserted into (479), the Laplacian of this V_{ex} exhibits third derivatives of the potential V, whereas the density contains only first and second derivatives, as we observed in the discussion of Eq.(210). If one wants, as we do, to maintain the basic simplicity of the TF approach, one ingredient of which is the low order of the differential equation, then one should aim at a second-order differential equation and look for a sensible approximation of $\nabla^2 V_{ex}$.

Since we must preserve the divergence property of $\nabla^2 V_{ex}$ in order to not destroy the boundary conditions of V, we really need an approximation of the gradient of the exchange potential. This is achieved by treating $\vec{\nabla}V$ like a constant when evaluating $\vec{\nabla}V_{ex}$,

$$\vec{\nabla}V_{ex} \cong \frac{\partial V_{ex}}{\partial V}\vec{\nabla}V = 2|2\vec{\nabla}V|^{-1/3} F_0(V,|\vec{\nabla}V|)\vec{\nabla}V \quad . \tag{4-496}$$

In passing we remark that one can do slightly better if V is spherically symmetric, details being discussed in Ref.47; here we are satisfied with (496). Upon taking the divergence of this equation we get, with the aid of (493) for m=0,

$$\begin{aligned}\nabla^2 V_{ex} &\cong 2|2\vec{\nabla}V|^{-1/3} \nabla^2 V\, F_0 - |2\vec{\nabla}V|\, F_{-1} \\ &\quad + \frac{4}{3}|2\vec{\nabla}V|^{-7/3}\, \vec{\nabla}V\cdot\vec{\nabla}\vec{\nabla}V\cdot\vec{\nabla}V\, F_{-3} \quad ,\end{aligned} \tag{4-497}$$

which is by construction an exact total divergence and contains only first and second derivatives of the effective potential.

<u>New differential equation.</u> With this Laplacian of the exchange potential and with the density of (198) and (210), the Poisson equation (479) appears as

$$-\frac{1}{4\pi}\nabla^2(V+\frac{Z}{r}) = n_{IME} + \tilde{n} - \frac{1}{4\pi}\nabla^2 V_{ex} =$$

$$= n_{IME} + \frac{1}{2\pi}|2\vec{\nabla}V|F_2 - \frac{2}{3\pi}|2\vec{\nabla}V|^{-1/3}\nabla^2V\,F_0$$

$$+ \frac{1}{4\pi}|2\vec{\nabla}V|F_{-1}$$

$$- \frac{1}{18\pi}|2\vec{\nabla}V|^{-5/3}[(\nabla^2V)^2 - \vec{\nabla}\vec{\nabla}V\cdot\cdot\vec{\nabla}\vec{\nabla}V]F_{-2} \qquad (4\text{-}498)$$

$$+ \frac{1}{9\pi}|2\vec{\nabla}V|^{-7/3}[\nabla^2V(\nabla V)^2 - 4\vec{\nabla}V\cdot\vec{\nabla}\vec{\nabla}V\cdot\vec{\nabla}V]F_{-3}$$

$$- \frac{1}{27\pi}|2\vec{\nabla}V|^{-11/3}[\nabla^2V\ \vec{\nabla}V\cdot\vec{\nabla}\vec{\nabla}V\cdot\vec{\nabla}V - \vec{\nabla}V\cdot\vec{\nabla}\vec{\nabla}V\cdot\vec{\nabla}\vec{\nabla}V\cdot\vec{\nabla}V]F_{-5} \quad .$$

This second-order differential equation for V is as a matter of fact linear in the second derivative if one adopts a coordinate system such that V depends only on one of the (orthogonal) coordinates. We illustrate this for the situation of spherical symmetry, $V=V(r)$, as is fitting for the application to an isolated atom for which (498) is actually written.

In spherical coordinates, we have for $V=V(r)$,

$$\vec{\nabla}V = \frac{\partial V}{\partial r}\frac{\vec{r}}{r} \quad ,$$

$$\vec{\nabla}\vec{\nabla}V = \frac{\partial^2 V}{\partial r^2}\frac{\vec{r}}{r}\frac{\vec{r}}{r} + (\overleftrightarrow{1} - \frac{\vec{r}}{r}\frac{\vec{r}}{r})\frac{1}{r}\frac{\partial V}{\partial r}$$

$$= \nabla^2V\frac{\vec{r}}{r}\frac{\vec{r}}{r} + (\overleftrightarrow{1} - 3\frac{\vec{r}}{r}\frac{\vec{r}}{r})\frac{1}{r}\frac{\partial V}{\partial r} \quad , \qquad (4\text{-}499)$$

$$\nabla^2V = \frac{\partial^2 V}{\partial r^2} + \frac{2}{r}\frac{\partial V}{\partial r} = \frac{1}{r}\frac{\partial^2}{\partial r^2}(rV) \quad ,$$

from which we deduce

$$(\nabla^2V)^2 - \vec{\nabla}\vec{\nabla}V\cdot\cdot\vec{\nabla}\vec{\nabla}V = \frac{4}{r}\frac{\partial V}{\partial r}\nabla^2V - 2(\frac{1}{r}\frac{\partial V}{\partial r})^2 \quad ,$$

$$\nabla^2V(\vec{\nabla}V)^2 - 4\vec{\nabla}V\cdot\vec{\nabla}\vec{\nabla}V\cdot\vec{\nabla}V = -3(\frac{\partial V}{\partial r})^2\nabla^2V + \frac{8}{r}(\frac{\partial V}{\partial r})^3 \quad , \qquad (4\text{-}500)$$

$$\nabla^2V\ \vec{\nabla}V\cdot\vec{\nabla}\vec{\nabla}V\cdot\vec{\nabla}V - \vec{\nabla}V\cdot\vec{\nabla}\vec{\nabla}V\cdot\vec{\nabla}\vec{\nabla}V\cdot\vec{\nabla}V$$

$$= \frac{2}{r}(\frac{\partial V}{\partial r})^3\nabla^2V - \frac{4}{r^2}(\frac{\partial V}{\partial r})^4 \quad .$$

Indeed, these right-hand sides are linear in ∇^2V.

At this stage, we have

$$-\frac{1}{4\pi}\nabla^2(V + \frac{Z}{r}) = \{n_{IME} + \frac{1}{2\pi}|2\vec{\nabla}V|F_2 + \frac{1}{4\pi}|2\vec{\nabla}V|F_{-1} +$$

$$
\begin{aligned}
&+ \frac{1}{36\pi}\frac{1}{r^2}|2\vec{\nabla}V|^{1/3}F_{-2} \pm \frac{1}{9\pi}\frac{1}{r}|2\vec{\nabla}V|^{2/3}F_{-3} \\
&+ \frac{1}{108\pi}\frac{1}{r^2}|2\vec{\nabla}V|^{1/3}F_{-5}\} \\
+ \frac{1}{4\pi}\nabla^2 V\{&-\frac{8}{3}|2\vec{\nabla}V|^{-1/3}F_0 \mp \frac{4}{9}\frac{1}{r}|2\vec{\nabla}V|^{-2/3}F_{-2} \\
&-\frac{1}{3}|2\vec{\nabla}V|^{-1/3}F_{-3} \mp \frac{1}{27}\frac{1}{r}|2\vec{\nabla}V|^{-2/3}F_{-5}\} ,
\end{aligned}
\tag{4-501}
$$

where the upper (lower) sign refers to $\partial V/\partial r > 0\,(<0)$ and originates in

$$\frac{\partial V}{\partial r} = \pm\frac{1}{2}|2\vec{\nabla}V| . \tag{4-502}$$

(Under the standard circumstances only the upper sign will occur. Since, however, these equations could be applied to negative ions, where $\partial V/\partial r$ changes sign beyond the atomic edge, we keep the two signs for the sake of completeness.) The factor multiplying $\nabla^2 V$ on the right-hand side of (501) vanishes thoroughly at r=0 which permits the replacement

$$\nabla^2 V \rightarrow \nabla^2\left(V + \frac{Z}{r}\right) . \tag{4-503}$$

After solving for this Laplacian, the <u>new differential equation</u> is obtained in its final form:

$$-\nabla^2\left(V + \frac{Z}{r}\right) = \mathit{Num}/\mathit{Den} , \tag{4-504a}$$

where the numerator and denominator are

$$
\begin{aligned}
\mathit{Num} = 4\pi n_{IME} &+ |2\vec{\nabla}V|(2F_2 + F_{-1}) \\
&+ \frac{1}{27}\frac{1}{r^2}|2\vec{\nabla}V|^{1/3}(3F_{-2} + F_{-5}) \\
&\pm \frac{4}{9}\frac{1}{r}|2\vec{\nabla}V|^{2/3}F_{-3} ,
\end{aligned}
\tag{4-504b}
$$

and

$$
\begin{aligned}
\mathit{Den} = 1 &- \frac{1}{3}|2\vec{\nabla}V|^{-1/3}(8F_0 + F_{-3}) \\
&\mp \frac{1}{27}\frac{1}{r}|2\vec{\nabla}V|^{-2/3}(12F_{-2} + F_{-5}) .
\end{aligned}
\tag{4-504c}
$$

One must certainly admit that this, although being an extension of the TF differential equation (2-48), no longer has any striking resemblance to it. However, there are common features: first, it is a second-order differential equation; second, it is one equation for all systems, as compared to HF formulations, where going from N to N+1 changes the number of functions to be found; third, different N and Z enter the problem via the boundary conditions, without direct effect on the differential equation.

There are enormous differences, too. But they are of a more technical natur. In the case of the TF equation, the numerical challenge was merely to find V and ζ such that the differential equation along with the boundary conditions

$$r\,V \rightarrow \begin{cases} -Z & \text{as} \quad r \rightarrow 0 \\ -(Z-N) & \text{as} \quad r \rightarrow \infty \end{cases} \qquad (4\text{-}505)$$

was satisfied. Now, we encounter additional complications because of the special treatment of the strongly bound electrons. The new parameters ζ_j and Q_j, that are implicit in n_{IME}, are given in terms of integrals involving the potential. These are Eqs.(3-50) and (197), respectively. The numerical procedure for handling these parameters was already described in Chapter Three in the context of the TFS model where the same complication is present. The main change from what is done there comes from the abundance of Airy functions in the new differential equation (504). That makes it numerically more involved (and more expensive), but again this is not a fundamental departure from the TF equation.

Before proceeding with the discussion of Eq.(504) and its implications, I must point out that this equation is not identical with the one obtained and studied in Ref.47, which produced the numbers of Table 7. The differences between these two equations arise primarily from the use of the energy functional (186) instead of (183) in Ref.8. A minor change originated in additional terms in the approximation to $\vec{\nabla} V_{ex}$, Eq.(496), that are made use of in Ref.47. The numerical results obtained with Eq.(504) do not differ substantially from the ones of Ref. 47 (the diamagnetic susceptibilities are somewhat larger here), but the time needed for the numerical computation is about one third less for the new differential equation than it is for the older one of Ref.47. [Since from the numerical point of view the main difference consists in the reduction of the number of F_m's from 14 in Ref.47 to 9 in (504), one can safely infer that a substantial amount of the computation time is spent on the evaluation of the F_m functions.] The following discussion will focus upon the new differential equation (504); the reader interested in a comparison with the older one is referred to Ref.47 for details.

Small distances. When approaching the site of the nucleus at r=0, we encounter the strong cancellations that are inherent in the structure (488). In the F_m's, y and the y_j's are for small r given by

$$y, y_j \cong 2\left(-\frac{Z}{r}\right)\left(\frac{2Z}{r^2}\right)^{-2/3} = -(2Zr)^{1/3} \tag{4-506}$$

[cf. Eq.(173)], their difference being

$$\begin{aligned} \Delta y_j &\equiv y_j - y = 2(\zeta_j-\zeta)\left(\frac{2Z}{r^2}\right)^{-2/3} \\ &= \frac{\zeta_j-\zeta}{2Z^2}(2Zr)^{4/3} \quad . \end{aligned} \tag{4-507}$$

This implies that the F_m's, for $r \to 0$, behave like

$$\begin{aligned} F_m &= \sum_j w_j\left[F_m(y_j-\Delta y_j) - F_m(Y_j) - \Delta y_j F_{m-1}(y_j)\right] \\ &= \sum_j w_j\left[\frac{1}{2!}F_{m-2}(y_j)(\Delta y_j)^2 + \frac{1}{3!}F_{m-3}(y_j)(\Delta y_j)^3 + \dots\right] \\ &\cong \frac{1}{8}(2Zr)^{8/3}\, F_{m-2}(0) \sum_j w_j\left(\frac{\zeta_j-\zeta}{Z^2}\right)^2 \quad . \end{aligned} \tag{4-508}$$

We insert this into (504b) and (504c) and learn that the denominator approaches unity as $r \to 0$, whereas the numerator has contributions from n_{IME} and from the term which possesses the factor $(1/r^2)\,|2\vec{\nabla}V|^{1/3} \sim r^{-8/3}$. Thus

$$\begin{aligned} &-\nabla^2\left(V + \frac{Z}{r}\right) \\ &\cong 4\pi n_{IME}(0) + \left(\frac{Z}{3}\right)^3\left(3F_{-4}(0)+F_{-7}(0)\right) \sum_j w_j\left(\frac{\zeta_j-\zeta}{Z^2}\right)^2 \quad , \end{aligned} \tag{4-509}$$

for $r \to 0$. This tells us that the potential is perfectly well behaved in the vicinity of the nucleus.

Correspondingly, the density at r=0 has contributions from n_{IME} and from the term in (210) involving F_{-2} and F_{-4}. These are

$$\begin{aligned} n_o &= n(r=0) = n_{IME}(0) + \tilde{n}(0) \\ &= n_{IME}(0) + \frac{(2Z)^3}{4\pi}\left[\frac{1}{72}F_{-4}(0) + \frac{1}{216}F_{-7}(0)\right] \sum_j w_j\left(\frac{\zeta_j-\zeta}{Z^2}\right)^2 \quad . \end{aligned} \tag{4-510}$$

Clearly, the (small) term $\tilde{n}(0)$ represents the s-electrons in the outer atomic shells. Here, $n_{IME}(0)$ is given by the expression found in Eq. (3-77),

$$n_{IME}(0) = \frac{(2Z)^3}{4\pi}\left[\sum_{n'=1}^{n_{s-}} \left(\frac{1}{n'}\right)^3 + \sum_j w_j \frac{Q_j}{2n_j^{\,5}}\right] , \tag{4-511}$$

and in $\tilde{n}(0)$ we can employ the recurrence relation (157) for y=0,

$$F_{m-3}(0) = (4m-2)F_m(0) , \tag{4-512}$$

to establish the identity

$$\begin{aligned} \frac{1}{72} F_{-4}(0) + \frac{1}{216} F_{-7}(0) &= \frac{5}{12} F_{-1}(0) = \frac{5\sqrt{3}}{36\pi} \\ &= 0.1326\ldots , \end{aligned} \tag{4-513}$$

which uses[52]

$$\begin{aligned} F_{-1}(0) &= 2[\mathrm{Ai}(0)]\times[-\mathrm{Ai}'(0)] \\ &= 2\left[\frac{3^{-1/6}}{2\pi}\left(-\frac{2}{3}\right)!\right]\left[\frac{3^{1/6}}{2\pi}\left(-\frac{1}{3}\right)!\right] \\ &= \frac{1}{2\pi}\,\frac{\left(-\frac{2}{3}\right)!\left(-\frac{1}{3}\right)!}{\pi} = \frac{1}{2\pi}\,\frac{1}{\sin(\pi/3)} \\ &= \frac{1}{\pi\sqrt{3}} . \end{aligned} \tag{4-514}$$

Upon inserting (511) and (513) into (510), we have

$$n_o \Big/ \frac{(2Z)^3}{4\pi} = \sum_{n'=1}^{n_{s-}} \left(\frac{1}{n'}\right)^3 + \sum_{j=1}^{J} w_j\left[\frac{Q_j}{2n_j^{\,5}} + \frac{5\sqrt{3}}{36\pi}\left(\frac{\zeta_j-\zeta}{Z^2}\right)^2\right] , \tag{4-515}$$

which is the statistical-model prediction for the electron density at the site of the nucleus.

One could now improve the TFS prediction (3-166) by inserting into (515) what here replaces Eqs.(3-135) and (3-139). This has not been done as yet.

Large distances. We turn to the region of large distances where all terms that refer to the innermost electrons are effectively zero. Thus, all $\mathcal{F}_m$'s are now just $F_m(y)$'s, and n_{IME} is absent. By "large r" we mean distances sufficiently far beyond the edge. That becomes more concrete by stating that the potential V is small compared to ζ, allowing the approximation

$$y = 2(V+\zeta)\,|2\vec{\nabla}V|^{-2/3} \cong 2\zeta\,|2\vec{\nabla}V|^{-2/3} \quad . \tag{4-516}$$

In conjunction with the asymptotic form (166), this produces

$$F_m(y) \cong \frac{1}{2\pi}(8\zeta)^{-(m+1)/2}\,|2\vec{\nabla}V|^{(m+1)/3} \times \exp\left(-\frac{2}{3}\,\frac{(2\zeta)^{3/2}}{|\vec{\nabla}V|}\right) \quad . \tag{4-517}$$

The large-r behavior of V, as enforced by (505), implies that, for ions, $|\vec{\nabla}V|$ approaches zero like $1/r^2$ and even faster for neutral atoms. Consequently, in a sum of F_m's the one with the most negative subscript dominates.

When applying this observation to the new differential equation (504), we find that, for $r \to \infty$, the numerator and denominator are, respectively, given by

$$\begin{aligned} Num &\cong \frac{32}{27\pi}\,\zeta^2\,\frac{1}{r^2}|2\vec{\nabla}V|^{-1}\,\exp\left(-\frac{2}{3}(2\zeta)^{3/2}/|\vec{\nabla}V|\right) \\ Den &\cong 1 \quad , \end{aligned} \tag{4-518}$$

so that the asymptotic shape of the potential is determined by

$$-\nabla^2 V \cong \frac{32}{27\pi}\,\zeta^2\,\frac{1}{r^2}|2\vec{\nabla}V|^{-1}\,\exp\left(-\frac{2}{3}(2\zeta)^{3/2}/|\vec{\nabla}V|\right) \quad . \tag{4-519}$$

For an ionized system, where $V(r) \to -(Z-N)/r$, the additional information thus obtainted is

$$V(r) \cong -\frac{Z-N}{r}\left[1 + \frac{1}{24\pi}\,\frac{1}{r}\,\frac{1}{\zeta}\,\exp\left(-\frac{2}{3}(2\zeta)^{3/2}\,\frac{r^2}{Z-N}\right)\right] \quad , \tag{4-520}$$

which is not particularly remarkable. For a neutral atom, the unknown asymptotic behavior of V(r) emerges from (519), at least in principle, if not in practice. To see what is involved, let us write

$$|\vec{\nabla}V| \equiv \frac{2}{3}(2\zeta)^{3/2}\,h(\rho)/\rho^2 \quad , \quad \rho \equiv 6\pi\zeta r \quad , \tag{4-521}$$

with the consequences

$$\begin{aligned} \nabla^2 V &= \frac{1}{r^2}\frac{\partial}{\partial r}(r^2|\vec{\nabla}V|) = 2\pi(2\zeta)^{5/2}\,\frac{1}{\rho^2}\,\frac{\partial}{\partial\rho}\,h(\rho) \quad , \\ \frac{32}{27\pi}\,\zeta^2\,\frac{1}{r^2}\,|2\vec{\nabla}V|^{-1} &= 2\pi(2\zeta)^{5/2}\,\frac{1}{h(\rho)} \quad , \end{aligned} \tag{4-522}$$

which turn (519) into

$$\frac{d}{d\rho}\, h(\rho) \;=\; -\,\frac{\rho^2}{h(\rho)}\,\exp\left(-\,\frac{\rho^2}{h(\rho)}\right) \quad . \tag{4-523}$$

What we need to know is: <u>How</u> does the solution of (523), obeying the boundary condition $h(\rho\to\infty)=0$, approach zero as $\rho\to\infty$? This would tell us the asymptotic shape of the neutral-atom effective potential in the statistical model. Unfortunately, I have not been able to extract this valuable information. Without this knowledge, however, the differential equation cannot be easily integrated for neutral atoms. The standard inward integration is not feasible because of this lack of initial values. [This is different from the situation in the TFS model, where one knew that the neutral-atom potential at large distances is a rescaled TF potential; see Eq.(3-171).] On the other hand, a simple outward integration, beginning with a trial value for the additive constant in

$$V \to -\,\frac{Z}{r} + \text{const} \;, \quad \text{for} \quad r\to 0 \quad , \tag{4-524}$$

is unstable due to its sensitivity to round-off errors; and a mixed strategy of integrating in both directions from an intermediate point would introduce two more numerical parameters (such as the value of the potential and its gradient at the intermediate point). Therefore, we resort to extrapolating the N=Z data from results obtained for ions in the manner described in Chapter Three. For this extrapolation we use the three ions which have fixed N and these values of Z-N:

$$\nu \equiv Z-N = \frac{1}{2}, \frac{1}{5}, \frac{1}{10}, \equiv \nu_1, \nu_2, \nu_3 \quad . \tag{4-525}$$

Suppose the quantity to be calculated is denoted by $\mu(\nu)$ and we possess the three numbers $\mu_j = \mu(\nu_j)$. A reliable procedure for the extrapolation to $\nu=0$ is, as experience indicates, supplied by representing the data in terms of an algebraic function,

$$\mu(\nu) \;=\; \frac{\mu_o + \lambda_1\nu}{1+\lambda_2\nu} \quad . \tag{4-526}$$

The extrapolated value $\mu_o = \mu(\nu=0)$ is then

$$\mu_o \;=\; \frac{\mu_1\nu_2\nu_3(\mu_2-\mu_3)+\mu_2\nu_3\nu_1(\mu_3-\mu_1)+\mu_3\nu_1\nu_2(\mu_1-\mu_2)}{\nu_2\nu_3(\mu_3-\mu_3) + \nu_3\nu_1(\mu_3-\mu_1) + \nu_1\nu_2(\mu_1-\mu_2)} \tag{4-527}$$

or,

$$\mu_o \;=\; \frac{5\mu_1\mu_2 + 3\mu_2\mu_3 - 8\mu_3\mu_1}{-5\mu_3 - 3\mu_1 + 8\mu_2} \tag{4-528}$$

when the ν_j of (525) are inserted.

This discussion of the asymptotic shape of the effective potential assumed implicitly that $\zeta < 0$ for neutral atoms. We now demonstrate why it cannot be zero for $N = Z$. Suppose it were. Then y would certainly be negative for large r, with two possible limits:

$$y = 2V|2\nabla V|^{-2/3} \to \begin{cases} y_o \le 0, & (4\text{-}529a) \\ -\infty, & (4\text{-}529b) \end{cases} \quad \text{as } r \to \infty .$$

The exchange potential (495) is, outside the region of the strongly bound electrons, given by

$$V_{ex} = -|2\vec{\nabla}V|^{1/3} F_1(y) = V \frac{2}{|2\vec{\nabla}V|^{1/3}} (-\frac{1}{y}) F_1(y) \quad . \qquad (4\text{-}530)$$

Since $F_1(y)$ is positive for all y, the realization of (529a) would imply that V_{ex} becomes an arbitrarily large multiple of V as $r \to \infty$. This is physically unacceptable, so that (529a) cannot be the situation. If however, (529b) happens, then the asymptotic form of $F_o(y)$, (Eq.167), inserted into (496) produces

$$\vec{\nabla}V_{ex} = \vec{\nabla}V \frac{1}{\pi} (-2V)^{-1/2} [1+\sin(\tfrac{4}{3}(-y)^{3/2})] \quad . \qquad (4\text{-}531)$$

Here we see that $\vec{\nabla}V_{ex}$ is arbitrarily larger than $\vec{\nabla}V$, as $r \to \infty$; again we encounter an unphysical behavior. So (529b) is equally discredited.

The lesson learned here is that, indeed, ζ is positive for neutral atoms; and V, $\vec{\nabla}V$, and $\nabla^2 V$ exceed V_{ex}, $\vec{\nabla}V_{ex}$, and $\nabla^2 V$ by amounts that are basically controlled by the exponential factor in (517), as $r\to\infty$. Further, we observe that $\zeta=0$ applies to a negative ion, one with an excess of electrons. Here, for instance, Eq.(530) immediately predicts $|V_{ex}| \ll |V|$ for large r, because y is now a large positive number under these circumstances. Numerical solutions corresponding to this situation of negative ions have not been calculated as yet. Of course, the excess of charge must be at least equal to one for $\zeta=0$; otherwise the solutions with $\zeta < \zeta(Z=N)$ are physically meaningless. This is, indeed, what seems to happen, when the extrapolation to find $\zeta(Z=N)$ is extended to yield N-Z when $\zeta=0$. For example, one gets $N-Z \cong 0.035$ for N=36 and $\zeta=0$. We infer that negative ions very probably do not exist in the statistical model, as they do not in the TF and TFS models.

Numerical results. We shall now report numerical results for the inert gases Z=N = 18,36,54, and for the related ions with Z=N+1 and Z=N+2.

The numerical procedure described in Chapter Three requires a choice to be made for the number n_s of Bohr shells of strongly bound electrons which we want to treat in the special way. Going with it is the number J of representative ζ_j's and the weights w_j. As in the numerical study of the TFS model we are content with the simplest choice J=2, $w_1=w_2=1/2$. In principle, the value of n_s should be such that N_s, the number of specially handeled innermost electrons, obeys the relation (3-14)

$$1 \ll N_s \ll Z \quad . \tag{4-532}$$

We are, however, now dealing with realistic, and therefore rather modest values of Z and N, and it is quite impossible to take (532) very seriously. The best one can do is to opt for that N_s which, on a logarithmic scale, is halfway between 1 and Z. Another way of stating this is to say that we choose that N_s which is closest to the square root of Z(≅N). For N=18 and N=54, the answer is unambiguous: N_s=1 and 2, respectively. In the N=36 systems, both N_s=2 and N_s=10 are equally distant from 6, the square root of N (or Z); we vote for n_s=1, N_s=2 in order to avoid the danger of overcorrecting for the strongly bound electrons.

The results thus obtained are displayed in Table 9. The numbers in the column IS are the initial slopes $(V+Z/r)/Z^2$ essentially the additive constant of (524). It is not necessary to comment on these num-

Table 4-9. Values of the parameters ζ, ζ_1, ζ_2, Q_1, Q_2, and of the initial slope (IS) as obtained for N=18, 36, and 54, Z=N, N+1, and N+2 in the statistical model.

N	Z	ζ	ζ_1	ζ_2	Q_1	Q_2	IS
	18	0.0095	106.8	7.24	0.745	3.25	0.2239
18	19	0.35901	121.14	9.164	0.7517	3.316	0.21513
	20	0.84605	136.56	11.379	0.7582	3.375	0.20691
	36	0.0094	498.6	61.12	0.823	4.10	0.1430
36	37	0.33467	529.42	66.103	0.8251	4.137	0.14027
	38	0.76868	561.29	71.377	0.8276	4.168	0.13761
	54	0.0093	152.1	14.96	4.42	10.5	0.1176
54	55	0.31170	160.14	16.871	4.441	10.55	0.11598
	56	0.70229	168.50	18.939	4.463	10.62	0.11444

bers, so we only remind the reader of the fact that entries referring to neutral atoms are the outcome of the extrapolation procedure discussed above, and are therefore less reliable.

Next, we look at various plots of radial densities

$$D(r) = 4\pi r^2 n(r) \quad , \tag{4-533}$$

which are always presented with the abscissa linear in the square root of r, in order to stretch the small-r region where the density curves have most of their structure. We begin, in Fig.9, with the radial densities of the potassium (N=18=Z-1), the rubidium (N=36=Z-1), and the cesium (N=54=Z-1) ions. We see that the statistical model yields a variety of shapes for the electronic densities of systems with different

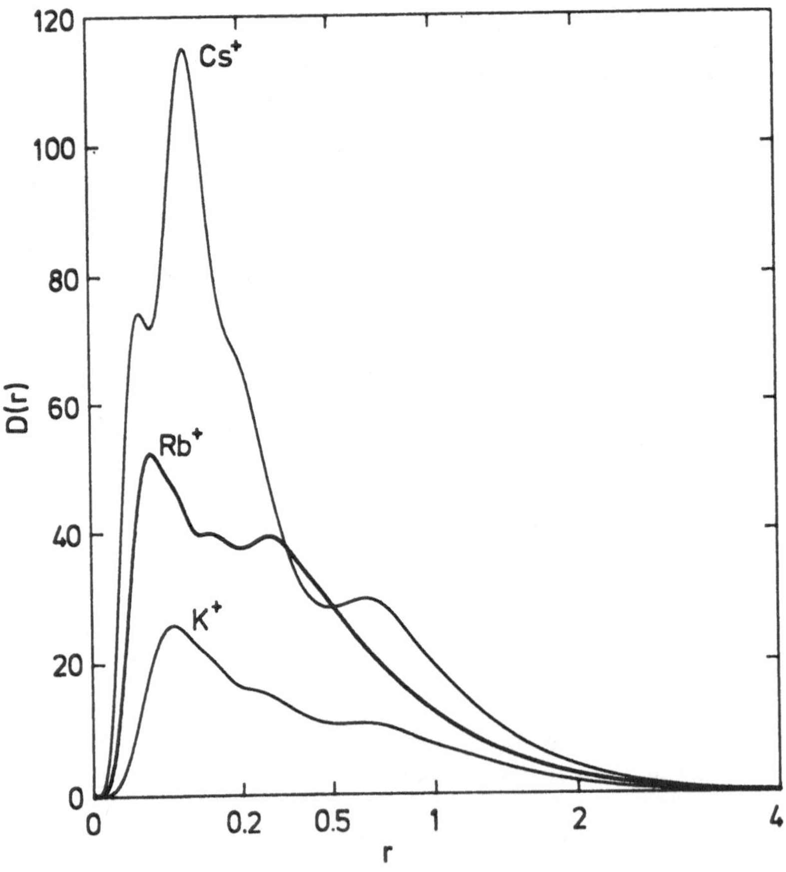

Fig.4-9. Radial electron densities of K^+, Rb^+, and Cs^+. Abscissa is linear in the square root of r.

N and Z; in contrast, the TF model gives a uniform look - now there is a lot of individuality. The K^+ ion has an almost structureless density spread out over a large volume. The density of Rb^+ is much more localized and has somewhat more structure. For Cs^+ we get a smooth, well-concentrated main peak accompanied by a smaller one which is farther away from the nucleus. The obvious question is now: How do these densities compare with those obtained by HF calculations?

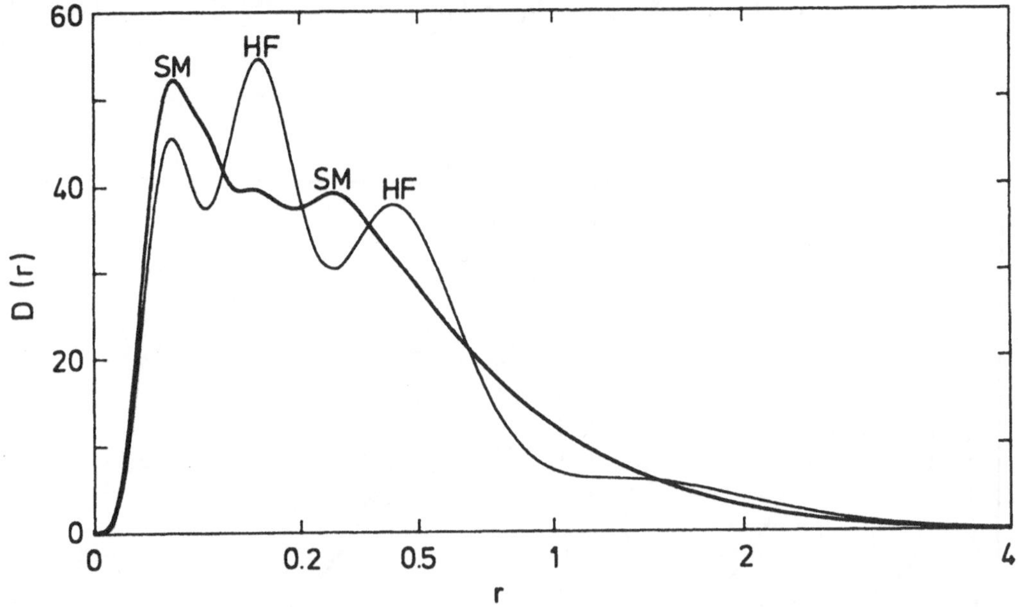

Fig.4-10. Radial electron density in Rb^+*. Comparison of HF prediction with that of the statistical model (SM). Abscissa is linear in the square root of* r.

For the comparison with the HF prediction, we pick the Rb^+ ion - it is the most striking example. Figure 10 shows the differences.[53] The two densities agree only in the domain of the strongly bound electrons. We observe quite different peak structures and notice that at large distances the HF density is significantly larger. Unfortunately, there is no simple way of telling which one is closer to reality because electron densities cannot be measured directly. One can, of course, compare derived quantities, such as $\overline{r^2}$, the expectation value of the square distance. We refer to Tables 3,6, and 7 (the latter reports the numbers obtained in Ref.47, but no matter).

One should not take the detailed structure of this statistical-model Rb^+ density too literally. In particular, it certainly contains some residual Bohr-shell artifacts that are not removed by the simple average over ζ_s employed in the computation. This is confirmed by Fig.11. It shows the decompositions

$$D = 4\pi r^2 n_{IME} + 4\pi r^2 \tilde{n} \equiv D_{IME} + \tilde{D} \tag{4-534}$$

and

$$D = (-r^2\nabla^2 V) - (-r^2\nabla^2 V_{ex}) \equiv D_V - D_{V_{ex}} \quad . \tag{4-535}$$

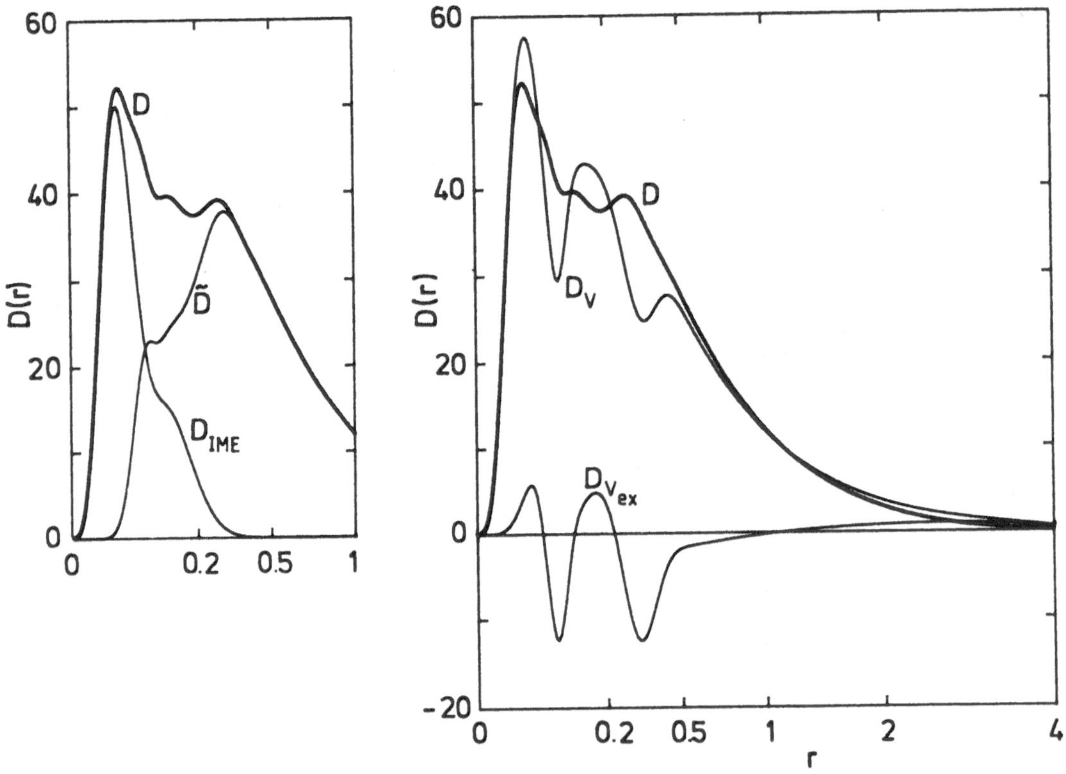

Fig.4-11. Radial electron density of Rb^+. *Left-hand side: decomposition* $D = D_{IME} + \tilde{D}$; *right-hand side: decomposition* $D = D_V - D_{V_{ex}}$. *Abscissa is linear in the square root of* r.

We note that the "fine structure" of the minimum between the two main peaks may well be spurious. Both D_{IME} and $\tilde{D}$ are much smoother, and only the rapid decrease of D_{IME} on top of the equally rapid increase of $\tilde{D}$ produces the wiggles. Likewise said fine structure is made by the interplay of D_V and $D_{V_{ex}}$ which are both smoother, though oscillating.

Figure 11 offers a remarkable observation. Near the edge of the atom ($V+\zeta=0$ at $r=3.34$, see below) the magnitudes of D_V and $D_{V_{ex}}$ are almost equal. Sufficiently beyond this edge, D_V is, of course, much larger than $D_{V_{ex}}$, inasmuch as we find, analogously to the derivation of (519),

$$\nabla^2 V_{ex} \cong -\frac{4}{3\pi}\zeta\frac{1}{r}\exp\left(-\frac{2}{3}(2\zeta)^{3/2}/|\vec{\nabla}V|\right) \quad , \tag{4-536}$$

for large r. This is smaller than $\nabla^2 V$ by a factor of

$$\frac{9}{16}\frac{r}{\zeta^2}|2\vec{\nabla}V| = \frac{9}{8}\frac{Z-N}{\zeta^2}\frac{1}{r} \quad , \tag{4-537}$$

so that D_V exceeds $D_{V_{ex}}$ substantially in the exterior of the atom.

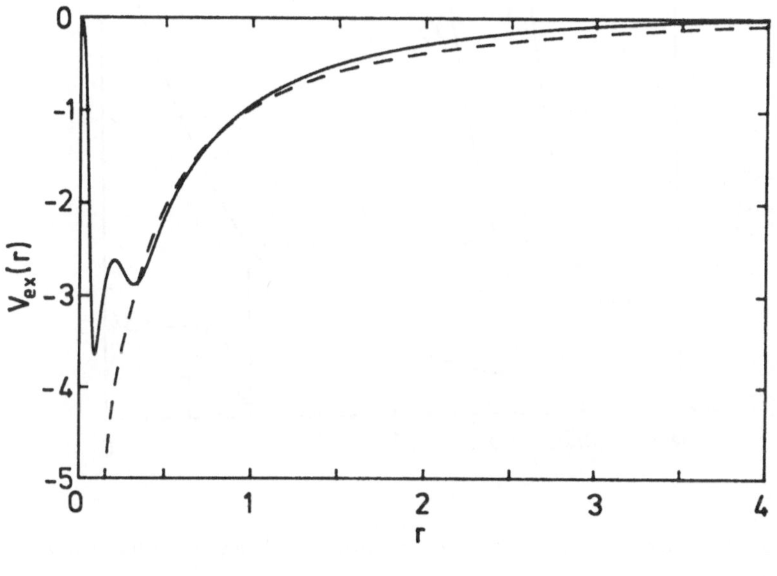

Fig.4-12. Exchange potential for Rb^+ *as a function of the distance* r *from the nucleus.*

This compensation of D_V by $D_{V_{ex}}$ at the edge is a manifestation of the attractive nature of the exchange potential: the density of electrons in the outer reaches of the atom is reduced to the benefit of the interior. Since we are at it, why not take a look at V_{ex} itself ?

In Fig.12, the exchange potential of Rb^+ is plotted as a function of r. It is, indeed, attractive. The dashed curve is the corresponding Dirac-Jensen exchange potential (486). It agrees with the actual one in the range $0.3 \lesssim r \lesssim 1$, which is the TF regime; at larger distances, it significantly exceeds the real V_{ex}, which, as we recall, is the origin of the trouble in the TFD and ES models. At small distances, we observe that V_{ex} tends to zero; this is very likely not a realistic behavior but a consequence of the way the strongly bound electrons are handled here. Following Eq.(495) there is the remark that our V_{ex} contains the typical cancellations near r=0 [cf. Eq.(508)], whereas the compensating term referring to the innermost electrons is missing. Future developments will tell to which extend a modification of V_{ex} is actually necessary.

A last plot is that of y, y_1, and y_2 for Rb^+ in Fig.13, which

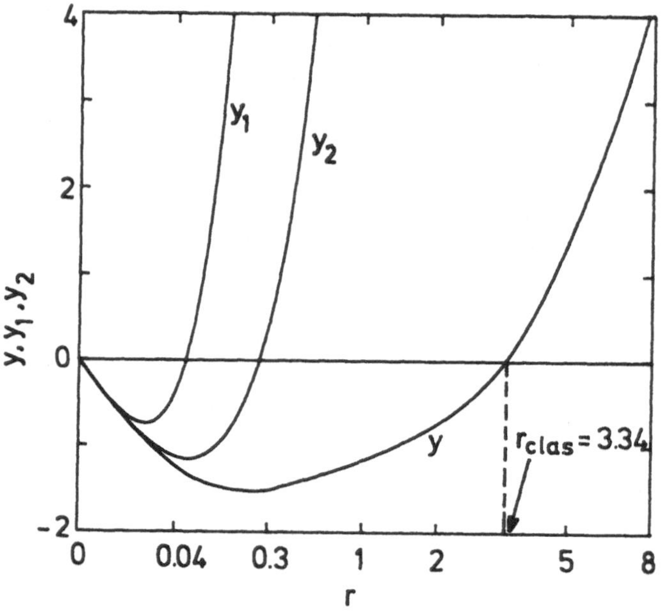

Fig.4-13. y, y_1, *and* y_2 *for* Rb^+ *as a function of* r. *Abscissa is linear in the cubic roof of* r.

should be compared with Fig.3. What was observed then, is confirmed here: y does not become a large negative number. We recall that one needs $y \lesssim -1.5$ for the asymptotic forms of the F_m's to be valid. From this point of view, the TF limit is not expected to be particularly accurate for real atoms; but it is, as the evidence of Fig.6 shows.

The sign change of $V+\zeta$, or equivalently of y, marks the edge of the atom. In Rb^+ it occurs at the distance 3.34 which we call the classical radius r_{clas} of this atomic system. Of course, y_1 and y_2 turn positive far inside the atom, at 0.055 and 0.26, respectively. Outside the domain of the strongly bound electrons, they are very large positive. In Fig.13, the abscissa is chosen linear in the cubic root of r, so that the y(r) functions are straight lines in the plot for $r \to 0$, as is implied by (506). As a consequence, the region of small r is enormously stretched in the plot, which creates the misleading impression, that the sign change of y_1 and y_2 happens at a distance that is a good fraction of r_{clas}. This is an optical illusion.

Problems

4-1. Derive the normalized wave-functions (143) of the constant-force potential from the known time transformation function $\langle\vec{r}',t|\vec{r}'',0\rangle$. [Recall that the approximations (45), (47), and (50) are exact for a linear potential.]

4-2. Find recurrence relations for the polynomials that are mentioned in connection with Eqs.(158) and (159).

4-3. Show that $F_m(y)$ obeys the differential equation

$$\left(\frac{d^3}{dy^3} - 4y\frac{d}{dy} + 4m-2\right)F_m(y) = 0 \quad ,$$

and demonstrate that it is consistent with the asymptotic forms (166), (161), and (170).

4-4. For which complex values of the integration variable x in Eq.(140) is the phase of the integrand stationary? Use this insight to deform the path of integration to contours appropriate for a stationary phase evaluation of the integral for $y \gg 1$ and $y \ll -1$, respectively. Confirm Eqs. (163) and (164).

4-5. Find the minimum of $y = 2V|2\vec{\nabla}V|^{-2/3}$ for the neutral-atom Tietz potential (222). Compare with the corresponding TF result (174).

4-6. Derive Eq.(109) from Eq.(219).

4-7. Demonstrate that $n_V^{(1)}$ and $n_V^{(2)}$, given explicitly in Eqs.(234) and

(236), and related to each other through (242), obey Eq.(231).

4-8. Confirm Eq.(280).

4-9. Expand the right-hand side of Eq.(374) in powers of $(\Psi(y)/y)^{1/2}$. What do you get if you keep only the leading term? Now, keep also the next-to-leading contribution. What is then the asymptotic form of $\Psi(y)$ as $y \to \infty$ for a neutral atom? [More about this in Ref.54.]

4-10. Show that

$$(n+1)F_{n+1}^2(y) = \frac{d}{dy}\left[-\frac{1}{4}F_{n+1}(y)F_{n-1}(y) + \frac{1}{8}F_n^2(y) + \frac{1}{2}yF_{n+1}^2(y)\right] .$$

For $n = -1$, this has the special implication that the contents of the square brackets is a constant. Find this constant. Then use the statement for n=0 to perform the integration over y' in Eq.(399). Demonstrate that the resulting exchange-energy density

$$\frac{V_o^2}{2\pi}\left(-\frac{1}{4}F_{-1}(y)F_1(y) + \frac{1}{8}F_o^2(y) + \frac{1}{2}yF_{n+1}^2(y)\right)$$

[with y determined by n through Eq.(400)] is equal to the Dirac-Jensen result

$$-\frac{1}{4\pi^3}(3\pi^2 n)^{4/3}$$

if n is sufficiently large, or V_o sufficiently small.

4-11. Show that a $\rho'(r)$ of the form (447), inserted into (426), produces a $v(r)$, for which the scale invariant expression (438) gives $\alpha_p = \infty$.

Chapter Five

SHELL STRUCTURE

In the preceding Chapter we obtained, in Eq.(4-267), the statistical-model prediction

$$-E_{stat} = 0.768745\ Z^{7/3} - \frac{1}{2} Z^2 + 0.269900\ Z^{5/3} \qquad (5\text{-}1)$$

for the binding energy of neutral atoms; it is compared with the corresponding integer-Z HF numbers in Fig.4-6. The deviations between E_{stat} and E_{HF}, indiscernible in that plot, shall be the subject of this Chapter.

We begin with a look at the relative deviation between the HF and the statistical-model predictions, presented in Fig.1 for $6 \leq Z \leq 120$

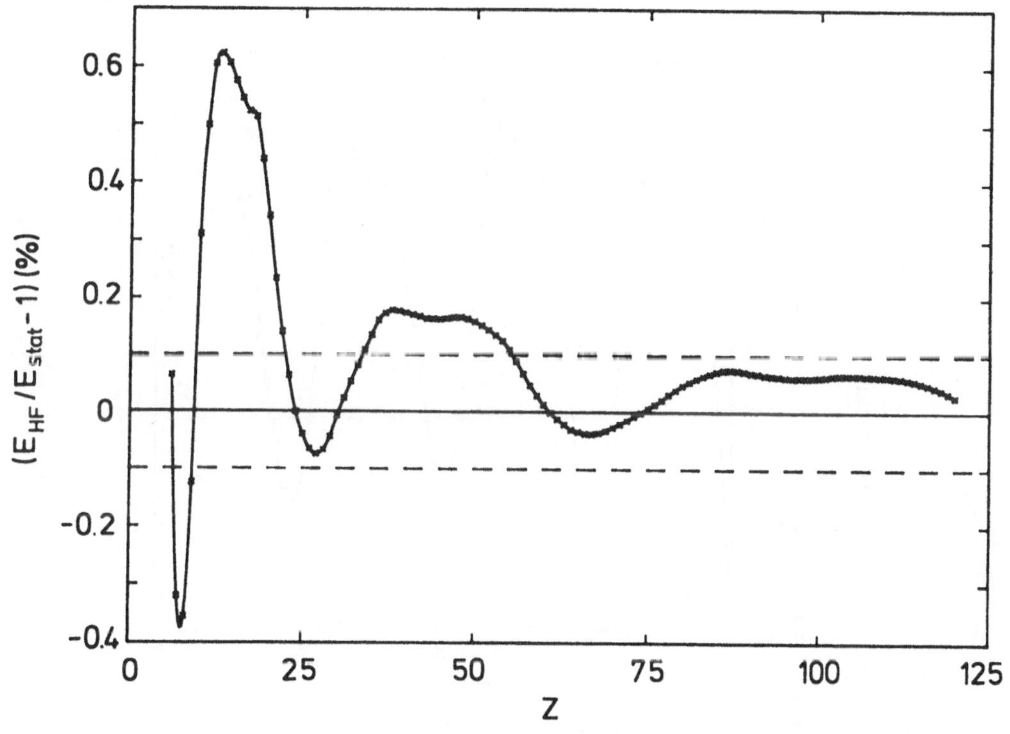

Fig.5-1. Relative deviation, in %, between HF binding energies and the statistical-model prediction (1), as a function of Z, for Z = 6,7,8,..., 120.

(for $Z = 1,2,3,4$, and 5 the respective deviations are -7.2, 4.8, 3.8, 2.3, and 0.9%). One sees that the relative difference between E_{HF} and E_{stat} is less than one percent for $Z \geq 5$, less than one-fifth of a percent for $Z \geq 22$, and less than one-tenth of a percent for $Z \geq 56$. We further observe that this deviation is oscillatory with a period that increases slowly with Z. Denoting the difference between the actual energy and E_{stat}, therefore, by E_{osc},

$$E = E_{stat} + E_{osc} \quad , \tag{5-2}$$

we expect that E_{osc} is an oscillatory function of $Z^{1/3}$ with an amplitude proportional to $Z^{4/3}$, so as to fit into the pattern laid out by Eq.(1). The HF prediction for E_{osc}, that is:

$$\left(E_{osc}\right)_{HF} = E_{HF} - E_{stat} \quad , \tag{5-3}$$

exhibits this anticipated Z-dependence, as demonstrated by Fig.2, which

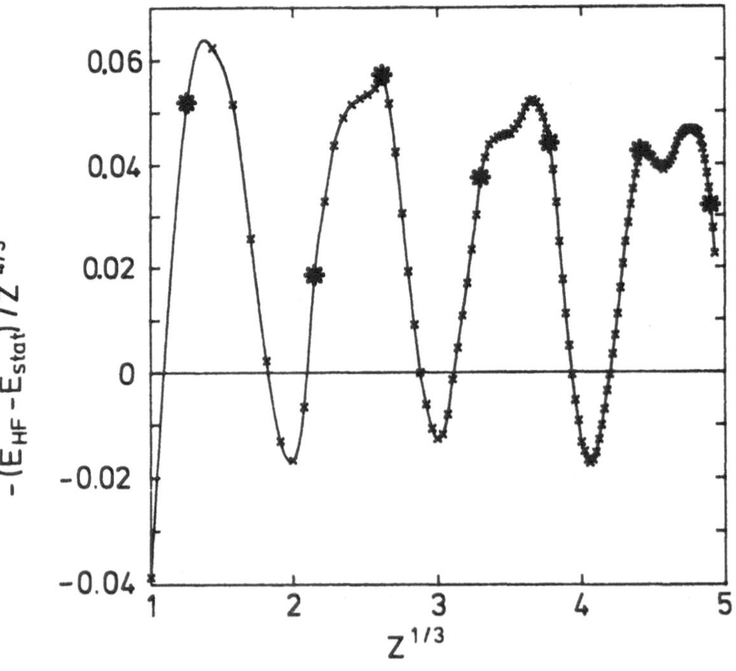

Fig.5-2. Absolute deviation between HF binding energies and the statistical-model prediction (1), divided by $Z^{4/3}$, as a function of $Z^{1/3}$. Stars mark inert-gas atoms.

shows a plot of $-\left(E_{osc}\right)_{HF}/Z^{4/3}$ as a function of $Z^{1/3}$.[1]

These atomic-binding-energy oscillations possess some very remarkable features. Certainly, their extremely high regularity is surprising; there is no essential difference between small and large values of $Z^{1/3}$; even hydrogen ($Z^{1/3}=1$) is no exception. Another particularly interesting property of E_{osc} is that the minima in Fig.2 are sharp and structureless in contrast to the broad maxima with their evolving double-peak structure.

Experimental evidence. When plotting Figs.1 and 2, the HF numbers are, of course, taken seriously. One can, I think, trust these numbers to the necessary five or six significant digits, but there is the possibility that the oscillations of Fig.2 are nothing more than a HF artifact.[2] To make sure that these binding-energy oscillations are a real physical phenomenon, we need the comparison with experimental neutral-atom binding energies. This knowledge has to be extracted from spectroscopic

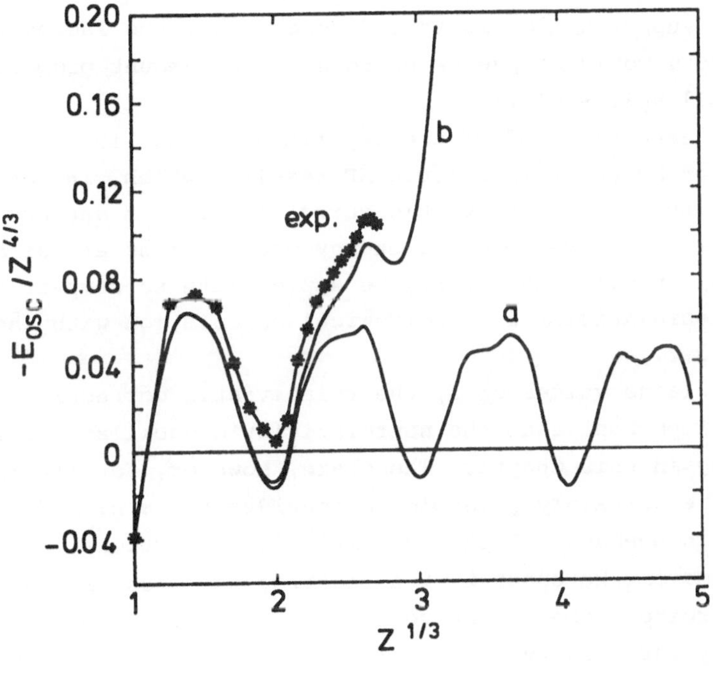

Fig.5-3. Binding energy oscillations. Stars are experimental values for Z = 1,...,20. Curve a shows the nonrelativistic HF oscillations of Fig.2. Curve b connects HF values with relativistic corrections.

data, the analysis of which supplies step-by-step ionization energies.[3] Unfortunately, this has produced binding energies only up to $Z = 20$. For more massive atoms, the ionization potentials after the first 20 electrons are rarely known. In short, we can compare E_{stat} of (1) with reality only for the first 20 members of the Periodic Table. This is done in Fig.3, which displays, in addition to the experimental data, also the nonrelativistic HF oscillations of Fig.2 and, for $Z \leq 31$, the results of HF calculations with relativistic corrections.[4] Please observe two things. First, the experimental values do confirm the existence of binding-energy oscillations. Second, there is, on this scale, a significant discrepancy between experiment and the HF values, even after including relativistic effects. This is a reminder that the HF model is not exact, the lion's share of the missing binding energy being the correlation energy of Eq.(4-248). In Fig.3 the difference between curve b and the stars representing experimental data is roughly a constant, implying the estimate

$$E_{corr} \cong -0.013\, Z^{4/3} \quad , \tag{5-4}$$

which somehow supports the widespread remark that the inclusion of correlation effects would change HF energies by an amount proportional to $Z^{4/3}$ (for small values of Z).

The general trend of the experimental data in Fig.3 is quite well reproduced by the relativistic HF results, which, in conjunction with the fact that the relativistic correction itself depends smoothly on Z, tells us that these binding-energy oscillations are of nonrelativistic origin. It will, therefore, be appropriate to compare the outcome of our nonrelativistic semiclassical calculation with the HF oscillations of Fig.2.

For large values of Z, the relativistic corrections are, evidently, more important than the nonrelativistic oscillations that we are addressing in this Chapter. These are, however, totally different problems. It is certainly possible to consider these binding-energy oscillations independent of the relativistic corrections. We shall therefore at this moment be content with fighting the wrong impression which Fig.3 might create, namely that relativistic effects dominate the total binding energy for large values of Z. As a matter of fact, even in uranium the relativistic correction amounts to less than 10% of the total binding energy;[4] more generally, a rule of thumb, given by Scott,[5] says that "the error in the total binding energy resulting from neglecting relativity is roughly $(Z/30)^2$%."

Qualitative arguments. There is a common reaction to Fig.2 that these oscillations have something to do with the filling of the atomic shells. This notion does, however, not explain a single quantitative detail. In fact, as a class, the inert-gas atoms [He(Z=2),Ne(10),Ar(18),Kr(36),Xe (54),Rn(86), and another one with Z=118, for which the chemists have not invented a name as yet] do not reside on prominent sites of the HF curve in Fig.2. They are, on the other hand, not randomly distributed over the oscillatory curve either, but show a clear tendency toward the maxima and away from the minima. We infer, therefore, that there is a connection between the energy oscillations and the existence of closed atomic shells, notwithstanding that their Z values cannot be predicted by simply looking at Fig.2. These two phenomena are manifestations of one physical effect.

To answer the question what effect that is, let us recall the origin of atomic shells. The reason for their being is the existence of quantum numbers in a spherically symmetric potential: angular quantum number ℓ, and radial quantum number n_r. But ℓ and n_r alone would not account for shells; we also need the fact of energetic degeneracy. States of differing quantum numbers may have almost the same binding energy. This is, of course, familiar for the Coulomb potential where the energies depend only on the principal quantum number $m' = n_r+\ell+1$ [a circumstance that we have made use of in Chapter One, see Eq.(1-10)] leading to the well-known $2m'^2$-fold degeneracy. Thus, in Bohr atoms the maximal radial quantum number equals the maximal angular quantum number. Such is the situation in a highly ionized atom where dynamics is dominated by the nucleus-electron interaction, the interelectronic forces being comparatively small. Not so for neutral atoms, where the ratio of the maximal values of n_r and ℓ is roughly 2:1, provided that Z is sufficiently large. For example, uranium possesses 7s electrons (n_r=6, ℓ=0) and 5f electrons (n_r=1,ℓ=3). The degeneracy of the weakly bound outermost electrons is certainly not of Coulombic type. We can learn more about it from another look at the Periodic Table, this time at the last row. There the 7s, 7p, 6d, and 5f electrons are filled in, but not in a given order, instead they compete with each other - a sure sign of degeneracy. In an ℓ-n_r diagram, Fig.4, these 4 states do not lie on a straight line; degenerate states are connected by bent curves which are the steeper, the larger ℓ is. Deep inside the atom, we expect Coulombic degeneracy for the strongly bound electrons. In this situation, states with the same binding energy do lie on a straight line in the ℓ-n_r diagram. In Fig.4 this is illustrated by the 2s and the 2p state.

It is clear that a theoretical description of the binding-

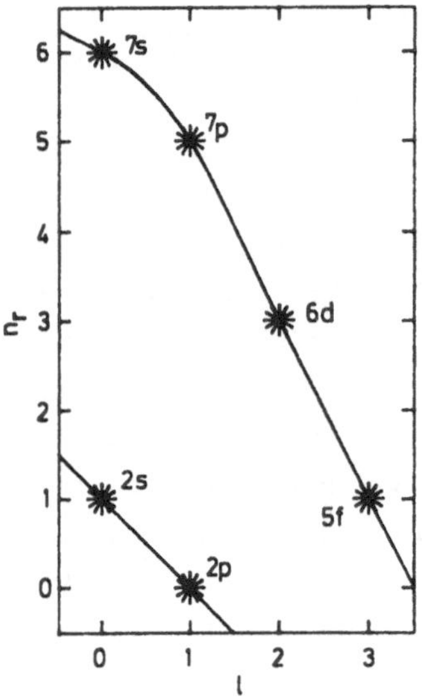

Fig.5-4. Energetic degeneracy in a large atom.

energy oscillations in Fig.2 must be based upon a detailed energetic treatment of those few electrons with least binding energy. This view is supported by the relative size of the effect we are looking for, which is of order $Z^{4/3}/Z^{7/3} = 1/Z$ as compared to the leading TF term in (1), or, like a few electrons compared to the totality of Z electrons in the neutral atom. This should be contrasted with the relative size of the corrections to the TF energy discussed in Chapters Three and Four, namely $Z^{6/3}/Z^{7/3} = Z^{2/3}/Z$ and $Z^{5/3}/Z^{7/3} = Z^{1/3}/Z$, respectively, which we now interpret as showing that large fractions of the total number of electrons contribute to these corrections.

Bohr atoms. As a first step toward a computation of the binding-energy oscillations, we study E_{osc} in the simple situation of NIE. In Chapter One we found that the total binding energy of a Bohr atom is given by

$$-E = Z^2\{y - \frac{1}{2} + \left(\langle y\rangle^2 - \frac{1}{4}\right)\frac{y - \frac{2}{3}\langle y\rangle}{(y-\langle y\rangle)^2}\} \tag{5-5}$$

[this is Eq.(1-20)], where y=y(N) is the solution of (1-12),

$$y^3 - \frac{1}{4}y = \frac{3}{2}N \quad , \tag{5-6}$$

and we recall that $\langle y\rangle$ differs from y by the integer part of $y + 1/2$ [Eq.

(1-14)],

$$\langle y\rangle = y - [y+1/2] \quad , \tag{5-7}$$

so that [Eq.(1-15)]

$$-\frac{1}{2} \le \langle y\rangle < \frac{1}{2} \quad . \tag{5-8}$$

The two-term approximation of (1-21),

$$y = \left(\tfrac{3}{2}N\right)^{1/3} + \frac{1}{12}\left(\tfrac{3}{2}N\right)^{-1/3} \tag{5-9}$$

is actually exact up to order $N^{-4/3}$ (see Problem 1). Consequently, we can exhibit the contributions to E in Eq.(5) up to this order. This calculation employs the identity

$$\frac{y - \frac{2}{3}\langle y\rangle}{(y-\langle y\rangle)^2} = \frac{1}{y}\sum_{k=0}^{\infty} \frac{k+3}{3}\left(\frac{\langle y\rangle}{y}\right)^k \tag{5-10}$$

and results in

$$-E/Z^2 = \left(\tfrac{3}{2}N\right)^{1/3} - \frac{1}{2} - E_{osc}/Z^2 \tag{5-11}$$

with

$$\begin{aligned} -E_{osc}/Z^2 = {} & \left(\tfrac{3}{2}N\right)^{-1/3}\left(\langle y\rangle^2 - \tfrac{1}{6}\right) \\ & + \left(\tfrac{3}{2}N\right)^{-2/3} \tfrac{4}{3}\langle y\rangle\left(\langle y\rangle^2 - \tfrac{1}{4}\right) \\ & + \left(\tfrac{3}{2}N\right)^{-3/3}\left(\langle y\rangle^2 - \tfrac{1}{4}\right)\left(\tfrac{5}{3}\langle y\rangle^2 - \tfrac{1}{12}\right) \\ & + \left(\tfrac{3}{2}N\right)^{-4/3} 2\langle y\rangle\left(\langle y\rangle^2 - \tfrac{1}{4}\right)\left(\langle y\rangle^2 - \tfrac{1}{9}\right) \\ & + \dots \quad , \end{aligned} \tag{5-12}$$

where y, as given in (9), is to be inserted. In $\langle y\rangle$ one must not neglect the second term of (9) because that would produce a wrong phase of the oscillations at smaller values of N.

In Fig.5 the exact amount of

$$\frac{-E_{osc}}{Z^2/N^{1/3}} = N^{1/3}\left(-E/Z^2 - \left(\tfrac{3}{2}N\right)^{1/3} + \tfrac{1}{2}\right) \tag{5-13}$$

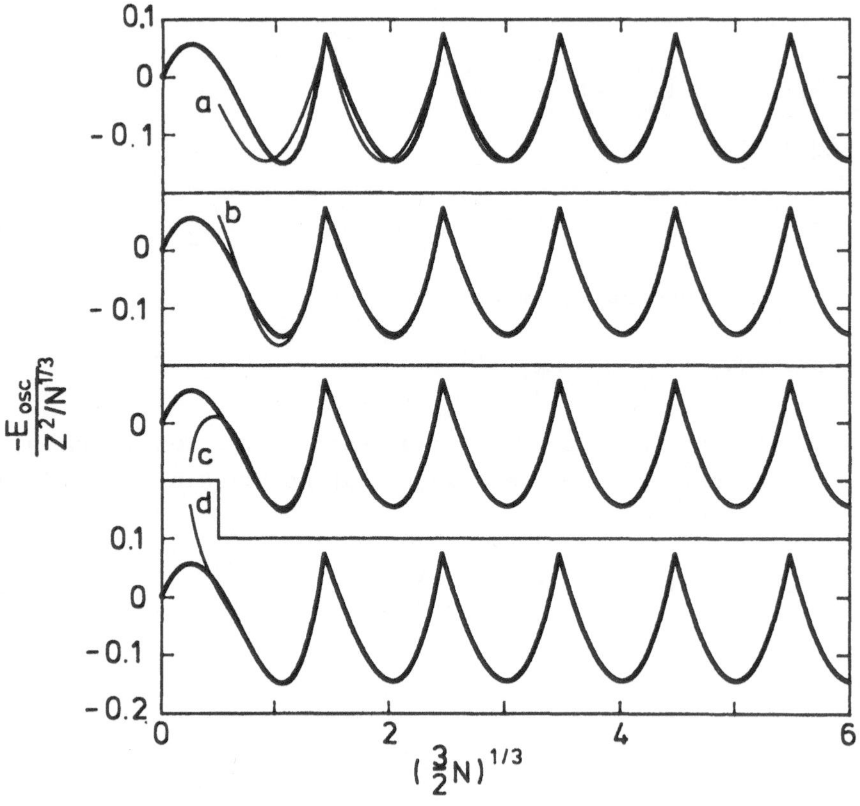

Fig.5-5. Binding-energy oscillation of Bohr atoms as a function of $(\frac{3}{2}N)^{1/3}$. Thick curve is the actual amount. Thinner curves are the successive approximations of Eq.(14): (a) leading term only; (b) terms up to order $N^{-1/3}$; (c) terms up to order $N^{-2/3}$; (d) terms up to order $N^{-3/3}$.

is compared to the successive approximations that result from the expansion (12),

$$\frac{-E_{osc}}{Z^2/N^{1/3}} = \left(\frac{2}{3}\right)^{1/3}\left(\langle y\rangle^2 - \frac{1}{6}\right) + \left(\frac{128}{81}\right)^{1/3}\left(\frac{3}{2}N\right)^{-1/3} \langle y\rangle\left(\langle y\rangle^2 - \frac{1}{4}\right) + \dots \quad . \tag{5-14}$$

During the filling of the first shell, that is $(\frac{3}{2}N)^{1/3} < 3^{1/3} = 1.44$, the exact E_{osc} naturally does not yet show the shape of the large-N oscillations. Note, in particular, how good already is the leading term of (14). Three terms are sufficient to make the difference between the two curves indiscernible for $N \gtrsim 1$, that is $\left(\frac{3}{2}N\right)^{1/3} \gtrsim 1.14$.

So far we have been filling the Coulomb potential with a certain number of electrons. Let us now shift our attention to a different, though related, problem. Consider the Coulombic potential

$$V(r) = -\frac{Z}{r} + E_o \tag{5-15}$$

with the constant E_o determined such that the TF count of negative energy states,

$$N_{TF} = \int (d\vec{r}) \, \frac{1}{3\pi^2} \, [-2V(r)]^{3/2} \quad , \tag{5-16}$$

equals a certain, given multiple of Z,

$$N_{TF} = \kappa Z \quad , \quad \kappa > 0 \quad . \tag{5-17}$$

The integral in (16) is of the structure encountered in (1-34), so that, as in (1-35),

$$N_{TF} = \frac{2}{3}\left(\frac{Z^2}{2E_o}\right)^{3/2} \quad , \tag{5-18}$$

which produces

$$E_o = Z^{4/3}/(18\kappa^2)^{1/3} \quad , \tag{5-19}$$

showing that E_o is proportional to $Z^{4/3}$. The Z-dependence of the potential is, therefore, somewhat more complicated now. It is reminiscent of that of the small-r TF potential; actually both (15) and the TF potential are of the form "$Z^{4/3}$ times a function of $Z^{1/3}r$." More about this in Problem 2.

Let us now study the Z-dependence of the energy when all negative energy states are occupied in the Coulombic potential (15). In this situation, the number of electrons is

$$\begin{aligned} N &= \sum_{m'=1}^{\infty} 2m'^2 \, \eta\left(\frac{Z^2}{2m'^2} - E_o\right) \\ &= \frac{2}{3}\left([\lambda_o] + \frac{1}{2}\right)^3 - \frac{1}{6}\left([\lambda_o] + \frac{1}{2}\right) \quad , \end{aligned} \tag{5-20}$$

where we have introduced

$$\lambda_o \equiv \frac{Z}{\sqrt{2E_o}} = \left(\frac{3}{2}\kappa Z\right)^{1/3} \quad . \tag{5-21}$$

Obviously, the principal quantum number of the last Bohr shell with negative energy is the integer part $[\lambda_o]$ of λ_o. Likewise, the binding energy is

$$-E = \sum_{m'=1}^{\infty} 2m'^2\left(\frac{Z^2}{2m'^2} - E_o\right)\eta\left(\frac{Z^2}{2m'^2} - E_o\right)$$

$$= Z^2[\lambda_o] - E_o N \tag{5-22}$$

$$= Z^2\{[\lambda_o] - \frac{1}{2\lambda_o^2} N\} \quad .$$

Smooth and oscillatory contributions are exhibited after employing

$$[\lambda_o] = \lambda_o - \frac{1}{2} - <\lambda_o + \frac{1}{2}> \tag{5-23}$$

in Eqs. (20) and (22). The outcome is

$$-E/Z^2 = \frac{2}{3}\lambda_o - \frac{1}{2} - E_{osc}/Z^2 \quad , \tag{5-24}$$

where

$$-E_{osc}/Z^2 = -\frac{1}{\lambda_o}\left(<\lambda_o + \frac{1}{2}>^2 - \frac{1}{12}\right)$$
$$+ \frac{1/3}{\lambda_o^2} <\lambda_o + \frac{1}{2}>\left(<\lambda_o + \frac{1}{2}>^2 - \frac{1}{4}\right) \quad . \tag{5-25}$$

Since λ_o is proportional to $Z^{1/3}$, this E_{osc} is proportional to $Z^{5/3}$.

Before proceeding, a remark concerning partly filled shells is in order. Whenever λ_o is an integer, say $\lambda_o = m_o'$, the energy of this m_o'-th Bohr shell equals zero. We can describe it as partly filled if we assign the whole range of values between $m_o'-1$ and m_o to $[\lambda_o]$, when $\lambda_o = m_o'$. Equivalently, the step function $\eta(x)$ equals any number between 0 and 1 for $x=0$. This way the number of occupied states, as calculated in (20), can equal any desired amount. As far as the binding energy (22) is concerned, these subtleties do not matter, because the step function is multiplied by its argument; this product, $x\eta(x)$, equals zero for $x=0$ independent of the value given to $\eta(x)$. Indeed, the resulting energy, displayed in Eqs. (24) and (25), is a continuous function of λ_o.

Let us be somewhat more specific and consider $\kappa=1$, that is neutral atoms in the TF limit. Then focusing on the leading term, Eq. (25) gives

$$-E_{osc}/Z^{5/3} = -\left(\frac{2}{3}\right)^{1/3}\left(<\left(\frac{3}{2}Z\right)^{1/3} + \frac{1}{2}>^2 - \frac{1}{12}\right)$$
$$+ \ldots \quad , \tag{5-26}$$

which we contrast with the N=Z version of (14),

$$-E_{osc}/Z^{5/3} = \left(\frac{2}{3}\right)^{1/3}\left(<\left(\frac{2}{3}Z\right)^{1/3} + \frac{1}{12}\left(\frac{2}{3}Z\right)^{-1/3}>^2 - \frac{1}{6}\right) + \ldots \quad . \tag{5-27}$$

These oscillations are markedly different, in particular in sign and phase, as illustrated by Fig.6. As a consequence, the closed-shell values

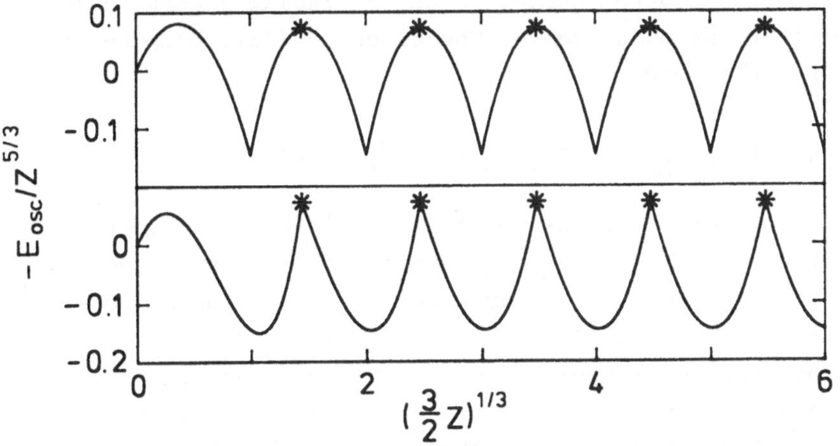

Fig.5-6. Binding-energy oscillations of Eq.(26) (upper curve) and Eq.(27) (lower curve), as functions of $(\frac{3}{2}Z)^{1/3}$. Stars mark the Z values of closed Bohr shells.

of Z[$(\frac{3}{2}Z)^{1/3}$ = 1.44, 2.47, 4.38, 4.48, and 5.48] mark the sharp maxima on the lower curve of Fig.6, which refers to Eq.(27), whereas they sit close to the tops of broad maxima on the upper curve referring to Eq.(26). The latter situation is reminiscent of Fig.2. This is, of course, not accidental, as we shall learn in the sequel. Despite this similarity, there are essential differences which one must not forget: in Fig.2, the oscillations are of order $Z^{4/3}$, not $Z^{5/3}$ as in Fig.6; and the Z values of closed-shell atoms, as a rule, do not coincide with the maxima of the energy oscillations plotted in Fig.2. One can state that the closing of a shell in real atoms is a much less dramatic event than in Bohr atoms.

TF quantization. The qualitative discussion given above and the study of energy oscillations in Bohr atoms indicate that a quantitative treatment of E_{osc} has to proceed from an improved evaluation of

$$E_1 = \mathrm{tr}(H+\zeta)\eta(-H-\zeta) \quad , \tag{5-28}$$

an evaluation that goes beyond the phase space integrals that we have been using so far. In particular, the discreteness of the (relevant part of the) spectrum of H must be taken into account.

The effective potential $V(\vec{r})$ is spherically symmetric, $V=V(r)$, in the situation of an isolated atom. We confine the discussion to potentials with this property. The eigenvalues E_{ℓ,n_r} of H are then labelled by the angular quantum number ℓ and the radial quantum number n_r, both being non-negative integers. The trace in (28), computed as a sum over ℓ and n_r, now reads

$$E_1 = 2\sum_{\ell,n_r=0}^{\infty}(2\ell+1)(E_{\ell,n_r}+\zeta)\eta(-E_{\ell,n_r}-\zeta) \quad , \tag{5-29}$$

where the factors of 2 and $2\ell+1$ reflect the spin multiplicity and the angular multiplicity, respectively. The latter one is, of course, due to the independence, of the energy eigenvalues, of the magnetic quantum number $m=\ell,\ell-1,\ldots,-\ell$.

The functional dependence of (29) upon $V(r)$ and ζ can only be investigated if we have an explicit expression that relates E_{ℓ,n_r} to both the quantum numbers and the potential. At the present stage $V(r)$ is quite arbitrary, and therefore such a relation can obviously be only an approximate one. Such an approximation is supplied by a semiclassical argument of remarkable simplicity.

For a start, consider a one-dimensional Hamilton operator $H_{1D}(x,p_x)$, for which the time independent Schrödinger equation

$$H_{1D}\left(x',\frac{1}{i}\frac{\partial}{\partial x'}\right)\chi_E(x') = E\,\chi_E(x') \tag{5-30}$$

tells us the eigenvalues $E=E_o,E_1,E_2,\ldots$. About H_{1D} we shall only assume that its spectrum contains a discrete part $-\infty<E_o<E_1<E_2<\ldots$ and that a possible continuous part consists of E values larger than the discrete eigenvalues. In other words: H_{1D} is supposed to belong to a reasonable physical system which possesses a ground state, some excited bound states, and, possibly but not necessarily, scattering states. We are only interested, here, in the discrete part, so that the wavefunctions $\chi_E(x')$ are normalizable,

$$\int_{-\infty}^{\infty} dx'\,|\chi_E(x')|^2 = 1 \quad . \tag{5-31}$$

The quantity

$$n_E = \mathrm{tr}\ \eta\left(E-H_{1D}(x,p_x)\right)$$
$$= \sum_j \eta(E-E_j) \tag{5-32}$$

is the count of eigenvalues of H_{1D} less than E; or: n_E equals the number of bound states with energy below E. Thus, n_E is piece-wise constant,

$$n_E = n \quad \text{for} \quad E_{n-1} < E < E_n \quad . \tag{5-33}$$

In a plot of n_E versus E, Fig.7, we see its staircase character. Obvi-

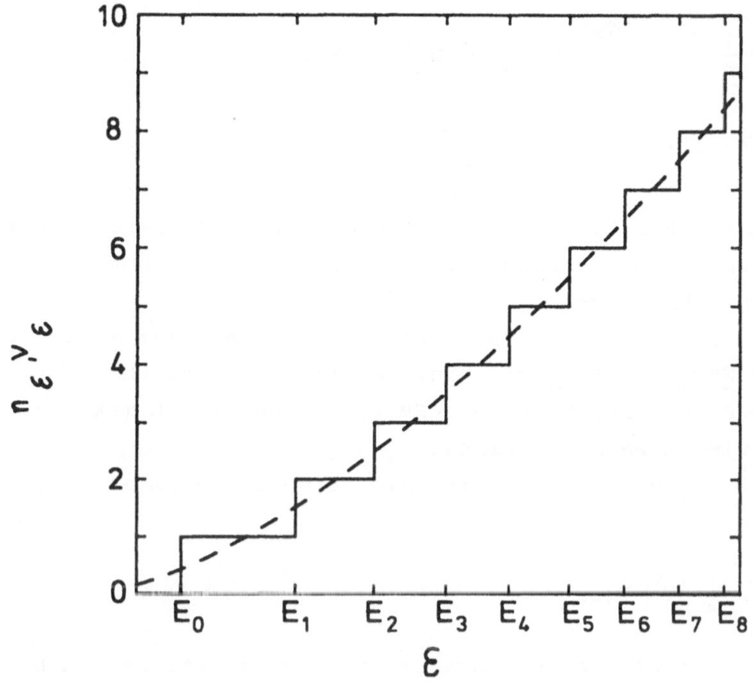

Fig.5-7. Sketch of n_E *(solid-line staircase) and* ν_E *(dashed smooth curve) as a function of* E.

ously, n_E can be regarded as consisting of a continuous smooth part ν_E plus an oscillatory (and discontinuous) supplement $n_E-\nu_E$. We can find ν_E by evaluating the trace in (32) TF wise,

$$\nu_E = \int\frac{dx'dp_x'}{2\pi}\ \eta\left(E-H_{1D}(x',p_x')\right) \quad , \tag{5-34}$$

as in Eq.(1-43). (Again, we use primed quantities x',p_x' to distinguish numbers from operators x,p_x.) Now, if ν_E does indeed smooth out n_E, it will equal some number between n and n+1 for $E= E_n$. It is natural to expect this number to be about n+1/2. Consequently, an implicit and approximate way of calculating E_n is to find the value of E for which ν_E equals n+1/2. In short:

$$\nu_{E_n} \cong n + 1/2 \tag{5-35}$$

is the TF quantization condition in one dimension. In particular, if H_{1D} has the standard structure

$$H_{1D}(x,p_x) = \frac{1}{2} p_x^2 + V(x) \quad , \tag{5-36}$$

this is

$$\begin{aligned} n + \frac{1}{2} &\cong \int \frac{dx'dp_x'}{2\pi} \, \eta\left(E_n - \frac{1}{2} p_x'^2 - V(x')\right) \\ &= \frac{1}{\pi} \int dx' \sqrt{2(E_n - V(x'))} \quad , \end{aligned} \tag{5-37}$$

where the range of integration extends over the classically allowed region inside which the argument of the square root is non-negative. (As always, we use the convention that $\sqrt{z} = 0$ if $z < 0$.)

The reader, so I trust, recognizes that (37) is identical with the familiar WKB quantization rule, and also notices that our derivation of (37) is much simpler than the WKB reasoning, which makes extensive use of approximate wave functions.

Now we turn to the three-dimensional spherically symmetric Hamilton operator

$$H = \frac{1}{2} \vec{p}^{\,2} + V(r) \quad . \tag{5-38}$$

The radial and angular dependence of the wave functions can be separated,

$$\langle \vec{r}\,' | E \rangle = \frac{1}{r'} R_{\ell,E}(r') \; Y_{\ell m}(\theta',\phi') \quad , \tag{5-39}$$

with the aid of the spherical harmonics $Y_{\ell m}$. The Schrödinger equation then implies a differential equation for $R(r')$,

$$\left[-\frac{1}{2}\frac{d^2}{dr'^2} + V(r') + \frac{\ell(\ell+1)}{2r'^2}\right] R_{\ell,E}(r') = E \, R_{\ell,E}(r') \quad , \tag{5-40}$$

which determines the energy eigenvalues E.

This differential equation cannot, as it stands, be read as the Schrödinger equation associated with a one-dimensional Hamilton operator $H_{1D}(x,p_x)$. The identification of r' with x and $-id/dr'$ with p_x

does not work, because the range of r' is restricted to positive values, as is illustrated by the normalization integral

$$\int_0^\infty dr' \, |R(r')|^2 = 1 \quad . \tag{5-41}$$

Therefore, we must, during an intermediate stage, introduce a new variable, x', that ranges from $-\infty$ to ∞. The physical distance r' is now expressed as a function of this auxiliary variable: $r' = r'(x')$. It is certainly reasonable to restrict ourselves do such transformations for which $dr'/dx' > 0$ and $r'(x' \to \infty) = 0$, $r'(x' \to \infty) = \infty$. In other words: as x' grows from $-\infty$ to ∞, $r'(x')$ increases monotonically from 0 to ∞. The normalization integral (41) is then

$$\int_{-\infty}^{\infty} dx' \, \frac{dr'(x')}{dx'} \, |R(r'(x'))|^2 = 1 \tag{5-42}$$

so that

$$X(x') = \left[\frac{dr'(x')}{dx'}\right]^{1/2} R(r'(x')) \tag{5-43}$$

is going to be the wave function in the new variable x'. The differential equation for $X(x')$ is

$$\left(\frac{dr'}{dx'}\right)^{1/2} \left[-\frac{1}{2}\left(\frac{dx'}{dr'}\,\frac{d}{dx'}\right)^2 + V(r') + \frac{\ell(\ell+1)}{2r'^2}\right] \left(\frac{dr'}{dx'}\right)^{-1/2} X(x') = E\,X(x') \quad . \tag{5-44}$$

Now we can identify $x' \to x$ and $-id/dx' \to p_x$ as coordinate and momentum of an effective one-dimensional description. The associated Hamilton operator is

$$H_{1D}(x,p_x) = \frac{1}{2}\,\frac{\sqrt{r(x)}}{s(x)}\left(\frac{s^2(x)}{r(x)}\,p_x\right)^2 \frac{s(x)}{\sqrt{r(x)}} + V(r(x)) + \frac{\ell(\ell+1)}{2r^2(x)} \,, \tag{6-45}$$

where s(x) is defined by

$$s(x) = \left[\frac{r(x)}{dr(x)/dx}\right]^{1/2} \quad . \tag{5-46}$$

The identity

$$\frac{\sqrt{r(x)}}{s(x)}\left(\frac{s^2(x)}{r(x)}\,p_x\right)^2 \frac{s(x)}{\sqrt{r(x)}} = \frac{s^2(x)}{r(x)}\,p_x^2\,\frac{s^2(x)}{r(x)} + \frac{1}{4r^2(x)}\left[1+4s(x)\,\frac{d^2s(x)}{dx^2}\right] \tag{5-47}$$

simplifies (45) to

$$H_{1D}(x,p_x) = \frac{1}{2}\left[\frac{dr(x)}{dx}\right]^{-1} p_x^2 \left[\frac{dr(x)}{dx}\right]^{-1} + V\big(r(x)\big)$$
$$+ \frac{(\ell+1/2)^2}{2r^2(x)} + \frac{s^3(x)}{2r^2(x)} \frac{d^2s(x)}{dx^2} \quad . \tag{5-48}$$

If we insert this H_{1D} into the TF quantization rule (35), we find that the radial motion is (approximately) quantized according to

$$n_r + \frac{1}{2} = \int \frac{dx'dp_x'}{2\pi} \eta\big(E_{\ell,n_r} - H_{1D}(x',p_x')\big)$$
$$= \frac{1}{\pi} \int dx' \frac{dr}{dx'} \frac{1}{r} \sqrt{2r^2\big(E_{\ell,n_r} - V(r)\big) - (\ell+1/2)^2 - s^3 \frac{d^2s}{dx'^2}} \ , \tag{5-49}$$

or since $dx' \frac{dr}{dx'} = dr$,

$$n_r + 1/2 = \frac{1}{\pi} \int \frac{dr}{r} \sqrt{2r^2\big(E_{\ell,n_r} - V(r)\big) - (\ell+1/2)^2 - s^3 \frac{d^2s}{dx'^2}} \ . \tag{5-50}$$

We compare this with what we would have obtained by wrongly taking (40) as a one-dimensional Schrödinger equation and observe that $\ell(\ell+1)$ is replaced by $(\ell+1/2)^2 + s^3 \, d^2s/dx'^2$. The latter term is dimensionless by construction, inasmuch as (46) shows that s^2 has the dimension of x. Consequently, as a function of r this term has to be a constant because elsewise we would be forced to introduce an artificial scale for r. The functions $s(x')$, for which $s^3 \, d^2s/dx'^2$ is independent of x', are given by

$$s(x') = \left[s_o + s_1 (x'-x_o)^2\right]^{1/2} \ , \tag{5-51}$$

where $s_o > 0$, $s_1 \geq 0$, and x_o are constants. Now recall that $r' \to 0; \infty$ for $x' \to -\infty; \infty$, so that $\log r' \to \pm\infty$ for $x' \to \pm\infty$. In terms of $s(x')$ this means

$$\infty = \int_{-\infty}^{\infty} dx' \log r'(x') = \int_{-\infty}^{\infty} dx' \left[\frac{1}{s(x')}\right]^2$$
$$= \int_{-\infty}^{\infty} \frac{dx'}{s_o + s_1 (x'-x_o)^2} \ . \tag{5-52}$$

The latter integral, however, results in a finite number unless s_1 equals zero. Then $s(x') = \text{const.}$, and (46) implies

$$r(x) = r_o \, e^{x/s_o} \quad , \quad r_o > 0 \ . \tag{5-53}$$

The term $s^3\, d^2s/dx^2$ in H_{1D} of (48) vanishes accordingly.

The particular choice for r_o and s_o is irrelevant, we can put both equal to unity. Then

$$H_{1D}(x,p_x) = \frac{1}{2}\,\frac{1}{r}\left[p_x^2 + (\ell + \tfrac{1}{2})^2\right]\frac{1}{r} + V(r)\,,\; r = e^x \quad , \tag{5-54}$$

and the resulting TF quantization rule

$$n_r + \frac{1}{2} = \frac{1}{\pi}\int dr\left[2\left(E_{\ell,n_r} - V(r) - \frac{(\ell+1/2)^2}{2r^2}\right)\right]^{1/2} \tag{5-55}$$

has the familiar WKB form. This is how we shall relate E_{ℓ,n_r} to both the quantum numbers and the potential, for eventual use in Eq.(29).

Before proceeding with this development, I would like to offer a remark. The necessity of replacing the naively expected "centrifugal barrier" $\ell(\ell+1)/(2r^2)$ by $(\ell+1/2)^2/(2r^2)$ is, of course, well-known since Kramers[6] observed, in 1926, that the correct behavior of the WKB wavefunctions near r=0 is not obtained otherwise. The reasoning given above, however, is quite different from Kramers' argument. Instead, it is in the spirit of Langer's[7] derivation in 1937. The main difference is that Langer settles for $r=e^x$ right away, whereas it is shown above that this is, in some sense, the only reasonable mapping of $0 \le r < \infty$ to $-\infty < x < \infty$. So much about the justification of $\ell(\ell+1) \to (\ell+1/2)^2$. But this is not the only reason for being so explicit. It is clear that in evaluating the trace of $\eta\left(E-H_{1D}(x,p_x)\right)$ one wants eventually to go beyond the TF approximation (34) by, for instance, introducing the relevant Airy averages. For that purpose, one must deal with the H_{1D} given in (54). Nothing of this has been worked out as yet, and it would certainly be interesting to find the resulting modifications of the quantization rules (37) and (55). For our present objective of studying E_{osc} it suffices to employ Eq.(55).

For large quantum numbers ℓ and n_r quantization according to the TF (or WKB) rule (55) is highly accurate by construction. What can we say about the small quantum numbers associated with the strongly bound electrons? For these the potential is basically a Coulombic potential (15) with a (small) additive constant E_o. Inserted into (55) it produces (see Problem 4)

$$n_r + \frac{1}{2} = \frac{Z}{\sqrt{2(E_o - E_{\ell,n_r})}} - (\ell + 1/2) \tag{5-56}$$

or

$$E_{\ell,n_r} = E_o - \frac{Z^2}{2(n_r+\ell+1)^2} \tag{5-57}$$

This is the <u>exact</u> answer with its well-known dependence on the principal quantum number $(n_r+1/2) + (\ell+1/2) = n_r+\ell+1$. Thus, TF quantization (55) is also very good for small quantum numbers.

Here is an additional pay-off. Since (55) gives the correct energies for the strongly bound electrons, the special treatment discussed in Chapter Three is no longer asked for. In particular, when the E_{ℓ,n_r} values of (55) are used to evaluate the spectral sum of E_1 in Eq. (29), the Scott correction to the energy is there without further ado. More about this later in this Chapter.

<u>Fourier formulation.</u> The new expression for $E_1(V+\zeta)$, Eq. (29), which is a sum over the quantum numbers ℓ and n_r with E_{ℓ,n_r} approximated by (55), is not well suited for a practical calculation as it stands. We therefore rewrite (29) in a few steps.

First, we recall that we can shift the emphasis from the energy E_1 to $N(E)$, the count of states with energy less than E,

$$N(E) = \text{tr}\ \eta(E-H) \quad , \tag{5-58}$$

because [see Eqs. (209) and (2-20), and note that $N(E)=N(\zeta=-E)$]

$$E_1 = \text{tr}(H+\zeta)\eta(-H-\zeta) = - \int_{-\infty}^{-\zeta} dE\ N(E) \quad , \tag{5-59}$$

so that information about $N(E)$ is immediately turned into knowledge of E_1.

Then, instead of summing over ℓ and n_r, we equivalently integrate over λ and ν, given by

$$\lambda \equiv \ell + 1/2 \quad , \quad \nu \equiv n_r + 1/2 \quad , \tag{5-60}$$

and introduce

$$E_{\lambda,\nu} \equiv E_{\ell,n_r} \quad , \tag{5-61}$$

which, according to (55), is related to λ and ν by

$$\nu = \frac{1}{\pi}\int\frac{dr}{r}\ [2r^2\left(E_{\lambda,\nu}-V(r)\right)-\lambda^2]^{1/2} . \tag{5-62}$$

After the introduction of Delta functions to select the discrete quantum numbers, we arrive at

$$
\begin{aligned}
N(E) &= 2 \sum_{\ell,n_r=0}^{\infty} (2\ell+1)\,\eta(E-E_{\ell,n_r}) \\
&= 4 \int_0^{\infty} d\lambda\, \lambda \sum_{\ell=-\infty}^{\infty} \delta(\ell+\tfrac{1}{2}-\lambda) \\
&\times \int_0^{\infty} d\nu \sum_{n_r=0}^{\infty} \delta(n_r+\tfrac{1}{2}-\nu)\,\eta(E-E_{\lambda,\nu}) \quad .
\end{aligned}
\tag{5-63}
$$

Now, the twofold application of the Poisson identity

$$
\begin{aligned}
\sum_{\ell=-\infty}^{\infty} \delta(\ell+\tfrac{1}{2}-\lambda) &= \sum_{k=-\infty}^{\infty} (-1)^k\, e^{i2\pi k\lambda} \quad , \\
\sum_{n_r=-\infty}^{\infty} \delta(n_r+\tfrac{1}{2}-\nu) &= \sum_{j=-\infty}^{\infty} (-1)^j\, e^{i2\pi j\nu}
\end{aligned}
\tag{5-64}
$$

produces

$$
N(E) = 4 \sum_{k,j=-\infty}^{\infty} (-1)^{k+j} \int_0^{\infty} d\lambda\lambda e^{i2\pi k\lambda} \int_0^{\infty} d\nu e^{i2\pi j\nu}\, \eta(E-E_{\lambda,\nu}) \quad . \tag{5-65}
$$

So far we have been reading Eq.(62) [and likewise (55) before] as implicitly defining $E_{\lambda,\nu}$ for given λ and ν. However, another view is more useful. It understands ν as a function of λ and E, $\nu=\nu_E(\lambda)$,

$$
\nu_E(\lambda) = \frac{1}{\pi} \int \frac{dr}{r} [2r^2\big(E-V(r)\big)-\lambda^2]^{1/2} \quad , \tag{5-66}
$$

which for each E defines a "line of degeneracy" in a λ,ν-diagram. The term degeneracy is appropriate here because such lines connect (λ,ν) values belonging to the same energy E. If it should happen that several (ℓ,n_r) pairs of (integer) quantum numbers refer to (λ,ν)'s on the same line of degeneracy, then there is more than one (orbital) state with the corresponding energy; these states are degenerate (in addition to the general spin and angular momentum multiplicity). This is certainly possible among the lines of degeneracy that are straight, but it can also occur for bent ones.

The domain of integration in (65) consists of all λ,ν below the line of degeneracy $\nu_E(\lambda)$. For a fixed value of λ, the step function in (65) selects, therefore, $\nu\leq\nu_E(\lambda)$,

$$
\eta(E-E_{\lambda,\nu}) = \eta\big(\nu_E(\lambda)-\nu\big) \quad . \tag{5-67}
$$

On the other hand, if λ exceeds λ_E, defined by

$$\lambda_E \equiv \underset{r}{\text{Max}}\,[2r^2(E-V(r))]^{1/2} \quad , \tag{5-68}$$

then the argument of the square root in (66) is negative for all r, implying

$$\nu_E(\lambda \geq \lambda_E) = 0 \quad . \tag{5-69}$$

Consequently, we now have

$$N(E) = 4\sum_{k,j=-\infty}^{\infty}(-1)^{k+j}\int_0^{\lambda_E} d\lambda\,\lambda\,e^{i2\pi k\lambda}\int_0^{\nu_E(\lambda)} d\nu\;e^{i2\pi j\nu} \quad , \tag{5-70}$$

which, inserted into (59), enables one to compute E_1.

<u>Isolating the TF contribution.</u> The j=k=0 term in (70) gives the result of integrating of λ and ν, <u>without</u> reference to the Delta functions that enforce the integral nature of λ-1/2 and ν-1/2. Therefore, we expect it to reproduce the TF version of N(E),

$$\Big(N(E)\Big)_{TF} = \int(d\vec{r})\,\frac{1}{3\pi^2}[2(E-V)]^{3/2} \quad . \tag{5-71}$$

Indeed, this happens when $\nu_E(\lambda)$ of (66) is put into the j=k=0 term of (70):

$$\begin{aligned}\Big(N(E)\Big)_{j=k=0} &= 4\int_0^{\lambda_E} d\lambda\,\lambda\,\nu_E(\lambda) \\ &= \frac{4}{\pi}\int\frac{dr}{r}\int_0^{\lambda_E} d\lambda\lambda[2r^2(E-V)-\lambda^2]^{1/2} \\ &= \frac{4}{3\pi}\int\frac{dr}{r}\,[2r^2(E-V)]^{3/2} \quad ,\end{aligned} \tag{5-72}$$

which, in view of $(d\vec{r}) = 4\pi r^2 dr$, agrees with (71).

This observation implies the decomposition of N(E) into its TF part and a supplement that represents quantum corrections,

$$N(E) = \Big(N(E)\Big)_{TF} + N_{qu}(E) \quad , \tag{5-73}$$

where the term "quantum corrections" is used with the meaning given to it in the paragraph after Eq.(1-43). In the present context we approximate $N_{qu}(E)$ by the right-hand side of Eq.(70) without the j=k=0 term. Quite analogously, E_1 is split into the TF expression plus a quantum correction[8] (not to be confused with $\Delta_{qu}F_1$ of Chapter Four). We shall see that this correction is usually small compared to the TF part, allowing its perturbative evaluation. As a preparation we first collect information about the lines of degeneracy $\nu_E(\lambda)$ which are the basic ingredients in (70).

Lines of degeneracy. The maximum of Eq.(68) is located at the distance r_E,

$$\lambda_E^2 = 2r_E^2[E-V(r_E)] \quad . \tag{5-74}$$

This maximum property implies that r_E obeys

$$V(r_E) + \frac{d}{dr_E}\left(r_E V(r_E)\right) = 2E \quad , \tag{5-75}$$

which has the consequence

$$\lambda_E^2 = r_E^2\left(\frac{d}{dr_E} - \frac{1}{r_E}\right)[r_E V(r_E)] \quad . \tag{5-76}$$

Another immediate implication is

$$\frac{d}{dE}\lambda_E^2 = 2r_E^2 \quad . \tag{5-77}$$

This shows that if we were given λ_E for some range of E, we could calculate r_E and then employ (74) to find $V(r)$ for r in the corresponding range of r_E. As an illustration hereof, consider

$$\lambda_E = Z/\sqrt{2(E_o-E)} \quad , \quad E < E_o \quad , \tag{5-78}$$

for which (77) gives

$$r_E = \frac{Z/2}{E_o-E} \quad , \quad r_E > 0 \quad , \tag{5-79}$$

so that

$$E = -\frac{Z/2}{r_E} + E_o \quad , \quad \lambda_E^2 = Z\,r_E \quad , \tag{5-80}$$

and, using Eq.(74),

$$V(r_E) = E - \frac{\lambda_E^2}{2r_E^2} = -Z/r_E + E_o \quad , \tag{5-81}$$

the Coulombic potential (15) emerges. Indeed, Eq.(56) is equivalent to

$$\nu_E(\lambda) = Z/\sqrt{2(E_o-E)} - \lambda = \lambda_E-\lambda \quad , \tag{5-82}$$

with λ_E from (78), which is to be supplemented by

$$\nu_E(\lambda) = 0 \quad \text{for} \quad \lambda > \lambda_E \tag{5-83}$$

as required by (69).

It is an important lesson that λ_E, in its dependence on E, contains a lot of information about the potential V(r). Since the range of λ_E is possibly limited (short-range potentials do not bind states with very large angular momenta), the corresponding range of r_E need not extend to infinity. Then λ_E determines V(r) within a sphere, of finite radius, around r=0.

For values of λ close to λ_E, the domain of integration in (66) is a small neighborhood of r_E. There one can approximate the argument of the square root by a quadratic polynomial in $r-r_E$,

$$2r^2\left(E-V(r)\right)-\lambda^2 \cong \lambda_E^2-\lambda^2-\tfrac{1}{2}\omega_E^2(r-r_E)^2 \quad , \tag{5-84}$$

with

$$\begin{aligned}\omega_E^2 &= \frac{d^2}{dr^2}\,[2r^2\left(V(r)-E\right)]\,\Big|_{r=r_E} \\ &= 2\,[r_E\frac{d^2}{dr_E^2}+\frac{d}{dr_E}-\frac{1}{r_E}]\,[r_E V(r_E)] \quad .\end{aligned} \tag{5-85}$$

The relations

$$\omega_E^2 = \frac{4r_E}{dr_E/dE} = 8\,\frac{d(\lambda_E^2)/dE}{d^2(\lambda_E^2)/dE^2} \tag{5-86}$$

indicate that, again, knowledge of λ_E is sufficient.

The integral that results from inserting (84) into (66),

$$\nu_E(\lambda) \cong \frac{1}{\pi}\int\frac{dr}{r}\,[\lambda_E^2-\lambda^2-\frac{1}{2}\,\omega_E^2(r-r_E)^2]^{1/2} \quad , \tag{5-87}$$

has the structure of the one producing the $\nu_E(\lambda)$ associated with the Coulomb potential:

$$\frac{1}{\pi}\int\frac{dr}{r}\,[2r^2(E+Z/r)-\lambda^2]^{1/2} = \frac{Z}{\sqrt{-2E}}-\lambda \quad . \tag{5-88}$$

The evaluation of (87) is, therefore, immediate. We find, for $\lambda \lesssim \lambda_E$,

$$\nu_E(\lambda) \cong \sqrt{2}\,\frac{\lambda_E}{\omega_E r_E}\,(\lambda_E-\lambda) \quad . \tag{5-89}$$

In the limit $\lambda \to \lambda_E$ this is exact, so that

$$\nu_E' \equiv -\frac{\partial}{\partial\lambda}\,\nu_E(\lambda)\Big|_{\lambda=\lambda_E} = \sqrt{2}\,\frac{\lambda_E}{\omega_E r_E} \quad . \tag{5-90}$$

Please note that, via Eqs.(77) and (86), the E dependence of λ_E determines the λ dependence of $\nu_E(\lambda)$ near $\lambda=\lambda_E$.

Higher derivatives of $\nu_E(\lambda)$ at $\lambda=\lambda_E$ can be calculated by a method (described in the Appendix to Ref.9) which improves upon the approximation (84). For example, the second derivative

$$\nu_E'' \equiv -\frac{\partial^2}{\partial\lambda^2}\,\nu_E(\lambda)\Big|_{\lambda=\lambda_E} \tag{5-91}$$

equals

$$\nu_E'' = \frac{\nu_E'}{\lambda_E}[1-(\nu_E')^2 + \tfrac{3}{2}\nu_E' v_3 - \tfrac{15}{8}v_3^2 + \tfrac{3}{2}v_4] , \tag{5-92}$$

where the coefficients

$$v_k \equiv \frac{(-1)^k}{k!}\,\frac{2}{\lambda_E^2}(\nu_E' r_E)^k \left(\frac{d}{dr_E}\right)^k [r_E^2 V(r_E)] \tag{5-93}$$

depend on E, and could be expressed in terms of E derivatives of λ_E.

The recognition that, for a spherically symmetric potential,

$$\begin{aligned} \frac{1}{r}\frac{d^2}{dr^2}(rV) &= \nabla^2 V \quad , \\ \left(\frac{d}{dr}-\frac{1}{r}\right)(rV) &= \vec{r}\cdot\vec{\nabla}V \quad , \end{aligned} \tag{5-94}$$

enables one to rewrite the right-hand side of (90), the outcome being

$$\nu_E' = \left[1 + \frac{r^2\nabla^2 V}{\vec{r}\cdot\vec{\nabla}V}\Big|_{r=r_E}\right]^{-1/2} \quad . \tag{5-95}$$

The force $-\vec{\nabla}V$ is towards the center at $r=0$, so that $\vec{r}\cdot\vec{\nabla}V$ is positive. On the other hand, the Laplacian of an atomic potential is related to the density ($-\nabla^2 V = 4\pi n$, in the simplest approximation) and is therefore negative. As a consequence, the contents of the square brackets in (95) are less than one, implying

$$\nu_E' = -\frac{\partial\nu_E(\lambda)}{\partial\lambda}\Big|_{\lambda=\lambda_E} > 1 \, . \tag{5-96}$$

The limit of unity is approached for large binding energies $-E$ belonging to strongly bound electrons for which $V \cong -Z/r$. Note that (96) is not true for any potential because $\nabla^2 V$ can be positive. An example is the oscillator potential $V=(\kappa/2)r^2$ where $\nu'_E=1/2$ for all energies $E>0$. Our interest, however, is in atomic potentials which approach $-Z/r$ as $r \to 0$ and vanish for $r \to \infty$. For these, Eq.(96) holds.

The derivation of (90) can also be done by using the general expression

$$-\frac{\partial}{\partial\lambda}\nu_E(\lambda) = \frac{1}{\pi}\int\frac{dr}{r}\frac{\lambda}{[2r^2(E-V(r))-\lambda^2]^{1/2}} , \qquad (5\text{-}97)$$

together with the approximation (84) (see Problem 5). Let us now employ (97) to find $\partial\nu_E/\partial\lambda$ for $\lambda \to 0$. No, the answer is not zero, for in this limit the integration reaches down to $r=0$ from which neighborhood a finite contribution arises. We isolate that part of the integral by introducing an upper limit $\bar{r}$, independent of λ and so small that $V \cong -Z/r$ already is a good approximation. At this stage we have

$$'\nu_E \equiv -\frac{\partial\nu_E(\lambda)}{\partial\lambda}\Big|_{\lambda=0} = \frac{1}{\pi}\int_{\lambda^2/(2Z)}^{\bar{r}} dr\, \frac{\lambda/r}{\sqrt{2Zr-\lambda^2}}\Big|_{\lambda\to 0} \qquad (5\text{-}98)$$

Now the substitution $2Zr = \lambda^2(1+x^2)$ yields

$$'\nu_E = \frac{2}{\pi}\int_0^\infty \frac{dx}{1+x^2} = 1 \quad . \qquad (5\text{-}99)$$

This statement holds for all E, except $E=0$, where there is the possibility of an additional contribution from the upper limit of the integral. This is the situation if $V \sim r^{-m}$ for $r \to \infty$, with $m>2$. Potentials with $m \leq 2$ are long-range potentials, of which the important example is the effective potential of an ion where $V \cong -(Z-N)/r$ for large r. In such a long-range potential there is no limit to the quantum numbers, and the entire line of degeneracy $\nu_o(\lambda)$ is infinitely distant from the origin in the λ,ν-diagram. On the other hand, for $m>2$, we have a short-range potential with a limit to the possible quantum numbers.

Again we isolate this upper part of the integral, now by a λ-independent lower limit $\bar{r}$, large enough to justify $V(r) \cong -c/r^m$ $(c>0, m>2)$:

$$\frac{1}{\pi}\int_{\bar{r}}^{(2c/\lambda^2)^{1/(m-2)}} dr\, \frac{\lambda/r}{[2cr^{2-m}-\lambda^2]^{1/2}}\Big|_{\lambda\to 0} =$$

$$= \frac{1}{m-2}\,\frac{2}{\pi}\int_0^\infty \frac{dx}{1+x^2} = \frac{1}{m-2} \quad , \tag{5-100}$$

where the substitution $2cr^{2-m} = \lambda^2(1+x^2)$ has been made. Thus, for potentials with $V(r\to\infty) \sim -1/r^m$, $m>2$ we have

$$'\nu_E = -\frac{\partial\nu_E(\lambda)}{\partial\lambda}\Big|_{\lambda=0} = \begin{cases} 1 & \text{for } E<0 \quad , \\ 1+\dfrac{1}{m-2} = \dfrac{m-1}{m-2} > 1 & \text{for } E=0. \end{cases} \tag{5-101}$$

For given λ, $\nu_E(\lambda)$ increases continuously with growing E. Therefore, the sudden increase of the initial slope at $E=0$ for short-range potentials must be accompanied by a rapid change of $\nu_E(\lambda=0)$ as E approaches zero. This is confirmed by the evaluation of

$$\frac{\partial}{\partial E}\,\nu_E(\lambda) = \frac{1}{\pi}\int dr\, \frac{r}{[2r^2(E-V)-\lambda^2]^{1/2}} \tag{5-102}$$

for $\lambda=0$, performed analogously to Eq.(100), which produces

$$\begin{aligned} \dot{\nu}_E(0) &\equiv \frac{\partial}{\partial E}\,\nu_E(0) \\ &= \frac{c^{1/m}}{\sqrt{2\pi}}\,\frac{(1/m-1/2)!}{(1/m-1)!}\left(\frac{1}{-E}\right)^{(2+m)/(2m)} \end{aligned} \tag{5-103}$$

for $E \lesssim 0$. This has the consequence

$$\nu_E(0) \cong \nu_o(0) + \frac{c^{1/m}}{\sqrt{2m}}\,\frac{(1/m-3/2)!}{(1/m-1)!}(-E)^{(m-2)/(2m)} \tag{5-104}$$

for $E \lesssim 0$. Please note that, because $m>2$, the numerical coefficient in (103) is positive, whereas it is negative in (104), and that the exponent of $(-E)$ in (104) is a positive number less than 1/2. Indeed, $\nu_E(0)$ grows rapidly as $E\to 0$.

Equations (103) and (104) are illustrations of the fact that $\nu_E(0)$ for $E \lesssim 0$ tests the outer reaches of the potential. However, the E-dependence of $\nu_E(0)$ is not converted into knowledge about $V(r)$ as easily as the E-dependence of λ_E [recall the remark after Eq.(83)].

In Eq.(103) we introduced a dot symbolizing the derivative with respect to E. This notational simplification will prove useful in the sequel. With this convention, Eqs.(77) and (86) can be written as

$$\begin{aligned} r_E^2 &= \lambda_E\dot{\lambda}_E \quad , \\ \omega_E^2 &= 4r_E/\dot{r}_E = 8\lambda_E\dot{\lambda}_E/(\lambda_E\ddot{\lambda}_E+\dot{\lambda}_E^2) \quad , \end{aligned} \tag{5-105}$$

and (90) implies

$$\frac{\lambda_E \ddot{\lambda}_E}{\dot{\lambda}_E^2} = 4(\nu_E')^2 - 1 \quad . \tag{5-106}$$

Differentiation of (85) gives

$$\begin{aligned} \frac{d}{dE}\omega_E^2 &= \dot{r}_E \frac{d}{dr_E}\omega_E^2 = \frac{4}{\omega_E^2} r_E \frac{d}{dr_E}\omega_E^2 \\ &= \frac{8}{\omega_E^2}\left[r_E^2 \frac{d^3}{dr_E^3} + 2r_E \frac{d^2}{dr_E^2} - \frac{d}{dr_E} + \frac{1}{r_E}\right][r_E V(r_E)] \quad , \end{aligned} \tag{5-107}$$

of which a more useful form is

$$\frac{1}{4}\frac{d}{dE}\,\omega_E^2 = 2\,\frac{r_E^2}{\omega_E^2}\left(3+r_E\frac{d}{dr_E}\right)\nabla^2 V(r_E) - (\nu_E')^2 \quad . \tag{5-108}$$

This will later be needed in

$$\frac{\lambda_E \dot{\nu}_E'}{\dot{\lambda}_E \nu_E'} = 1 - 2(\nu_E')^2 - \frac{1}{4}(\nu_E')^2 \frac{d}{dE}\,\omega_E^2 \quad , \tag{5-109}$$

the derivation of which I leave to the reader.

Classical orbits. Some of the equations of the preceding section possess an elementary significance when interpreted as referring to classical orbits of a particle in the spherically symmetric, attractive potential $V(r)$. For instance, the velocity in a circular orbit of energy E and radius r_E is determined by the kinetic energy

$$\frac{1}{2}\,v^2 = E - V(r_E) \quad , \tag{5-110}$$

which combined with the statement that the gradient of V must supply the necessary centripal force,

$$-\frac{d}{dr_E}\,V(r_E) = -\frac{v^2}{r_E} \tag{5-111}$$

reproduces (75). In other words: r_E, as obtained from (75), is the radius of the classical circular orbit with energy E. Further, insertion of (110) into (74) identifies λ_E as the classical angular momentum in this circular orbit. It is, indeed, well known that of all orbits to a certain energy the circular one has the maximal angular momentum.[10]

If the angular momentum λ is less than λ_E, the classical orbit is of the kind sketched in Fig.8. The radial motion is an oscillation between two distances r_1 and r_2 which define the classically allowed domain. These distances are the limits of integration in Eqs.(66), (97), and (102). The differential equation of the orbit is (the sign changes whenever $r=r_1$ or $r=r_2$)

$$d\phi = \pm \frac{dr}{r} \frac{\lambda}{\sqrt{2r^2\left(E-V(r)\right)-\lambda^2}} \quad ; \tag{5-112}$$

it determines r as a function of the azimuth ϕ. In particular, the angular period Φ of the orbit, $r(\phi+\Phi) = r(\phi)$, is twice the difference in

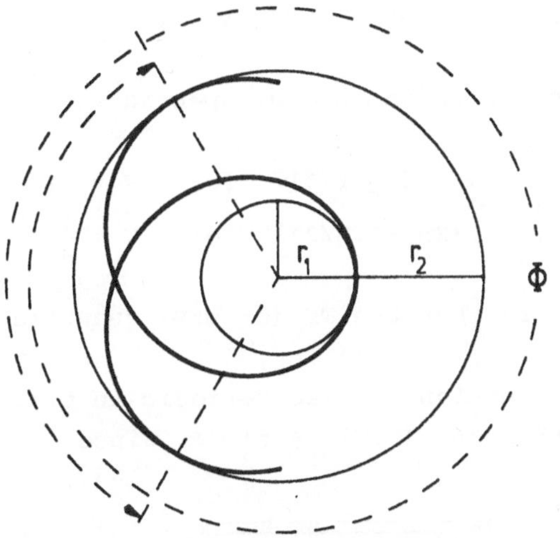

Fig.5-8: Sketch of typical classical trajectory in a spherically symmetric, attractive potential.

azimuth corresponding to $r = r_1 \rightarrow r = r_2$:

$$\Phi = 2 \int_{r_1}^{r_2} d\phi = 2 \int_{r_1}^{r_2} \frac{dr}{r} \frac{\lambda}{\sqrt{2r^2(E-V)-\lambda^2}} \quad . \tag{5-113}$$

Comparison with (97) establishes

$$\Phi/2\pi = -\frac{\partial}{\partial\lambda} \nu_E(\lambda) \quad ; \tag{5-114}$$

the slope of the lines of degeneracy is the angular period of the corresponding classical orbit, measured in units of 2π.

This insight can be used to find $\nu_E(\lambda)$ for both the Coulomb potential and the harmonic oscillator potential. For the first Φ is 2π, for the second it is π. Accordingly, we have

$$-\frac{\partial}{\partial\lambda}\,\nu_E(\lambda) = \begin{cases} 1 & \text{for} \quad V = -Z/r\ , \\ 1/2 & \text{for} \quad V = \frac{1}{2}\kappa r^2\ , \end{cases} \tag{5-115}$$

or,

$$\nu_E(\lambda) = \begin{cases} \lambda_E - \lambda & \text{for} \quad V = -Z/r\ , \\ \frac{1}{2}(\lambda_E-\lambda) & \text{for} \quad V = \frac{1}{2}\kappa r^2\ . \end{cases} \tag{5-116}$$

In conjunction with

$$\lambda_E = \begin{cases} Z/\sqrt{-2E} & \text{for} \quad V = -Z/r\ , \\ E/\kappa & \text{for} \quad V = \frac{1}{2}\kappa r^2\ , \end{cases} \tag{5-117}$$

this leads to the corresponding energy spectra

$$E_{\ell,n_r} = \begin{cases} -\frac{1}{2}Z^2/(n_r+\ell+1)^2 & \text{for} \quad V = -Z/r\ , \\ \kappa(2n_r+\ell+3/2) & \text{for} \quad V = \frac{1}{2}\kappa r^2\ , \end{cases} \tag{5-118}$$

demonstrating once more that the TF (or WKB) quantization is exact for these potentials.

The orbital motion is also periodic in time t, the period T being related to Φ through $\phi(t+T) = \phi(t) + \Phi$. Since

$$dt = \frac{r^2}{\lambda}d\phi = \pm\, dr\,\frac{r}{\sqrt{2r^2\big(E-V(r)\big)-\lambda^2}}\ , \tag{5-119}$$

we find

$$T = 2\int_{r_1}^{r_2} dr\,\frac{r}{\sqrt{2r^2(E-V)-\lambda^2}}\ , \tag{5-120}$$

so that, in view of Eq.(102), the analog to (113) is

$$T/2\pi = \frac{\partial}{\partial E}\,\nu_E(\lambda)\ . \tag{5-121}$$

The energy derivative of $\nu_E(\lambda)$, multiplied by 2π, equals the orbit's period in time. Note how Eqs.(113) and (121) pair angular momentum and angle as well as energy and time.

Applying (121) to (116) with (117), we obtain

$$T = \begin{cases} 2\pi Z/(-2E)^{3/2} & \text{for} \quad V = -Z/r\ , \\ \pi/\kappa & \text{for} \quad V = \frac{1}{2}\kappa r^2\ , \end{cases} \tag{5-122}$$

which are familiar results.

Degeneracy in the TF potential. After this excursion into the realm of classical mechanics we now return to the quantum world of atomic physics.

Inserting the neutral-atom TF potential $V = -(Z/r)F(x)$ into (66) produces

$$\frac{\nu_E(\lambda)}{Z^{1/3}} = \frac{1}{\pi}\int \frac{dx}{x}\left[2ax\left(F(x) + \frac{E}{Z^{4/3}}\,a\,x\right) - \left(\frac{\lambda}{Z^{1/3}}\right)^2\right]^{1/2} \quad . \tag{5-123}$$

This way of writing it makes explicit that $\nu_E(\lambda)/Z^{1/3}$ is a Z-independent function of both $\lambda/Z^{1/3}$ and $E/Z^{4/3}$. These lines of degeneracy are plotted in Fig.9, for the binding energies $-E/Z^{4/3} = 10, 1, 10^{-1}, \ldots, 10^{-5}, 0$. Please observe that $\nu_E(\lambda)$ is steeper for larger values of λ; in particular, note that the slope is -1 for $\lambda=0$ (and $E < 0$) whereas it exceeds

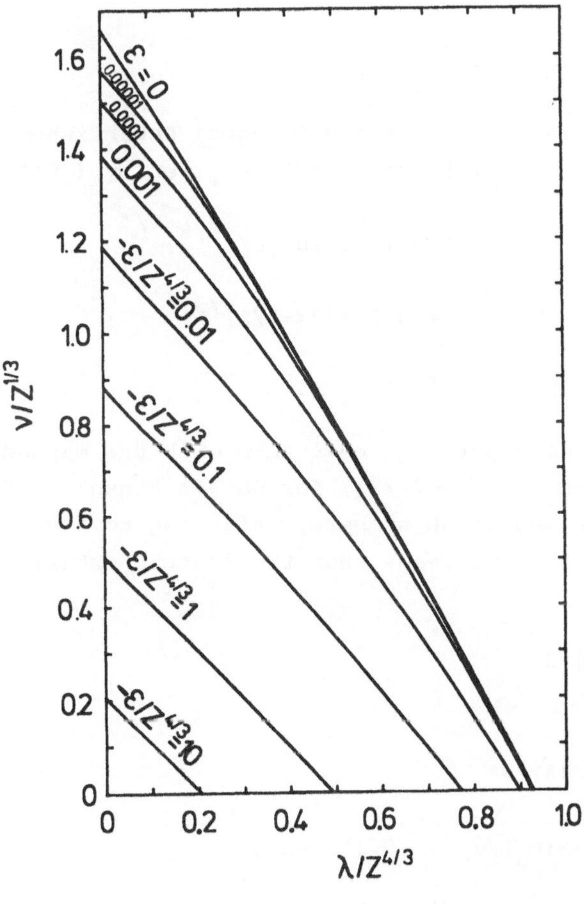

Fig.5-9. Lines of degeneracy for the neutral-atom TF potential.

unity, in magnitude, for $\lambda=\lambda_E$, where $\nu_E=0$. This is the message of Eqs. (96) and (101). The latter states in addition, that

$$'\nu_o = \frac{3}{2} \tag{5-124}$$

for the TF potential [m = 4 in (101)]. Indeed, this is the initial slope of the $E=0$ curve in Fig.9. Further, this figure confirms Eq.(104) (here with m=4 and $c=144a^3=81\pi^2/8$),

$$\frac{\partial}{\partial E}\nu_E(\lambda=0)/Z^{-3/3} \cong 0.4263(-E/Z^{4/3})^{-3/4} \quad , \tag{5-125}$$

for $E \lesssim 0$, inasmuch as $\nu_E(0)$ does change very rapidly as $E\to 0$. The maximum of $\nu_E(\lambda)$ is

$$\begin{aligned} \nu_o(0) &= \frac{1}{\pi}\sqrt{2a}\, Z^{1/3}\int_0^\infty dx\,\sqrt{F(x)/x} \\ &= 1.65865\, Z^{1/3} \quad . \end{aligned} \tag{5-126}$$

which uses the numerical value of the integral established in Problem 2-3.[11] The deviation of $\nu_E(0)$ from $\nu_o(0)$ is given by (104),

$$\begin{aligned} \nu_E(0)/Z^{1/3} &= 1.65865 - 1.70528(-E/Z^{4/3})^{1/4} \\ &\quad + 0.33172(-E/Z^{4/3})^{(1+\gamma)/4} \\ &\quad + \ldots \quad , \end{aligned} \tag{5-127}$$

for $E\lesssim 0$, where we also report the next term with the exponent $(1+\gamma)/4 = 0.443000$ [recall that $\gamma=(\sqrt{73}-7)/2$]; for detail consult Ref.9.

Some other important numbers, referring to $E=0$, can be computed starting from the knowledge that the function $xF(x)$ has its maximum at

$$x_o = 2.10403 \quad , \tag{5-128}$$

where

$$\begin{aligned} F(x_o) &= 0.231151 \ , \\ \frac{dF}{dx}(x_o) &= -F(x_o)/x_o = -0.109862 \ , \\ \frac{d^2F}{dx^2}(x_o) &= [F(x_o)]^{3/2}/x_o^{1/2} = 0.0766160 \ ; \end{aligned} \tag{5-129}$$

see Problem 2-4. We express r_o and λ_o in terms of x_o and $F(x_o)$,

$$r_o = a\,x_o\,Z^{-1/3}\ ,\qquad \lambda_o = \sqrt{2a\,x_o\,F(x_o)}\ Z^{1/3}\ . \tag{5-130}$$

Next, Eq.(85) appears as

$$\begin{aligned}\omega_o^2 &= \frac{2}{a}\left(x\frac{d}{dx^2}+\frac{d}{dx}-\frac{1}{x}\right)\left(-F(x)\right)Z^{4/3}\Big|_{x=x_o}\\ &= \frac{4}{a}\,\frac{F(x_o)}{x_o}\ [1-\tfrac{1}{2}x_o\sqrt{x_oF(x_o)}]\ Z^{4/3}\ ,\end{aligned} \tag{5-131}$$

so that (90) provides

$$\nu_o' = [1-\tfrac{1}{2}x_o\sqrt{x_oF(x_o)}]^{-1/2}\ . \tag{5-132}$$

With the aid of the differential equation obeyed by the TF potential for $r>0$, $\nabla^2V = (-4/3\pi)(-2V)^{3/2}$, Eq.(108) is simplified to

$$\begin{aligned}\frac{1}{4}\frac{d}{d\mathcal{E}}\,\omega_{\mathcal{E}}^2 &= -\frac{8}{3\pi}\,\frac{r_{\mathcal{E}}^2}{\omega_{\mathcal{E}}^2}\left(3+r_{\mathcal{E}}\frac{d}{dr_{\mathcal{E}}}\right)\left(-2V(r_{\mathcal{E}})\right)^{3/2}-(\nu_{\mathcal{E}}')^2\\ &= \frac{8}{\pi}\,\frac{r_{\mathcal{E}}^2}{\omega_{\mathcal{E}}^2}[-2V(r_{\mathcal{E}})]^{1/2}\left(\frac{1}{r_{\mathcal{E}}}+\frac{d}{dr_{\mathcal{E}}}\right)\left(r_{\mathcal{E}}V(r_{\mathcal{E}})\right)-(\nu_{\mathcal{E}}')^2\ .\end{aligned} \tag{5-133}$$

Now Eqs.(74) and (75) are employed to arrive at

$$\frac{1}{4}\frac{d}{d\mathcal{E}}\,\omega_{\mathcal{E}}^2 = \frac{16}{\pi}\,\mathcal{E}\left(\frac{r_{\mathcal{E}}}{\omega_{\mathcal{E}}}\right)^2\sqrt{(\lambda_{\mathcal{E}}/r_{\mathcal{E}})^2-2\mathcal{E}}-(\nu_{\mathcal{E}}')^2\ , \tag{5-134}$$

which inserted into (109) produces

$$\frac{\lambda_{\mathcal{E}}\dot{\nu}_{\mathcal{E}}'}{\dot{\lambda}_{\mathcal{E}}\nu_{\mathcal{E}}'} = [(\nu_{\mathcal{E}}')^2-1]^2+\frac{16}{\pi}(-\mathcal{E})(r_{\mathcal{E}}/\omega_{\mathcal{E}})^2\sqrt{(\lambda_{\mathcal{E}}/r_{\mathcal{E}})^2-2\mathcal{E}}\ . \tag{5-135}$$

In particular, for $\mathcal{E}=0$ we get

$$\begin{aligned}\dot{\omega}_o &= -2(\nu_o')^2/\omega_o = -4\,\frac{\lambda_o^2/r_o^2}{\omega_o^3}\\ &= -\sqrt{ax_o/F(x_o)}\ [1-\tfrac{1}{2}\,x_o\sqrt{x_oF(x_o)}]^{-3/2}\ Z^{-2/3}\ ,\end{aligned} \tag{5-136}$$

and

$$\frac{\lambda_o\dot{\nu}_o'}{\dot{\lambda}_o\nu_o'} = [(\nu_o')^2-1]^2 = \frac{1}{8}\,x_o^3F(x_o)\,[1-\tfrac{1}{2}\,x_o\sqrt{x_oF(x_o)}]^{-2}\ . \tag{5-137}$$

The corresponding numerical statements are

$$
\begin{aligned}
r_o/Z^{-1/3} &= 1.86278 \;, \\
\lambda_o/Z^{1/3} &= 0.927992 \;, \\
\omega_o/Z^{2/3} &= 0.363593 \;, \\
\nu'_o &= 1.93768 \;(\cong 2-\tfrac{1}{16}) \;, \\
-\dot{\omega}_o/Z^{-2/3} &= 20.6527 \;, \\
\frac{\lambda_o\dot{\nu}'_o}{\dot{\lambda}_o\nu'_o} &= 7.58781 \;.
\end{aligned}
\qquad (5\text{-}138)
$$

In addition, from Eqs.(105) and (106) we obtain

$$
\begin{aligned}
\dot{\lambda}_o/Z^{-3/3} &= (r_o/Z^{-1/3})^2/(\lambda_o/Z^{1/3}) = 3.73920 \;, \\
\frac{\lambda_o\ddot{\lambda}_o}{\dot{\lambda}_o^2} &= 4(\nu'_o)^2 - 1 = 14.0184 \;,
\end{aligned}
\qquad (5\text{-}139)
$$

and (92) leads to (see Problem 7)

$$
\begin{aligned}
\lambda_o\nu''_o &= \frac{1}{24}\,\nu'_o[(\nu'^2_o)-1][-5(\nu'_o)^4 + 23(\nu'_o)^2 - 15] \\
&= 0.193647 \;.
\end{aligned}
\qquad (5\text{-}140)
$$

These numbers will be used below.

For the purpose of illustration, we present, in Fig.10, various quantities as a function of $E/Z^{4/3}$. Observe in particular that for large binding energies, that is $-E \gg Z^{4/3}$, $\nu_E(0)$ equals λ_E. This is typical for Coulombic potentials, for which these relations hold:

$$
\begin{aligned}
\nu_E(0)/Z^{1/3} &= \lambda_E/Z^{1/3} \;, \\
r_E/Z^{1/3} &= (\lambda_E/Z^{1/3})^2 \;, \\
\omega_E/Z^{2/3} &= \sqrt{2}\,/\,(\lambda_E/Z^{1/3}) \;;
\end{aligned}
\qquad (5\text{-}141)
$$

see Eqs.(78), (79), and (82). Of course, Coulombic degeneracy for strongly bound electrons is not unexpected; recall the discussion around Eqs.

(56) and (57).

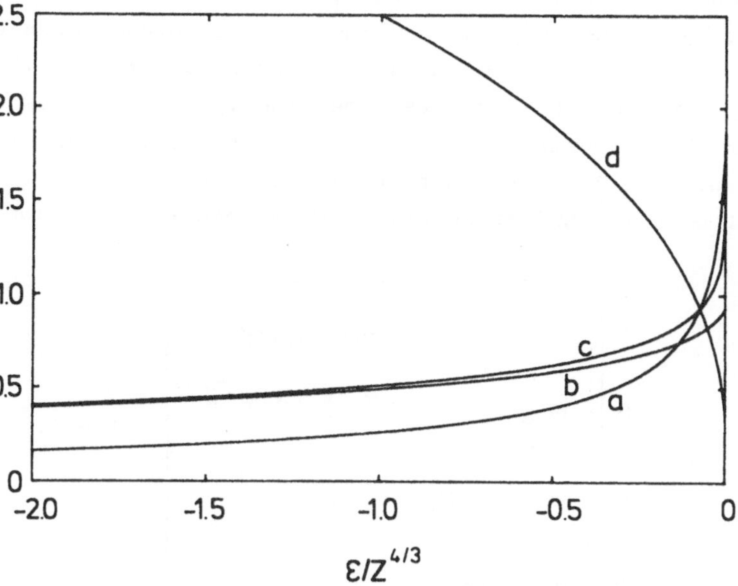

Fig.5-10. As a function of $E/Z^{4/3}$ are shown: (a) $r_E/Z^{-1/3}$, (b) $\lambda_E/Z^{1/3}$, (c) $\nu_E(0)/Z^{1/3}$, (d) $\omega_E/Z^{2/3}$; for the neutral-atom TF potential.

TF degeneracy and the systematics of the Periodic Table. Is there any reality to the energetic degeneracy as predicted via the neutral-atom TF potential?

Our affirmative answer begins with pointing out the similarity between Figs.4 and 9. In quantitative terms we note that the curve connecting 7s with 5f in Fig.4 has terminal slopes of about -1 and -2, which agrees with those of $\nu_E(\lambda)$ for $E \lesssim 0$; in particular, $\partial\nu_E/\partial\lambda$ at $\lambda=\lambda_E$ for $E \lesssim 0$ is practically equal to $-\nu_o'$, which according to (138) differs from -2 by a small amount. Further, the maximal values of ν and λ in the Periodic Table are $\nu = 6 + 1/2$ and $\lambda = 3 + 1/2$, referring to $n_r=6$ and $\ell=3$ (7s and 5f, respectively), whose ratio

$$\frac{6+1/2}{3+1/2} = 1.86 \tag{5-142}$$

does not differ much from the TF number

$$\frac{\nu_o(0)}{\lambda_o} = \frac{1.659\ Z^{1/3}}{0.928\ Z^{1/3}} = 1.79\ . \tag{5-143}$$

Next, consider a certain value of Z, say Z = 88, which is the atomic number of radium. For this Z, the TF line of degeneracy $\nu_o(\lambda)$ is plotted in Fig.11, where the physical values of λ and ν are marked on the axes and labelled by the corresponding values of the integer quantum numbers n_r and ℓ. The lattice points thus defined mark the orbital states specified by these quantum numbers. The line $\nu_o(\lambda)$ separates those available in the TF potential from the ones that are unavailable. For the chosen value of Z we see that the TF prediction about the occupied orbital states is in perfect agreement with spectroscopic observations.

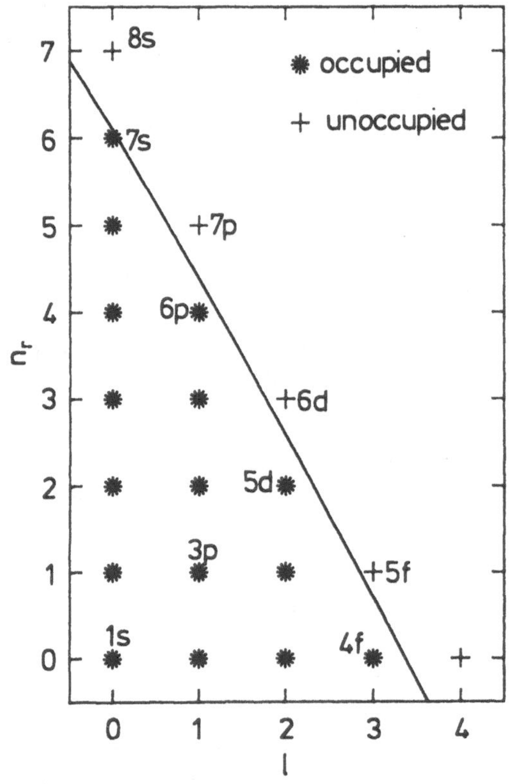

Fig.5-11. TF prediction for occupied states in radium (Z=88).

Let us now imagine that we increase Z. Then the curve of $\nu_o(\lambda)$ will move away from $\lambda=\nu=0$ in Fig.11, thereby keeping its shape. It is merely stretched proportional to $Z^{1/3}$. The next states that become available are the 5f and the 6d states which are crossed by $\nu_o(\lambda)$ for practically the same value of Z. For an even larger Z we obtain the 8s electrons, and so on. Whenever a new pair of integer quantum numbers ℓ, n_r lies below the $\nu_o(\lambda)$ curve, the TF potential can bind $2(2\ell+1)$ electrons more. Obviously, the total number of bound states does not always agree with Z; as a rule, it is a little bit less or more than Z. When picking Z=88 we made a lucky choice in Fig.11. However, this figure speaks in favor of the TF potential not because the number of bound states agrees with Z, but because the states selected by $\nu_o(\lambda)$ are the ones that are spectroscopically known to be occupied. No doubt, there is reality in the TF potential. A precise statement is the following: the TF potential reproduces the correct order in which the orbital states are filled as Z increases. This order is, both as derived from the TF potential and as known from the systematics of the Periodic Table: 1s, 2s, 2p, 3s, 3p, 4s, 3d, 4p, 5s, 4d, 5p, 6s, 4f, 5d, 6p, 7s, 5f, 6d, ... The corresponding values of $Z^{1/3}$ are listed in Table 1 below. It is remarkable that the ratio of $Z^{1/3}$ values of two successive orbital states in this sequence differ the least for 4f and 5d, and for 5f and 6d (differences: 1.5% and 1.7%, respectively). This is consistent with the strong competition between these states known from the electronic structure of the lanthanides and the actinides.

General features of N_{qu}. We shall now learn more about $N_{qu}(E) = N(E) - (N(E))_{TF}$ for the TF potential. For simplicity we confine ourselves to $E = 0$, for a start, and ask the question: how does $N(E=0)$ depend on Z? In other words: how many electrons can the TF potential bind for a given value of Z? In the limit of $Z \to \infty$, N_{qu} is small, so that $N(E=0) \cong (N(E=0))_{TF}$ $= Z$. Consequently, $N_{qu}(E=0)$ describes the deviation of $N(E=0)$ from its asymptotic value, Z.

A detailed answer is somewhat elaborate, and we present it in a later section. However, some general qualitative features of N_{qu} can be demonstrated without great effort. This is our objective here.

Let us first consider the sequence of states 1s, 3p, 5d, ... which is characterized by the constant value of the ratio

$$\frac{\nu}{\lambda} = \frac{n_r+1/2}{\ell+1/2} = \frac{0+1/2}{0+1/2}\ ;\ \frac{1+1/2}{1+1/2}\ ;\ \frac{2+1/2}{2+1/2}\ ;\ \ldots = 1\ . \tag{5-144}$$

In the ℓ, n_r-diagram Fig.11 these states are on the straight line through $\nu=\lambda=0$ with unit slope. Now observe that the respective distances from $\nu=\lambda=0$ are in proportions of 1:3:5:... which is an immediate consequence of the circumstance that the physical values of λ and ν are all odd multiples of $\frac{1}{2}$. Also recall that the line of degeneracy $\nu_o(\lambda)$ stretches proportional to $Z^{1/3}$. This implies that the $Z^{1/3}$ values at which the successive states of this sequence become available are in proportions of 1:3:5:... as well. The contribution to the number of available states made by said sequence of states is, therefore, given by

$$N_{(1s)}(Z) = \begin{cases} 0 & \text{for} \quad (Z/Z_{1s})^{1)3} < 1 \quad , \\ 2 & \text{for} \quad 1 < (Z/Z_{1s})^{1/3} < 3 \quad , \\ 2+6 & \text{for} \quad 3 < (Z/Z_{1s})^{1/3} < 5 \quad , \\ 2+6+10 & \text{for} \quad 5 < (Z/Z_{1s})^{1/3} < 7 \quad , \\ \vdots & \\ 2m^2 & \text{for} \quad 2m-1 < (Z/Z_{1s})^{1/3} < 2m+1 \quad , \end{cases} \tag{5-145}$$

where Z_{1s} stands for the minimal value of Z required to bind the 1s state, the first state in this sequence. In (145), m is the integer part of $\frac{1}{2}(Z/Z_{1s})^{1/3} + 1/2$, so that we can write, with the aid of Eq. (7),

$$\begin{aligned} N_{(1s)}(Z) &= 2\left(\tfrac{1}{2}(Z/Z_{1s})^{1/3} - \langle\tfrac{1}{2}(Z/Z_{1s})^{1/3}\rangle\right)^2 \\ &\equiv 2\,\mathcal{N}(Z/Z_{1s}) \quad . \end{aligned} \tag{5-146}$$

The function $\mathcal{N}$ thus defined is universal, which is to say: it is the same for each such sequence of states characterized by a common ratio $\nu/\lambda = (n_r+1/2)/(\ell+1/2)$. This is so because in each sequence the distances from $\lambda=\nu=0$ are in proportions of the odd integers, 1:3:5:... . And this is the only ingredient in $\mathcal{N}$. For any sequence we have accordingly

$$N_{(\text{sequence})}(Z) = 2(2\ell_o+1)\,\mathcal{N}(Z/Z_{min}) \tag{5-147}$$

where $2(2\ell_o+1)$ is the multiplicity of the initial state of the sequence, and Z_{min} is the Z value at which this initial state becomes available. Table 1 gives the essential numbers for the first fifteen sequences of states, ordered by increasing Z_{min}.

The total number of occupied states, $N(E=0)$, is then given by the sum over all sequences

$$N(E=0) = \sum_{\text{sequences}} \text{multiplicity} \times \mathcal{N}(Z/Z_{min}) \quad . \tag{5-148}$$

It is technically impossible to perform this summation. Nevertheless, we can certainly use it to study the structure of $N(E=0)$ as a function of Z. Note that for large Z, $\mathcal{N}(Z/Z_{min})$ appears as

Table 5-1. Initial state (IS), characterizing ratio $\nu:\lambda$, multiplicity (MULT) of initial state, and minimal $Z^{1/3}$ of initial state, for the first 15 sequences of states (ordered by increasing $Z^{1/3}_{\min}$). The orbital states 3p, 5d, and 6p do not initialize a new sequence.

IS	$\nu:\lambda$	MULT	$Z^{1/3}_{\min}$
1s	1:1	2	0.822
2s	3:1	2	1.41
2p	1:3	6	1.90
3s	5:1	2	2.00
[3p	1:1	(1s sequence)	3 × 0.822 = 2.47]
4s	7:1	2	2.60
3d	1:5	10	2.97
4p	5:3	6	3.05
5s	9:1	2	3.19
4d	3:5	10	3.54
5p	7:3	6	3.63
6s	11:1	2	3.79
4f	1:7	14	4.05
[5d	1:1	(1s sequence)	5 × 0.822 = 4.11]
[6p	3:1	(2s sequence)	3 × 1.41 = 4.23]
7s	13:1	2	4.39
5f	3:7	14	4.61
6d	7:5	10	4.69

$$N(Z/Z_{\min}) = \frac{1}{4}(Z/Z_{\min})^{2/3} + (Z/Z_{\min})^{1/3} < \frac{1}{2}(Z/Z_{\min})^{1/3} > + \ldots \tag{5-149}$$

where the leading terms have been exhibited: a smooth term of order $Z^{2/3}$, and an oscillatory term of order $Z^{1/3}$. The sum over sequences has to turn the smooth term into the TF part $\big(N(E=0)\big)_{\mathrm{TF}} = Z$. Thus this smooth term gains a factor of $Z^{1/3}$ when all sequences are summed. This will not be equally true for the oscillatory terms. An individual one has the periodicity $Z^{1/3} \to Z^{1/3} + 2Z^{1/3}_{\min}$, but as Table 1 shows the various sequences have what looks like randomly assigned values of $Z^{1/3}_{\min}$. The amplitude of each oscillatory term is also determined by $Z^{1/3}_{\min}$ in conjunction with the multiplicity of the sequence. There is nothing regular about these amplitudes as well. Therefore, we have to sum oscillatory functions that all have the same shape but irregular amplitudes and periods. The interference of these oscillations cannot be constructive. We conclude that the resulting fluctuating function of $Z^{1/3}$ has an amplitude factor of $Z^{1/3}$ as do the individual oscillations, no enhancement takes place. What we have found is:

$$N_{qu}(E=0) = Z^{1/3} \times \{\text{fluctuating function of } Z^{1/3}\} \quad . \tag{5-150}$$

As a matter of fact, the periods of the oscillations of the various sequences are not really assigned randomly. They are all determined by the shape of $\nu_o(\lambda)$. Accordingly, there is a little bit of amplification of the amplitude. The detailed analysis given below shows that the leading oscillatory term in N_{qu} is of the order $Z^{1/2} = Z^{1/3} \times Z^{1/6}$. However, for it to really dominate the terms of order $Z^{1/3}$, one needs the enormous value of 5×10^{10} for Z. In the small-Z range of physical interest, this "leading" oscillation is utterly insignificant.

Our algebraic results about $N(E=0)$ and $N_{qu}(E=0)$ as a function of Z are confirmed by the plots presented in Figs.12 and 13, of which the first one shows the staircase shape of (145) and compares it to the straight-line TF result, and the second one illustrates (150).

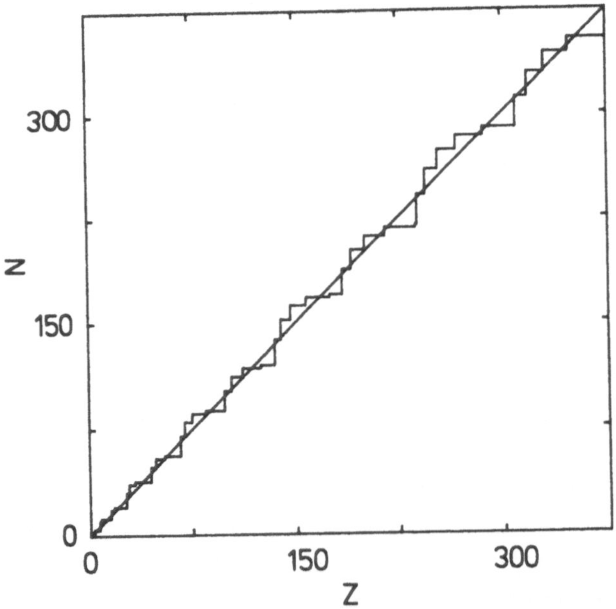

Fig.5-12. N(E=0) as a function of Z for the neutral-atom TF potential. The straight-line is $(N(E=0))_{TF} = Z$.

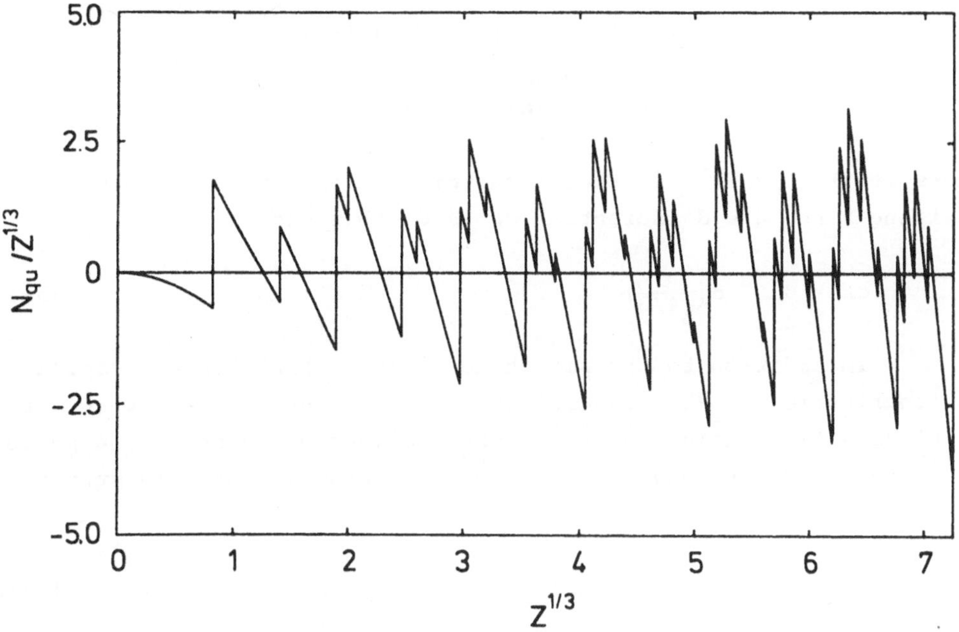

Fig. 5-13. $N_{qu}(E=0)/Z^{1/3} = [N(E=0)-Z]/Z^{1/3}$ *as a function of* $Z^{1/3}$ *for the neutral-atom TF potential.*

General features of E_{qu}. Armed with all this insight into the number of available states, we can now employ Eq. (59) and gain related information about the energy. What we have said about $N(E=0)$ holds also for $E<0$ if we keep the ratio $E/Z^{4/3}$ fixed when changing Z. This is an implication of the particular dependence of $\nu_E(\lambda)/Z^{1/3}$ on $E/Z^{4/3}$ and $\lambda/Z^{1/3}$, and therefore a consequence of the scaling properties of the TF model.

This means that $(N(E))_{TF}$ equals Z times a smooth function of $E/Z^{4/3}$ (for the TF potential, of course, or more generally for every potential of the form "$Z^{4/3}$ times a function of $Z^{1/3}r$"). Indeed, we know it does:

$$\begin{aligned}(N(E))_{TF} &= \int(d\vec{r})\ \frac{1}{3\pi^2}[2(E-V_{TF}(r))]^{3/2} \\ &= Z\int_0^\infty dx\ x^{1/2}[F(x)+ax\frac{E}{Z^{4/3}}]^{3/2} \quad .\end{aligned} \tag{5-151}$$

The integration thereof over E,

$$(E_1(\zeta))_{TF} = -\int_{-\infty}^{-\zeta} dE\ (N(E))_{TF} =$$

$$= \int (d\vec{r}) \left(-\frac{1}{15\pi^2}\right) [-2(V_{TF}(r)+\zeta)]^{5/2}$$
$$= -\frac{2}{5} \frac{Z^{7/3}}{a} \int_0^\infty dx\, x^{-1/2} [F(x) - ax \frac{\zeta}{Z^{4/3}}]^{5/2} \quad , \tag{5-152}$$

results in $(E_1)_{TF} \sim Z^{7/3}$ so that a factor $Z^{4/3}$ is acquired. This is not surprising since the differential dE is of this order:

$$dE = Z^{4/3}\, d\left(\frac{E}{Z^{4/3}}\right) \quad . \tag{5-153}$$

In addition to the smooth term (151), $N(E)$ has the oscillatory contribution $N_{qu}(E)$. Generalizing (150) we state that it equals $Z^{1/3}$ times a fluctuating function, the argument of which is the product of $Z^{1/3}$ and a smooth function of $E/Z^{4/3}$. An example for this structure is provided by

$$\text{"}N_{qu}(E)\text{"} = \lambda_E \sin(\lambda_E) \quad . \tag{5-154}$$

Upon performing successive partial integrations, this produces

$$\begin{aligned} \text{"}E_{qu}(\zeta)\text{"} &= -\int_{-\infty}^{-\zeta} dE\ \text{"}N_{qu}(E)\text{"} \\ &= -\int_{-\infty}^{-\zeta} dE \frac{d}{dE}\Big[-\frac{\lambda_E}{\dot{\lambda}_E}\cos(\lambda_E) \\ &\qquad + \frac{1}{\dot{\lambda}_E}\frac{d}{dE}\left(\frac{\lambda_E}{\dot{\lambda}_E}\right)\sin(\lambda_E) + \ldots\Big] \quad , \end{aligned} \tag{5-155}$$

where the dot represents differentiation with respect to E, and the ellipsis indicates further terms. The contents of the square brackets, evaluated at the upper limit, are oscillatory functions of $Z^{1/3}$. Their respective amplitudes are of the orders

$$\begin{aligned} &\lambda_E/\dot{\lambda}_E \sim Z^{1/3}/Z^{-3/3} = Z^{4/3} \quad , \\ &\frac{1}{\dot{\lambda}_E}\frac{d}{dE}\left(\frac{\lambda_E}{\dot{\lambda}_E}\right) \sim \frac{1}{Z^{-3/3}}\frac{1}{Z^{4/3}}\, Z^{4/3} = Z^{3/3} \quad , \end{aligned} \tag{5-156}$$

then $Z^{2/3}$, and so on. In short: when $N_{qu} \sim Z^{1/3}$, then the oscillatory part of E_{qu} is $\sim Z^{4/3}$. Oscillatory terms merely gain by a factor of $Z^{3/3}$.

We infer that the binding-energy oscillation $-E_{osc}$ is, for the TF potential, of amplitude $Z^{4/3}$ and (in some sense) periodic in $Z^{1/3}$. Indeed, as promised, this is small compared to the leading TF energy term ($\sim Z^{7/3}$) if only Z is sufficiently large. A pertubative treatment of these oscillations is fully justified. We shall arrive at quantitative statements about amplitude and period below, after a short detour.

We did not identify E_{qu} with E_{osc}, because E_{qu} also contains nonoscillatory contributions. Since the semiclassical spectral sum (29) handles the strongly bound electrons correctly, whereas they are misrepresented in $(E_1)_{TF}$, the Scott correction must be part of E_{qu}. Accordingly, N_{qu} possesses a related smooth term. The slight asymmetry in Fig.13, somewhat larger negative peaks than positive ones, is consistent with the presence of such a term. In the following section, we shall exhibit the Scott correction explicitly. Further, E_{qu} must contain (at least part of) the quantum corrections to E_1, that were discussed in Chapter Four. So far it has not been demonstrated how one can isolate them in E_{qu}.

Linear degeneracy. Scott correction. Let us briefly look back at the energy of Bohr atoms with filled shells, Eqs.(24) and (25). There the leading oscillatory term has an amplitude of order $Z^{5/3}$. Why is the chain of arguments that we applied to the TF potential not equally valid for a Coulombic potential? The reason is that for the TF potential the lines of degeneracy $\nu_E(\lambda)$ are bent, for $0 \lesssim -E/Z^{4/3} \ll 1$, not straight as for Coulombic potentials. A straight line results in sequences of states for which the respective $N(Z/Z_{min})$ are perfectly in phase. The random character of the periods, which we have observed for the TF potential in Table 1, is absent in the situation of linear degeneracy, that is when $\nu_E(\lambda)$ is a linear function of λ, a straight line in the λ,ν-diagram. In other words: the existence of a principal (or energy) quantum number is what distinguishes Coulombic potentials from the TF potential. As a matter of fact, one can demonstrate[12] that linear degeneracy near $E=-\zeta$ always leads to an energy oscillation of order $Z^{5/3}$.

The main example of a physical system displaying linear degeneracy throughout is a highly ionized atom, where the effective potential differs but little from a Coulombic potential. In neutral atoms, one has linear degeneracy for the strongly bound electrons. As in Chapter Three, we isolate these electrons by introducing a separating binding energy $\zeta_s = -E_s$ that selects the part of the spectrum with Coulombic degeneracy. Thus we write

$$E_1(\zeta) = [E_1(\zeta) - E_1(\zeta_s)] + E_1(\zeta_s) \quad . \tag{5-157}$$

According to (24) and (25), $E_1(\zeta_s)$ is given by

$$E_1(\zeta_s) = -Z^2\left[\frac{2}{3}\lambda_{E_s} - \frac{1}{2} - \frac{1}{\lambda_{E_s}}\left(\langle\lambda_{E_s} + \frac{1}{2}\rangle^2 - \frac{1}{12}\right) + \ldots\right] , \tag{5-158}$$

where [Eq.(78)]

$$\lambda_{E_s} = Z/\sqrt{2(E_o - E_s)} , \tag{5-159}$$

E_o being the additive constant of Eq.(15); for the TF potential it equals $Z^{4/3}B/a = 1.79374\ Z^{4/3}$. The leading term of $E_1(\zeta_s)$ is, as always, the TF contribution

$$\begin{aligned} -\frac{2}{3} Z^2\lambda_{E_s} &= -\frac{2}{3} Z^3/\sqrt{2(E_o - E_s)} \\ &= \int (d\vec{r})\left(-\frac{1}{15\pi^2}\right)\left[2\left(E_s + \frac{Z}{r} - E_o\right)\right]^{5/2} \\ &= \left(E_1(\zeta_s = -E_s)\right)_{TF} , \end{aligned} \tag{5-160}$$

when we insert (15) into (152) [see also (1-36)]. It combines with the TF part of the integral

$$E_1(\zeta) - E_1(\zeta_s) = -\int_{-\zeta_s}^{-\zeta} dE\ N(E) \tag{5-161}$$

to produce $(E_1(\zeta))_{TF}$. More interesting is the next-to-leading term of $E_1(\zeta_s)$. It equals $\frac{1}{2}Z^2$ and does not depend on E_s. Actually, being the only part of $E_1(\zeta_s)$ independent of $\zeta_s = -E_s$, this term is the only visible contribution of the strongly bound electrons to $E_1(\zeta)$. All the other terms in (158) cannot themselves be present in $E_1(\zeta)$ since $E_1(\zeta)$ does not depend on ζ_s.

We have thus identified the two leading contributions to $E_1(\zeta)$,

$$E_1(\zeta) = \left(E_1(\zeta)\right)_{TF} + \frac{1}{2}Z^2 + \ldots \tag{5-162}$$

thereby rediscovering the Scott correction to which Chapter Three is dedicated. Please note that at that earlier stage $E_{\zeta\zeta_s}$ was evaluated TF wise with the consequence that the Bohr shell oscillations of (158) [or (3-22)] had to be removed explicitly. In (157) we compute $E_{\zeta\zeta_s}$ with the aid of the semiclassical sum (29), so that all Bohr shell artifacts are taken care of automatically.

Perturbative approach to E_{osc}. After eliminating the density in favor of the effective potential from the energy functional (2-434) and separating the electron-electron interaction energy E_{ee} into its electrostatic part and a remainder E'_{ee}, as in (2-36), the potential functional of the energy (2-40)

$$E(V,\zeta) = E_1(V+\zeta) - \zeta N - \frac{1}{8\pi}\int(d\vec{r})\,[\vec{\nabla}(V+\frac{Z}{r}-V'_{ee})]^2 + \{E'_{ee}(n) - \int(d\vec{r})\,n\,V'_{ee}(n)\}_{n=n(V)} \tag{5-163}$$

emerges, for $V_{ext} = -Z/r$. We recall that the density is expressed in terms of the potential, symbolically: $n=n(V)$, by solving Eq.(2-432), in which $V_{ee} = V_{ee}(n)$, for n. The potential V'_{ee} is V_{ee} minus the electrostatic potential (2-28) [see (2-37)]; its lion's share is the exchange potential.

We exhibit the TF part of (163) by splitting E_1 into $(E_1)_{TF}$ and E_{qu},

$$\begin{aligned} E_1(V+\zeta) &= \int(d\vec{r})\left(-\frac{1}{15\pi^2}\right)[-2(V+\zeta)]^{5/2} + E_{qu}(V+\zeta) \\ &= \left(E_1(V+\zeta)\right)_{TF} + E_{qu}(V+\zeta) \quad , \end{aligned} \tag{5-164}$$

and by employing the identity

$$\begin{aligned} &-\frac{1}{8\pi}\int(d\vec{r})\,[\vec{\nabla}(V+\frac{Z}{r}-V'_{ee})]^2 - \int(d\vec{r})\,n\,V'_{ee} \\ &= -\frac{1}{8\pi}\int(d\vec{r})\,[\vec{\nabla}(V+\frac{Z}{r})]^2 + \frac{1}{8\pi}\int(d\vec{r})\,(\vec{\nabla}V'_{ee})^2 \quad . \end{aligned} \tag{5-165}$$

Thus,

$$\begin{aligned} E(V,\zeta) &= \left(E_1(V+\zeta)\right)_{TF} - \frac{1}{8\pi}\int(d\vec{r})\,[\vec{\nabla}(V+\zeta)]^2 - \zeta N \\ &\quad + E_{qu}(V+\zeta) + \left\{E'_{ee}(n) + \frac{1}{8\pi}\int(d\vec{r})\,[\vec{\nabla}V'_{ee}(n)]^2\right\}\Big|_{n=n(V)} \\ &= E_{TF}(V,\zeta) + E_{qu}(V+\zeta) \\ &\quad + \left\{E'_{ee}(n) + \frac{1}{8\pi}\int(d\vec{r})\,[\vec{\nabla}V'_{ee}(n)]^2\right\}\Big|_{n=n(V)} \end{aligned} \tag{5-166}$$

where we recognized the TF energy functional (2-45).

We evaluate

$$E_{qu} = -\int_{-\infty}^{-\zeta} dE\, N_{qu}(E) \tag{5-167}$$

approximately by inserting

$$N_{qu}(E) = 4 \sum_{k,j=\infty}^{\infty}{}' (-1)^{k+j} \int_0^{\lambda_E} d\lambda\, \lambda\, e^{i2\pi k\lambda} \int_0^{\nu_E(\lambda)} d\nu\, e^{i2\pi j\nu} \quad , \qquad (5\text{-}168)$$

where the primed sum does not include the j=k=0 term. This $N_{qu}(E)$ is, of course, the difference between N(E) of Eq. (70) and the TF term (71) or (72). Now we recall, that for given Z and N, the energy functional (166) is stationary for the actual potential V and the actual value of ζ. Also, E_{TF} alone has a stationary property; it is optimized, for neutral atoms, by $V = -(Z/r)F(x)$ and $\zeta = 0$. Above we have observed that, with the ζ and V, the leading oscillation in E_{qu} is relatively small at sufficiently large Z. All this means that we are justified in evaluating E_{osc} perturbatively by simply inserting $\zeta = 0$ and $V = V_{TF} = -(Z/r)F(x)$ into E_{qu}. In other words, we are going to extract E_{osc} out of (167) by using the TF lines of degeneracy $\nu_E(\lambda)$ in (168).

The terms in curly brackets in (166) are ignored in this process. This contribution to the energy is mainly exchange energy, which is smaller than the TF contribution by a factor of $Z^{-2/3}$. Consequently, the resulting modification of the effective potential is a small correction, and the energy oscillations that grow out of the exchange interaction are expected to be smaller than the leading oscillatory term by said factor of $Z^{-2/3}$. They are, therefore, negligible at the present level of accuracy.

<u>ℓ- quantized TF model.</u> Let us start our quantitative treatment of (168) by picking out the j=0 terms. With the disappearance of $\exp(i2\pi j\nu)$ all reference to the Delta functions, that initially enforced integral values for $n_r = \nu - 1/2$, is gone. We are thus, in effect, integrating over n_r instead of summing. Consequently, we are considering now the improvement obtained by quantization of angular motion <u>only</u>, without having radial motion also quantized. We call this the ℓTF model, short for ℓ-quantized Thomas-Fermi model.[13]

Equation (168) is here reduced to

$$\begin{aligned}\left(N_{qu}(E)\right)_{\ell TF} &= 4\sum_{k\neq 0}(-1)^k \int_0^{\lambda_E} d\lambda\, \lambda\, e^{i2\pi k\lambda}\, \nu_E(\lambda) \\ &= 8\sum_{k=1}^{\infty}(-1)^k \int_0^{\lambda_E} d\lambda\, \lambda\, \nu_E(\lambda)\cos(2\pi k\lambda) \quad .\end{aligned} \qquad (5\text{-}169)$$

Repeated partial integrations provide the identity

$$\lambda \nu_E(\lambda)\cos(2\pi k\lambda) = \frac{\partial}{\partial\lambda}\sum_{m=0}^{\infty}\frac{\cos(2\pi k\lambda + \frac{m-1}{2}\pi)}{(2\pi k)^{m+1}}\left(\frac{\partial}{\partial\lambda}\right)^m\left(\lambda\nu_E(\lambda)\right) , \tag{5-170}$$

which, inserted into (169), supplies oscillatory terms from the upper limit of integration. The smooth terms from the lower limit are of no interest to us here. Thus we found

$$\left(N_{osc}(E)\right)_{\ell TF} = 8\sum_{m=1}^{\infty}\left(\frac{\partial}{\partial\lambda}\right)^m\left(\lambda\nu_E(\lambda)\right)\Bigg|_{\lambda=\lambda_E}\sum_{k=1}^{\infty}\frac{(-1)^k}{(2\pi k)^{m+1}}\cos(2\pi k\lambda_E + \frac{m-1}{2}\pi), \tag{5-171}$$

where, because $\nu_E(\lambda_E) = 0$, no m=0 term is present. Since $\lambda \sim Z^{1/3}$ and $\nu_E \sim Z^{1/3}$, the m-th term in this series is of order $Z^{(2-m)/3}$; we exhibit the leading ones:

$$\begin{aligned}\left(N_{osc}(E)\right)_{\ell TF} &= -2\lambda_E\,\nu_E'\sum_{k=1}^{\infty}\frac{(-1)^k}{(\pi k)^2}\cos(2\pi k\lambda_E)\\ &\quad + (\lambda_E\nu_E''+2\nu_E')\sum_{k=1}^{\infty}\frac{(-1)^k}{(\pi k)^3}\sin(2\pi k\lambda_E)\\ &\quad + O(Z^{-1/3})\\ &= -2\lambda_E\nu_E'(\langle\lambda_E\rangle^2 - \tfrac{1}{12})\\ &\quad + \tfrac{2}{3}(\lambda_E\nu_E''+2\nu_E')\,\langle\lambda_E\rangle(\langle\lambda_E\rangle^2 - \tfrac{1}{4})\\ &\quad + O(Z^{-1/3}) \quad .\end{aligned} \tag{5-172}$$

The notations introduced in Eqs.(90) and (91) have been used; and the sums have been recognized to be the ones of Eqs.(3-29).

The corresponding ℓTF approximation to E_{osc} is obtained by inserting (172) into (167) and picking out the contributions from the upper limit (ζ=0 for neutral atoms) of the E integration. For this purpose, successive partial integrations are employed analogously to (155). For instance,

$$-\int^{0} dE\;\lambda_E\nu_E'\cos(2\pi k\lambda_E) \quad =$$

$$= \int^{0} dE \, \frac{d}{dE}\left[\lambda_E \nu'_E \, \frac{\sin(2\pi k\lambda_E)}{2\pi k\dot{\lambda}_E} + \frac{d}{dE}\left(\frac{\lambda_E\nu'_E}{2\pi k\dot{\lambda}_E}\right)\frac{\cos(2\pi k\lambda_E)}{2\pi k\dot{\lambda}_E} + \ldots\right]$$

$$= -\frac{1}{2}\,\frac{\lambda_o\nu'_o}{\dot{\lambda}_o}\,\frac{1}{\pi k}\,\sin(2\pi k\lambda_o) \tag{5-173}$$

$$-\frac{1}{4}\,\frac{\nu'_o}{\dot{\lambda}_o}\left(1+\frac{\lambda_o\nu'_o}{\dot{\lambda}_o\nu'_o} - \frac{\lambda_o\ddot{\lambda}_o}{\dot{\lambda}_o^2}\right)\left(\frac{1}{\pi k}\right)^2\cos(2\pi k\lambda_o)$$

$$+ \ldots \quad ,$$

where, once more, the dots symbolize differentiation with respect to E. The terms displayed are of the orders $Z^{4/3}$ and $Z^{3/3}$, the ellipsis indicates terms with amplitudes proportional to $Z^{2/3}$, $Z^{1/3}$, and so on.

This way we find

$$(-E_{osc})_{\ell TF} = -\frac{\lambda_o\nu'_o}{\dot{\lambda}_o}\sum_{k=1}^{\infty}\frac{(-1)^k}{(\pi k)^3}\sin(2\pi k\lambda_o) \tag{5-174}$$

$$-\frac{1}{2}\,\frac{\nu'_o}{\lambda_o}\left(3+\frac{\lambda_o\dot{\nu}'_o}{\dot{\lambda}_o\nu'_o} - \frac{\lambda_o\ddot{\lambda}_o}{\dot{\lambda}_o^2} + \frac{\lambda_o\nu''_o}{\nu'_o}\right)\sum_{k=1}^{\infty}\frac{(-1)^k}{(\pi k)^4}\cos(2\pi k\lambda_o)$$

$$+ O(Z^{2/3}) \quad .$$

After supplementing (3-29) by

$$\sum_{k=1}^{\infty}\frac{(-1)^k}{(\pi k)^4}\cos(2\pi k\lambda_o) = \frac{1}{90} - \frac{1}{3}\left(\langle\lambda_o\rangle^2 - \frac{1}{4}\right)^2 \quad , \tag{5-175}$$

and inserting the TF numbers reported in Eqs. (138), (139), and (140), this reads

$$\left(\frac{-E_{osc}}{Z^{4/3}}\right)_{\ell TF} = 0.320\,594\,\langle\lambda_o\rangle\left(\frac{1}{4} - \langle\lambda_o\rangle^2\right)$$

$$+ 0.287\,660\,Z^{-1/3}\left[\frac{1}{30} - \left(\frac{1}{4} - \langle\lambda_o\rangle^2\right)^2\right] \tag{5-176}$$

$$+ O(Z^{-2/3})$$

with $\lambda_o = 0.927\,992\,Z^{1/3}$. We see that λ_o determines the periodicity of the ℓTF energy oscillations, namely $Z^{1/3} \to Z^{1/3} + Z^{1/3}/\lambda_o = Z^{1/3} + 1.078$, which agrees quite well with the period of the HF oscillations of Fig. 5-2.

In Fig.14 the leading ℓTF contribution to E_{osc} is compared with the HF prediction. We see that the ℓTF model gives the correct period and phase, but accounts only for about half the amplitude; the other half is expected to be supplied by radial quantization. Also there is no sign of the intriguing double peak structure that the HF curve displays.

In plotting this figure, we extrapolated the large-$Z^{1/3}$ result

$$(-E_{osc})_{\ell TF} \cong 0.32\, Z^{4/3} \langle 0.928\, Z^{1/3}\rangle \left(\tfrac{1}{4} - \langle 0.928\, Z^{1/3}\rangle^2\right) \qquad (5\text{-}177)$$

down to small values of $Z^{1/3}$. This procedure needs justification. It is provided by Fig.15, where the next-to-leading ℓTF oscillation, that is the correction of relative size $Z^{-1/3}$ in (176), is recognized to be a small correction to the leading one. The sum of both has the overall characteristics of the leading ℓTF oscillation, the main difference ap-

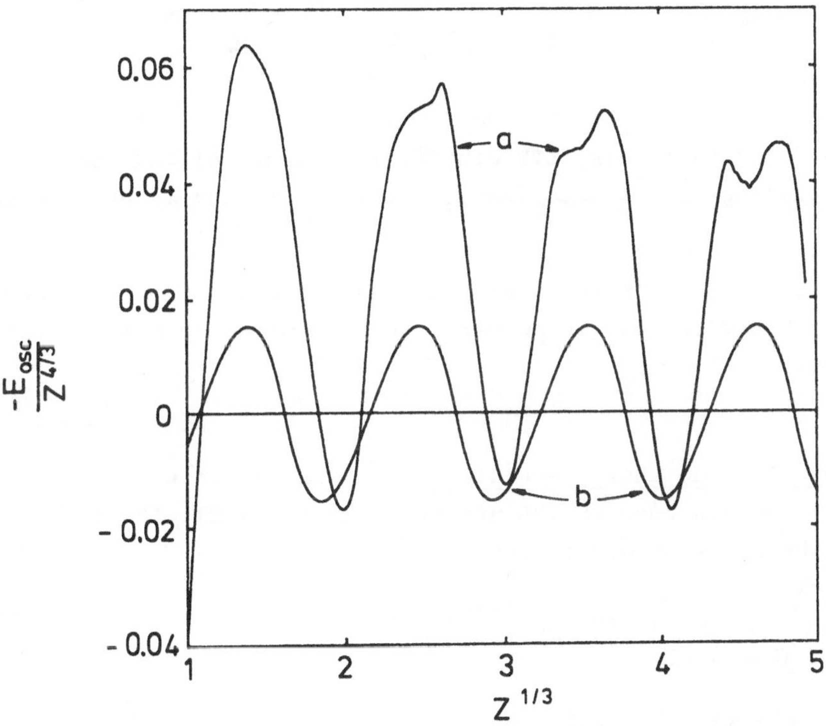

Fig.5-14. Comparison of the binding-energy oscillations as predicted by the HF method (curve a) with the leading ℓTF oscillation (curve b).

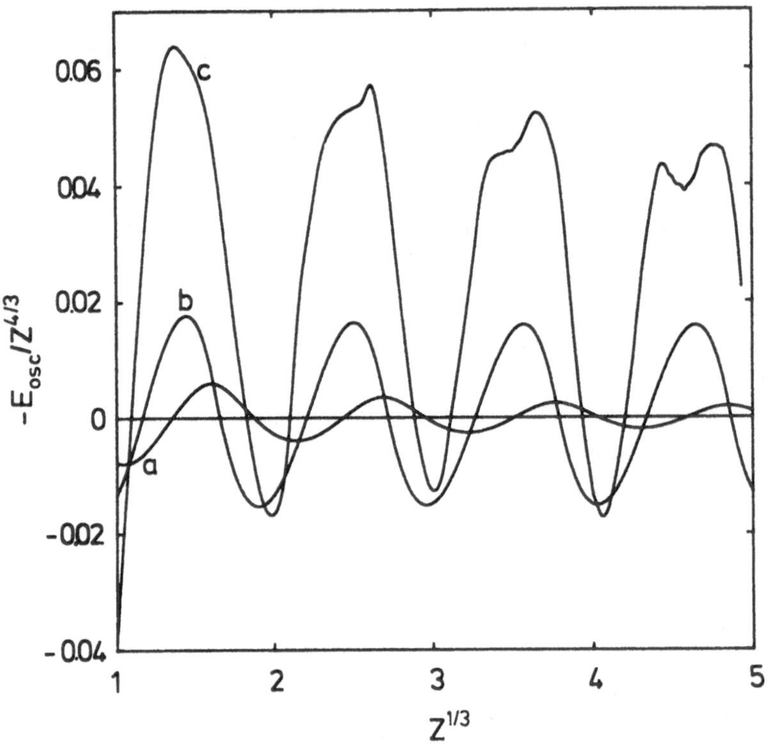

Fig.5-15. Next-to-leading ℓTF oscillation (curve a) and leading plus next-to-leading ℓTF oscillation (curve b), compared to HF prediction (curve c).

pearing at $Z^{1/3} \lesssim 2$. It is reassuring that in this range the agreement with the HF predictions is somewhat better for curve b in Fig.15 than in Fig.14.

<u>j≠0 terms. Leading energy oscillation.</u> Having dealt with the j=0 term in $N_{qu}(E)$, we now turn to the j≠0 terms in Eq.(168). It is expedient to rewrite the relevant double sum according to

$$\sum_{j=-\infty}^{\infty}{}' \sum_{k=-\infty}^{\infty} (-1)^{k+j} e^{i2\pi(k\lambda+j\nu)}$$

$$= \sum_{j=1}^{\infty} \sum_{k=-\infty}^{\infty} (-1)^{k+j} \left[e^{i2\pi(k\lambda+j\nu)} + e^{i2\pi(k\lambda-j\nu)} \right] \qquad (5\text{-}178)$$

$$= 2 \operatorname{Re} \sum_{j=1}^{\infty} \sum_{k=-\infty}^{\infty} (-1)^{k+j} e^{i2\pi(k\lambda+j\nu)} \quad .$$

The ν integration in (168) then produces

$$\left(N_{qu}(E)\right)_{j\neq 0} = 4\ \mathrm{Re} \sum_{j=1}^{\infty} \sum_{k=-\infty}^{\infty} \frac{(-1)^{k+j}}{i\pi j} \int_0^{\lambda_E} d\lambda\, \lambda\, e^{i2\pi k\lambda} \left(e^{i2\pi j\nu_E(\lambda)} - 1\right) \tag{5-179}$$

With the aid of Poisson's identity (64), we establish

$$\sum_{j=1}^{\infty} \frac{(-1)^j}{i\pi j} \sum_{k=-\infty}^{\infty} (-1)^k \int_0^{\lambda_E} d\lambda\, \lambda\, e^{i2\pi k\lambda} \tag{5-180}$$

$$= \frac{i}{\pi}(\log 2) \sum_{\ell=0}^{\infty} \left(\ell + \frac{1}{2}\right) \eta\left(\lambda_E - \frac{1}{2} - \ell\right) \quad ,$$

which - and this is the point - is purely imaginary. Therefore, the "-1" in (179) does not contribute, so that

$$\left(N_{qu}(E)\right)_{j\neq 0} = 4\ \mathrm{Re} \sum_{j=1}^{\infty} \sum_{k=-\infty}^{\infty} \frac{(-1)^{k+j}}{i\pi j} \int_0^{\lambda_E} d\lambda\, \lambda\, \exp\left[i2\pi\left(k\lambda + j\nu_E(\lambda)\right)\right] \ . \tag{5-181}$$

The required λ integration cannot be performed explicitly because of the complicated λ dependence of the exponent.

The integrals in (181) are largest for those j,k pairs for which the phase is stationary at some value of λ. For these values of j and k, the equation

$$k + j \frac{\partial \nu_E}{\partial \lambda}(\bar{\lambda}) = 0 \tag{5-182}$$

is obeyed by a $\bar{\lambda}$ in the range $0 < \bar{\lambda} \leq \lambda_E$. (For notational simplicity we leave the dependence of $\bar{\lambda}$ on E, j, and k implicit.) The possibility $\bar{\lambda}=0$ is only apparent because there is the weight λ in the integral. Now, since Fig.9 tells us that

$$'\nu_E = -\frac{\partial \nu_E}{\partial \lambda}(\lambda=0) \leq -\frac{\partial \nu_E}{\partial \lambda}(\lambda) \leq -\frac{\partial \nu_E}{\partial \lambda}(\lambda=\lambda_E) = \nu_E' \quad , \tag{5-183}$$

such a $\bar{\lambda}$ exists only if

$$j\ '\nu_E < k \leq j\nu_E' \quad . \tag{5-184}$$

For $E < 0$, $'\nu_E$ equals unity, see Eq.(101); and ν_E' does not exceed its $E=0$ value, which is $\nu_0' = 1.938 \cong 31/16$, see Eqs.(138). Thus, there is a point of stationary phase for the λ integration only if

$$j < k \lesssim \frac{31}{16}\, j < 2j \quad . \tag{5-185}$$

That is for j=2, k=3; j=3, k=4,5; and so on. These constitute a small fraction of all j,k pairs in (181)[14], but the corresponding λ integrals are particularly large and promise compensation for their small number.

For j,k obeying (184) we expand the exponent in (181) around the point of stationary phase,

$$k\lambda + j\nu_E(\lambda) \cong k\bar{\lambda} + j\nu_E(\bar{\lambda}) - \frac{1}{2}\left[-j\frac{\partial^2\nu_E}{\partial\lambda^2}(\bar{\lambda})\right](\lambda-\bar{\lambda})^2 \quad , \tag{5-186}$$

and obtain ($\tilde{\lambda} = \lambda - \bar{\lambda}$)

$$\begin{aligned}
&\int_0^{\lambda_E} d\lambda\, \lambda \exp\left[i2\pi\left(k\lambda + j\nu_E(\lambda)\right)\right] \\
&\quad \cong \bar{\lambda} \exp\left[i2\pi\left(k\bar{\lambda} + j\nu_E(\bar{\lambda})\right)\right] \int d\tilde{\lambda}\, \exp\left[-i\pi\left(-\frac{\partial^2\nu_E}{\partial\lambda^2}(\bar{\lambda})\right)\tilde{\lambda}^2\right] \\
&\quad = \bar{\lambda}\left[-j\frac{\partial^2\nu_E}{\partial\lambda^2}(\bar{\lambda})\right]^{-1/2} \exp\left[i2\pi\left(k\bar{\lambda} + j\nu_E(\bar{\lambda})\right) - i\pi/4\right] \quad .
\end{aligned} \tag{5-187}$$

The corresponding contribution to $N_{osc}(E)$ is then

$$\begin{aligned}
\left(N_{osc}(E)\right)_{\lambda,\nu} = -4\sum_{j=2}^{\infty}\sum_{k} \frac{(-1)^{k+j}}{\pi j}\, \bar{\lambda}\left[-j\frac{\partial^2\nu_E}{\partial\lambda^2}(\bar{\lambda})\right]^{-1/2} \\
\times \cos\left[2\pi\left(k\bar{\lambda} + j\nu_E(\bar{\lambda})\right) + \pi/4\right] \quad ,
\end{aligned} \tag{5-188}$$

where the range of the k summation is given by (184). The subscript λ,ν stands for "mixed λ,ν oscillations," which name will become plausible in the next section.

Since $\lambda \sim Z^{1/3}$ and $\nu_E \sim Z^{1/3}$, the amplitude of these oscillations is $\sim Z^{1/2}$. They constitute the leading contribution to N_{osc}. However, the difference between $Z^{1/2}$ and $Z^{1/3}$ is not large of $Z \lesssim 100$; it is at most a factor of about two. Therefore, the contribution to E_{osc} that results from (188) does not dominate the ℓTF contribution (and others to be found below) for the Z values of interest.

The dependence of the right-hand side of (188) upon E is complicated because of the implicit E dependence of $\bar{\lambda}$. This is the <u>technical</u> reason for which we approximate $\nu_E(\lambda)$ by

$$\nu_E(\lambda) \cong \nu_E'(\lambda_E - \lambda) - \frac{1}{2}\, \nu_E''(\lambda_E - \lambda)^2 \quad . \tag{5-189}$$

Used for the computation of the ℓTF contribution to E_{osc}, this approximation reproduces the leading and the next-to-leading terms correctly, as a glance at Eqs.(172) and (174) demonstrates. This is the principal justification for (189). In the present context, different from the ℓTF calculation, the slope of $\nu_E(\lambda)$ at $\lambda=0$ is crucial. In order to maintain its correct value,

$$1 = {}'\nu_E = -\frac{\partial \nu_E}{\partial \lambda}(\lambda=0) \cong \nu_E' - \nu_E''{}_E \quad , \tag{5-190}$$

we have to set

$$\nu_E'' = (\nu_E'-1)/\lambda_E \tag{5-191}$$

in (189), insteat of the actual second λ-derivative of $\nu_E(\lambda)$ at $\lambda=\lambda_E$. In particular, for $E=0$, we use

$$\nu_0'' = (\nu_0'-1)/\lambda_0 = 1.01044\ Z^{-1/3} \quad , \tag{5-192}$$

which is about five times the actual $\nu_0''=0.208674\ Z^{-1/3}$, see Eq.(140). contrary to the immediate expectation, this replacement does not cause a significant error in E_{osc}: the coefficient of the next-to-leading ℓTF oscillation in (176) is changed from 0.287660 to 0.254497, which difference is indiscernible in Fig.15. We should further remark that $'\nu_0$ equals 3/2 for the TF potential, whereas $'\nu_E=1$ for $E<0$. This abrupt change of $'\nu_E$, however, can hardly be taken seriously, since it refers to the unrealistically slow decrease of the potential at large distances. Any realistic potential has $'\nu_0=1$. With this justification we shall from now on adopt the approximation (189) with ν_E'' from (191).

Then we find from Eq.(182) that $\bar{\lambda}$ is given by

$$\bar{\lambda} = \lambda_E + \frac{1}{j\nu_E''}(k-j\nu_E') = \frac{1}{j\nu_E''}[k-j(\nu_E'-\lambda_E\nu_E'')] \quad . \tag{5-193}$$

In the latter version we recognize $'\nu_E$ of (190), so that $\bar{\lambda}$ and $\bar{\lambda}-\lambda_E$ are related to $'\nu_E$ and ν_E', respectively, in an identical manner:

$$\bar{\lambda} = \frac{1}{j\nu_E''}(k-j\,'\nu_E) \quad ,$$
$$\bar{\lambda}-\lambda_E = \frac{1}{j\nu_E''}(k-j\nu_E') \quad . \tag{5-194}$$

These equations translate the range $0<\bar{\lambda}\le\lambda_E$ into the range for k given in (184), which is, of course, the origin of our emphasis on correct terminal slopes of $\nu_E(\lambda)$.

We integrate (188) over E once by parts, as required by (167), and obtain the leading term of the corresponding contribution to E_{osc},

$$(-E_{osc})_{\lambda,\nu} = -\frac{2}{\pi^2}\sum_{j=2}^{\infty}\sum_{k}\frac{(-1)^{k+j}}{j}\,\bar{\lambda}\Big[-j\frac{\partial^2\nu_E}{\partial\lambda^2}(\bar{\lambda})\Big]^{-1/2} \tag{5-195}$$

$$\times\Big[\frac{\partial}{\partial E}\big(k\bar{\lambda}+j\nu_E(\bar{\lambda})\big)\Big]^{-1}\sin\big[2\pi\big(k\bar{\lambda}+j\nu_E(\bar{\lambda})\big)+\pi/4\big]\Big|_{E=0}+\dots\ .$$

With (189) we have

$$\big(k\bar{\lambda}+j\nu_E(\bar{\lambda})\big)\Big|_{E=0} = \frac{1}{2j\nu_o''}\big[\big((\nu_o')^2-1\big)j^2+(k-j)^2\big] \tag{5-196}$$

and

$$\frac{\partial}{\partial E}\big(k\bar{\lambda}+j\nu_E(\bar{\lambda})\big)\Big|_{E=0} = \frac{\dot{\nu}_o''}{2j(\nu_o'')^2}\big[\big(2\nu_o'\dot{\nu}_o'\nu_o''/\dot{\nu}_o''-(\nu_o')^2+1\big)j^2-(k-j)^2\big]\ . \tag{5-197}$$

Differentiation of (191) supplies

$$\dot{\nu}_o'' = \dot{\nu}_o'/\lambda_o - \nu_o''\dot{\lambda}_o/\lambda_o = 59.7680\ Z^{-5/3}\ , \tag{5-198}$$

so that the amplitude of an individual term of the double sum of (195) is given by[15]

$$-\frac{4}{\pi^2}\frac{(\nu_o'')^{1/2}}{\dot{\nu}_o''}(-1)^{k+j}\frac{k-j}{j^{3/2}}\big[\big(2\nu_o'\dot{\nu}_o'\nu_o''/\dot{\nu}_o''-(\dot{\nu}_o')^2+1\big)j^2-(k-j)^2\big]^{-1}$$

$$= -\ 0.006816\ Z^{3/2}\ (-1)^{k+j}\ \frac{(k-j)/j^{3/2}}{1.127j^2-(k-j)^2} \tag{5-199}$$

$$= Z^{4/3}\times\begin{cases} 0.000\,687\ Z^{1/6} & \text{for } j=2,k=3\ ,\\[1ex] 0.000\,144\ Z^{1/6} & \text{for } j=3,k=4\ ,\\ -0.000\,427\ Z^{1/6} & \text{for } j=3,k=5\ ,\\[1ex] 0.000\,050\ Z^{1/6} & \text{for } j=4,k=5\ ,\\ -0.000\,121\ Z^{1/6} & \text{for } j=4,k=6\ ,\\ 0.000\,283\ Z^{1/6} & \text{for } j=4,k=7\ ,\\[1ex] \dots\dots \end{cases}$$

Compared to the leading ℓTF oscillation, Eq.(174) and Fig.14, which has an amplitude of

$$\frac{1}{18\sqrt{3}}\frac{\lambda_o\nu'_o}{\dot{\lambda}_o} = 0.015425\ Z^{4/3} \quad , \qquad (5\text{-}200)$$

the mixed λ,ν oscillation of (195) is very small in the small-$Z^{1/3}$ range of interest. For large values of Z, unphysically large values, (195) is the dominant contribution, as anticipated, but it has practically no significance for $Z \lesssim 100$, when $Z^{1/6} \lesssim 2.2$.

For the sake of completeness, let us also report the periods of the terms in (195). The argument of the sine function changes by 2π if $Z^{1/3}$ increases by

$$\begin{aligned} &2j\nu''_o Z^{1/3}/[((\nu'_o)^2-1)j^2+(k-j)^2] \\ &= \frac{2.021\ j}{2.755j^2+(k-j)^2} \\ &= \begin{cases} 0.336 & \text{for} \quad j=2,k=3 \ , \\ 0.235 & \text{for} \quad j=3,k=4 \ , \\ 0.211 & \text{for} \quad j=3,k=5 \ , \\ 0.179 & \text{for} \quad j=4,k=5 \ , \\ 0.168 & \text{for} \quad j=4,k=6 \ , \\ 0.152 & \text{for} \quad j=4,k=7 \ , \\ \ldots & . \end{cases} \end{aligned} \qquad (5\text{-}201)$$

The amplitudes (199) and periods (201) are not well matched so that the terms of (195) will tend to interfere destructively, thereby reducing the size of these oscillations even more. We infer that what appears to be the leading contribution to E_{osc} is rather irrelevant for the Z values of physical interest.

Fresnel integrals. After utilizing $\bar{\lambda}$ and the approximation (189) in writing

$$\begin{aligned} k\lambda+j\nu_E(\lambda) &= k\bar{\lambda}+j\nu_E(\bar{\lambda}) - \frac{1}{2}j\nu''_E(\lambda-\bar{\lambda})^2 \\ &= k\lambda_E + \frac{1}{2}j\nu''_E(\lambda_E-\bar{\lambda})^2 - \frac{1}{2}j\nu''_E(\lambda-\bar{\lambda})^2 \quad , \end{aligned} \qquad (5\text{-}202)$$

Eq. (181) becomes

$$\left(N_{qu}(E)\right)_{j\neq 0} = 4\,\mathrm{Re}\sum_{j=1}^{\infty}\sum_{k=-\infty}^{\infty}\frac{(-1)^{k+j}}{i\pi j}\exp[i2\pi\left(k\lambda_E+\tfrac{1}{2}j\nu_E''(\lambda_E-\bar{\lambda})^2\right)]$$

$$\times\int_0^{\lambda_E}d\lambda\;\lambda\,\exp[-i\pi j\nu_E''(\lambda-\bar{\lambda})^2]\quad . \tag{5-203}$$

The weight factor λ in the integral can be equivalently replaced by

$$\lambda \to \bar{\lambda} - \frac{1}{i2\pi j\nu_E''}\frac{\partial}{\partial\lambda} \tag{5-204}$$

which allows an immediate partial integration. At this stage we have

$$\left(N_{qu}(E)\right)_{j\neq 0}$$

$$= 4\,\mathrm{Re}\sum_{j=1}^{\infty}\sum_{k=-\infty}^{\infty}\frac{(-1)^{k+1}}{i\pi j}\left\{\frac{\exp\left(i2\pi j\nu_E(0)\right)-\exp(i2\pi k\lambda_E)}{i\;2\pi j\nu_E''}\right. \tag{5-205}$$

$$\left.+\;\bar{\lambda}\exp[i2\pi\left(k\bar{\lambda}+j\nu_E(\bar{\lambda})\right)]\int_0^{\lambda_E}d\lambda\;\exp[-i\pi j\nu_E''(\lambda-\bar{\lambda})^2]\right\}\quad .$$

The remaining integral is of Fresnel type. Its standard form is[16]

$$E(z) = \int_0^z dt\;\exp(-i\tfrac{\pi}{2}t^2) = C(z)-iS(z)\quad , \tag{5-206}$$

where the letters E, C, and S refer to the exponential, the cosine, and the sine functions that are integrated. $C(z)$ and $S(z)$ are, of course, real functions.

In (205) we introduce $E(z)$ through the identity

$$\exp[-i\pi j\nu_E''(\lambda-\bar{\lambda})^2] = (2j\nu_E'')^{-1/2}\frac{d}{d\lambda}E\left((2j\nu_E'')^{1/2}(\lambda-\bar{\lambda})\right)\quad , \tag{5-207}$$

with the result

$$\left(N_{qu}(E)\right)_{j\neq 0}$$

$$= 4\,\mathrm{Re}\sum_{j=1}^{\infty}\sum_{k=-\infty}^{\infty}\frac{(-1)^{k+j}}{i\pi j}\left\{\frac{\exp\left(i2\pi j\nu_E(0)\right)-\exp(i2\pi k\lambda_E)}{i\;2\pi j\nu_E''}\right. \tag{5-208}$$

$$+\;\bar{\lambda}(2j\nu_E'')^{-1/2}\exp[i2\pi\left(k\bar{\lambda}+j\nu_E(\bar{\lambda})\right)]$$

$$\left.\times\;[E\left((2j\nu_E'')^{1/2}(\lambda_E-\bar{\lambda})\right)+E\left((2j\nu_E'')^{1/2}\bar{\lambda}\right)]\right\}\quad .$$

In order to reveal the significance of the individual terms here, we need to know some of the properties of E(z). Its oscillatory nature is made explicit by writing, for z≥0,

$$E(z) = \frac{1}{2}(1-i) + i\,h(z)\exp(-i\tfrac{\pi}{2}z^2) \quad . \qquad (5\text{-}209)$$

The slowly varying function h(z) obeys the differential equation

$$h'(z) = \frac{d}{dz}\,h(z) = i\pi z h(z) - i \qquad (5\text{-}210)$$

and is subject to

$$h(o) = \frac{1}{2}(1+i) \quad . \qquad (5\text{-}211)$$

In terms of the real and the imaginary parts of h(z) these two equations are expressed by

$$\begin{aligned} h(z) &= f(z) + ig(z) \quad , \\ f'(z) &= -\pi z g(z) \quad , \quad g'(z) = \pi z f(z) - 1 \quad , \\ f(0) &= g(0) = \frac{1}{2} \quad . \end{aligned} \qquad (5\text{-}212)$$

The asymptotic expansions of f(z) and g(z) can be obtained either by repeated partial integrations in (206) or, equivalently, by iterating (210), which for this purpose has to be solved for h(z). The outcome is

$$\begin{aligned} f(z) &\sim \frac{1}{\pi z} - \frac{3}{\pi^3 z^5} + \ldots \quad \text{for } z \gg 1 \quad , \\ g(z) &\sim \frac{1}{\pi^2 z^3} - \frac{15}{\pi^4 z^7} + \ldots . \quad \text{for } z \gg 1 \quad . \end{aligned} \qquad (5\text{-}213)$$

The leading asymptotic forms represent highly accurate approximations already for relatively small z. This is demonstrated in Fig.16, which also illustrates our statement that h(z) [that is: f(z) and g(z)] is a slowly varying function compared to the exponential in Eq.(209).[17]

As defined in (206), E(z) is an odd function of z. Therefore, we take h(z) to be an odd function,

$$h(z<0) = -h(-z) \quad , \qquad (5\text{-}214)$$

which is consistent with (210) (for z≠0). The extension of (209) to include negative values of z then reads

$$E(z) = \pm\,\frac{1}{2}(1-i) + ih(z)\exp(-i\tfrac{\pi}{2}z^2) \quad , \qquad (5\text{-}215)$$

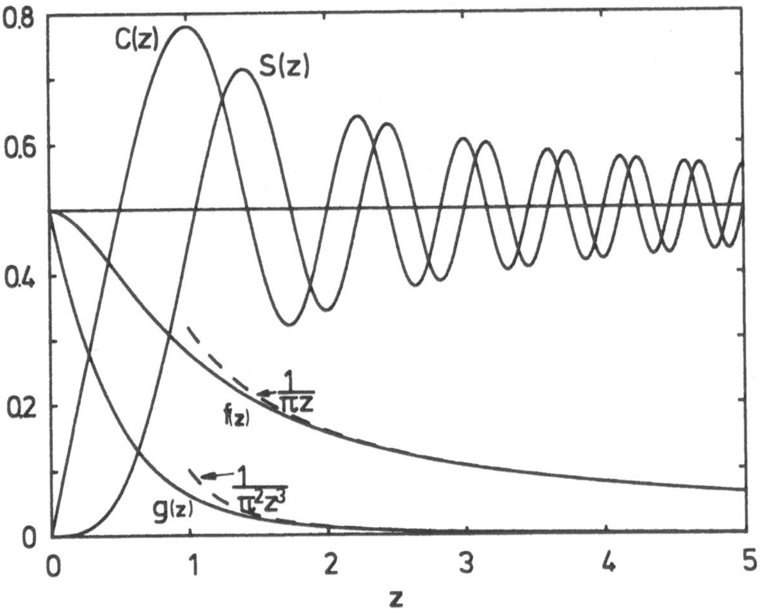

Fig.5-16. Fresnel integrals C(z) *and* S(z); *auxiliary functions* f(z) *and* g(z) *together with their leading asymptotic forms.*

where the lower sign applies for $z<0$. Note that $h(z)$ is discontinuous at $z=0$, where (210) does not apply. Of course, $E(z)$ itself is perfectly continuous.

The insertion of (215) into (208) now shows that $N_{qu}(E)$ consists of three distinct parts which are characterized by their oscillatory behavior, that is by the argument of the exponential. First, we have from the oscillatory part of the first E(...)

$$\begin{aligned}&\exp[i2\pi\left(k\bar{\lambda}+j\nu_E(\bar{\lambda})\right)-\tfrac{1}{2}j\nu_E''(\lambda_E-\bar{\lambda})^2]\\ &\quad=\exp(2\pi ik\lambda_E)\quad,\end{aligned} \tag{5-216}$$

where Eq.(202) for $\lambda=\lambda_E$ is used. The periodicity of these terms is given by the maximum value of λ; we call them <u>λ oscillations</u>. Second, analogously the second E(...) in (208) provides terms proportional to

$$\begin{aligned}&\exp[i2\pi\left(k\bar{\lambda}+j\nu_E(\bar{\lambda})-\tfrac{1}{2}j\nu_E''\bar{\lambda}^2\right)]\\ &\quad=\exp[i2\pi j\nu_E(0)]\end{aligned} \tag{5-217}$$

[Eq.(202) for $\lambda=0$ this time] which involve solely the maximum value of

ν_E; these are ν oscillations. Third, there are the constants of (215) in the E(...)'s of (208); because of the two signs in (215) these constants add up to a null result, unless the arguments of the two E(...)'s agree in sign. This is the situation if $0 \le \bar{\lambda} \le \lambda_E$, which is the requirement that $\bar{\lambda}$ lies within the range covered by the λ integration. Not surprisingly, we rediscover here the mixed λ,ν oscillations discussed in the last section. Indeed, the sum of the E(...)'s is simply $1-i = \sqrt{2}\exp(-i\pi/4)$ here, so that we are led immediately to (188).

Please note how the introduction of h(z) has enabled us to separate the rapidly oscillating circular functions from the slowly varying amplitudes. With this achievement we can again integrate over E (repeatedly) by parts to produce the various contributions to E_{osc}. The asymptotic forms of f(z) and g(z) determine the order of the respective contribution (since $z \sim Z^{1/6}$, see below), and the extrapolation down to the small-Z range of interest is done by the full z dependence of these auxiliary functions.

λ oscillations. To obtain the λ oscillations of $N_{qu}(E)$, we have to add the ℓTF oscillations (172) to those terms of (208) which exhibit the exponential (216). Thus

$$\left(N_{osc}(E)\right)_{\lambda} = \left(N_{osc}(E)\right)_{\ell TF} + 2\mathrm{Re}\sum_{j=1}^{\infty}\sum_{k\neq 0}\frac{(-1)^{k+j}}{(\pi j)^2\nu_E''}\, e^{i2\pi k\lambda_E}\left[1+\pi\bar{\lambda}\sqrt{2j\nu_E''}\,h(z_E)\right] , \tag{5-218}$$

where z_E has the significance

$$z_E \equiv \sqrt{2j\nu_E''}\,(\lambda_E-\bar{\lambda}) = \sqrt{2/(j\nu_E'')}\,(j\nu_E'-k) , \tag{5-219}$$

the last equality uses (194). The k=0 terms are omitted because they are nonoscillatory and of no interest to us in the present context. As a consequence of the jump of h(z) at z=0, the right-hand side of Eq. (218) is discontinuous for those values of E for which λ_E equals one of the $\bar{\lambda}$. Since the whole $N_{qu}(E)$ is certainly continuous, this discontinuity is not a physical effect, but rather a product of the mathematical separation into the three types of oscillations. Indeed, there is also a discontinuity in the λ,ν oscillations (188), which exactly compensates for the one in (218); see Problem 10. Therefore, we need not worry about the discontinuity in (218), and shall pretend that all arguments of h(z) (and its derivatives) are nonzero for $E \lesssim 0$.

The identity

$$\bar{\lambda}\sqrt{2j\nu_E''} = \lambda_E\sqrt{2j\nu_E''} - z_E \tag{5-220}$$

and (212) are employed in establsihing

$$\begin{aligned} &1 + \pi\bar{\lambda}\sqrt{2j\nu_E''}\, h(z_E) \\ &= \pi\lambda_E\sqrt{2j\nu_E''}\, f(z_E) + i\pi\bar{\lambda}\sqrt{2j\nu_E''}\, g(z_E) - \big(\pi z_E f(z_E) - 1\big) \\ &= \pi\lambda_E\sqrt{2j\nu_E''}\, f(z_E) + i\pi\bar{\lambda}\sqrt{2j\nu_E''}\, g(z_E) - g'(z_E) \quad . \end{aligned} \tag{5-221}$$

We use this decomposition for the evaluation of the real part in (218), with the outcome

$$\begin{aligned} \big(N_{osc}(E)\big)_\lambda = {} & \big(N_{osc}(E)\big)_{\ell TF} \\ & + 2\lambda_E \sum_{j=1}^{\infty}\sum_{k\neq 0} \frac{(-1)^{k+j}}{\pi j}\sqrt{2/(j\nu_E'')}\, f(z_E)\cos(2\pi k\lambda_E) \\ & - 2\sum_{j=1}^{\infty}\sum_{k\neq 0} \frac{(-1)^{k+j}}{\pi j}\bar{\lambda}\sqrt{2/(j\nu_E'')}\, g(z_E)\sin(2\pi k\lambda_E) \\ & - \frac{2}{\nu_E''}\sum_{j=1}^{\infty}\sum_{k\neq 0} \frac{(-1)^{k+j}}{(\pi j)^2}\, g'(z_E)\cos(2\pi k\lambda_E) \quad . \end{aligned} \tag{5-222}$$

These three double sums have differing large-Z behavior. The asymptotic forms of f(z) and g(z), given in (213), combined with the Z dependences

$$\begin{aligned} \lambda,\ \lambda_E &\sim Z^{1/3} \ , \\ \nu_E'' &\sim Z^{-1/3} \ , \\ z_E &\sim Z^{1/6} \ , \end{aligned} \tag{5-223}$$

imply that these three terms describe λ oscillations with amplitudes proportional to $Z^{1/3}$, Z^{o}, and $Z^{-1/3}$, respectively. These simple Z dependences, however, hold only for very large Z; more precisely: they hold when the asymptotic forms of f(z) and g(z) can be used for all j and k to be summed over. For the rather small values of Z we are interested in, there are j,k pairs (mainly the ones with k=2j) for which z_E is not in the asymptotic domain. In other words, while the asymptotics of f(z) and g(z) identify the double sums of (222) to belong to the leading λ oscillation, the next-to-leading one, ... , the extrapolation to the small-$Z^{1/3}$ range is not done correctly if one sticks to these asymptotic forms. Instead, as we stated already at the end of the preceding section, this extrapolation is supplied by the use of the full $f(z_E)$ and $g(z_E)$.

An important observation is that the ℓTF terms in (222) combine naturally with the contribution from the asymptotic forms of $f(z_E)$ and $g(z_E)$. We illustrate this unification for the leading λ oscillation. The relevant terms are here

$$- 2\lambda_E\nu_E' \sum_{k=1}^{\infty} \frac{(-1)^k}{(\pi k)^2} \cos(2\pi k\lambda_E)$$

$$+ 2\lambda_E \sum_{j=1}^{\infty} \sum_{k\neq 0} \frac{(-1)^{k+j}}{\pi j} \sqrt{2/(j\nu_E')}\, \frac{1}{\pi z_E} \cos(2\pi k\lambda_E) \qquad (5\text{-}224)$$

$$= - \lambda_E \sum_{k\neq 0} (-1)^k \cos(2\pi k\lambda_E) \left\{\frac{\nu_E'}{(\pi k)^2} - \sum_{j=1}^{\infty} \frac{(-1)^j}{\pi j} \frac{2}{\pi j\nu_E' - \pi k}\right\}$$

$$= - \lambda_E \sum_{k\neq 0} \frac{(-1)^k}{\pi k} \cos(2\pi k\lambda_E) \left\{\frac{\nu_E'}{\pi k} - \sum_{j\neq 0} (-1)^j \left(\frac{1}{-\pi j} - \frac{1}{\pi k/\nu_E' - \pi j}\right)\right\} ,$$

where the invariance of the summands under $j \to -j$, $k \to -k$ has been used for rewriting the expressions. Now we can employ the identity

$$\sum_{j\neq 0} (-1)^j \frac{1}{x-\pi j} = \sum_{j=-\infty}^{\infty} (-1)^j \frac{1}{x-\pi j} - \frac{1}{x} \qquad (5\text{-}225)$$

$$= \frac{1}{\sin x} - \frac{1}{x}$$

twice and equate (224) to

$$- \lambda_E \sum_{k\neq 0} \frac{(-1)^k}{\pi k} \frac{\cos(2\pi k\lambda_E)}{\sin(\pi k/\nu_E')} \qquad (5\text{-}226)$$

$$= - 2\lambda_E \sum_{k=1}^{\infty} \frac{(-1)^k}{\pi k} \frac{\cos(2\pi k\lambda_E)}{\sin(\pi k/\nu_E')} \quad .$$

The leading λ oscillation in (222) is therefore given by

$$\left(N_{osc}(E)\right)_\lambda = -2\lambda_E \sum_{k=1}^{\infty} \frac{(-1)^k}{\pi k} \frac{\cos(2\pi k\lambda_E)}{\sin(\pi k/\nu_E')}$$

$$+ 2\lambda_E \sum_{k\neq 0} \sum_{j=1}^{\infty} \frac{(-1)^{k+j}}{\pi k} \sqrt{2/(j\nu_E'')}\, \tilde{f}(z_E)\cos(2\pi k\lambda_E) \tag{5-227}$$

$$+ \ldots ,$$

where the ellipsis represents the nonleading λ oscillations, and $\tilde{f}(z)$ equals $f(z)$ without its leading asymptotic term

$$\tilde{f}(z) \equiv f(z) - \frac{1}{\pi z} , \quad \tilde{g}(z) \equiv g(z) - \frac{1}{\pi^2 z^3} , \tag{5-228}$$

likewise for $g(z)$ and $\tilde{g}(z)$. In passing, we remark that in the situation of linear degeneracy $[\nu_E'' \to 0,\ z_E \sim \nu_E''^{-1/2} \to \infty,\ \tilde{f}(z_E) \sim -z_E^{-3}]$ the second term in (227) vanishes, which identifies the first one as the leading λ oscillation for linear degeneracy. This can, of course, also be demonstrated directly. For detail consult Ref.12.

A first partial E integration of (227) supplies the leading λ oscillation of E_{osc}. It is given by

$$(-E_{osc}/Z^{4/3})_\lambda = \sum_{k=1}^{\infty} s_k \sin(2\pi k\lambda_o) + \ldots , \tag{5-229}$$

with the Z dependent coefficients s_k given by

$$s_k = \left(\frac{\lambda_o/Z^{1/3}}{\dot{\lambda}_o/Z^{-3/3}}\right)\left[-\frac{(-1)^k}{(\pi k)^2} \frac{1}{\sin(\pi k/\nu_o')}\right.$$

$$+ \frac{(-1)^k}{\pi k} \sum_{j=1}^{\infty} \frac{(-1)^j}{\pi j} \sqrt{2/(j\nu_o'')} \left\{\tilde{f}\left(\sqrt{2/(j\nu_o'')}\,(j\nu_o' - k)\right)\right. \tag{5-230}$$

$$\left.\left. - \tilde{f}\left(\sqrt{2/(j\nu_o'')}\,(j\nu_o' + k)\right)\right\}\right] .$$

Please note that, because $\nu_o' = 1.93\ldots$, very large values of j and k are required to obtain a vanishing argument of $\tilde{f}$. These terms do not contribute significantly to the Fourier sum of Eq.(229). Therefore, our disregarding of the consequences of the discontinuity of $h(z)$ at $z = 0$ is, for all practical purposes, harmless (not to mention the possibility that ν_o' is irrational); see also Problem 11.

For very large Z, the sum over j does not contribute to s_k in (230), so that with the numbers from (138) and (139)

$$\delta_k = \frac{(-1)^{k+1}}{k^2} \frac{0.02515}{\sin(1.621k)} \quad \text{for} \quad Z \gg 1 \quad . \tag{5-231}$$

The first few of the δ_k's are thus

$$\begin{aligned} \delta_1 &= 0.02518 \;, & \delta_2 &= 0.06232 \;, \\ \delta_3 &= -0.00283 \;, & \delta_4 &= -0.00783 \;, \\ \delta_5 &= 0.00104 \;, & \delta_6 &= 0.00234 \;, \\ \delta_7 &= -0.00055 \;, & \delta_8 &= -0.00100 \;. \end{aligned} \tag{5-232}$$

These large-Z values cannot be used for the small-Z range of physical interest. The numbers listed in Table 2 show that the δ_k with even k do not change much with $Z^{1/3}$, whereas those with odd k change markedly and differ, for $Z^{1/3}$=1...5, substantially from their asymptotic values (232). Another reason why (232) cannot be used in the small-Z range is the subject of Problem 12.

This essential difference between $\delta_1, \delta_3, \delta_5, \ldots$ and $\delta_2, \delta_4, \delta_6, \ldots$ is understood upon recalling that $\tilde{f}(z)$, the difference between f(z) and its asymptotic form, is large only for small arguments z. In Eq.(230) this requires $j\nu_o' \cong k$. Now, $\nu_o' \cong 2 - 1/16$, so that the term with $2j = k$ is picked out, which happens only for even k, of course. Let us use this insight to find an approximation for δ_k. If k is odd, (231) will do. If k is even, we add the j=k/2 term of the sum in (230), where we evaluate the relevant $\tilde{f}(z)$ according to

$$\tilde{f}\left(\sqrt{2/(j\nu_o'')}\,(j\nu_o'-k)\right) = \tilde{f}\left(-\sqrt{k/\nu_o''}\,(2-\nu_o')\right)$$
$$= f\left(-\sqrt{k/\nu_o''}\,(2-\nu_o')\right) + \frac{1}{\pi\sqrt{k/\nu_o''}\,(2-\nu_o')} \quad \cong .$$

Table 5-2. Coefficients $\delta_1, \delta_2, \delta_3$, and δ_4 for $Z^{1/3} = 1, 1.5, \ldots, 5$.

$Z^{1/3}$	δ_1	δ_2	δ_3	δ_4
1	0.02467	0.00683	-0.00248	-0.00132
1.5	0.02490	0.00876	-0.00261	-0.00166
2	0.02500	0.01035	-0.00268	-0.00193
2.5	0.02506	0.01174	-0.00272	-0.00216
3	0.02509	0.01297	-0.00275	-0.00236
3.5	0.02511	0.01409	-0.00277	-0.00255
4	0.02513	0.01512	-0.00279	-0.00271
4.5	0.02514	0.01607	-0.00279	-0.00287
5	0.02515	0.01696	-0.00279	-0.00301

$$\cong -\frac{1}{2} + \frac{1}{\pi}\sqrt{\nu_o''/k}\,\frac{1}{2-\nu_o'} \quad . \tag{5-233}$$

In addition to the definition of $\tilde{f}(z)$ in (228), we made use of

$$f(z\lesssim 0) = -\frac{1}{2} + O(z^2) \cong -\frac{1}{2} \quad , \tag{5-234}$$

which is an immediate consequence of Eqs. (212) and (214). The correction to (231) for even k is, therefore, given by

$$\left(\frac{\lambda_o/Z^{1/3}}{\dot{\lambda}_o/Z^{-3/3}}\right)\frac{(-1)^k}{\pi k}\,\frac{(-1)^{k/2}}{\pi k/2}\,\sqrt{\frac{2}{k\nu_o''/2}}\left[-\frac{1}{2}+\frac{1}{\pi}\sqrt{\frac{\nu_o''}{k}}\,\frac{1}{2-\nu_o'}\right]$$

$$= \left(\frac{\lambda_o/Z^{1/3}}{\dot{\lambda}_o/Z^{-3/3}}\right) 2\,\frac{(-1)^{k/2}}{(\pi k)^2}\left[-\frac{1}{\sqrt{k\nu_o''}}+\frac{1}{\pi k}\,\frac{2}{2-\nu_o'}\right] \quad . \tag{5-235}$$

We insert the numbers given in (138), (139), and (192), and summarize in stating

$$\delta_k \cong \frac{0.02515}{k^2} \times \begin{cases} 1/\sin(1.621k) & \text{for } k \text{ even} \quad , \\ -1/\sin(1.621k) + (-1)^{k/2}\left(\dfrac{20.430}{k} - \dfrac{1.990Z^{1/6}}{\sqrt{k}}\right) & \text{for } k \text{ odd.} \end{cases} \tag{5-236}$$

For most applications this approximation suffices. One must be aware of its limitations, however. Evidently, the arguments that led us from (230) to (236) are such that (236) is reliable only if neither k nor $Z^{1/6}$ is large. [The failure for large Z is also demonstrated by the fact that the exact δ_k become Z independent for $Z \gg 1$, whereas the even-k ones do not in (236).] Fortunately, the Z values of interest are not large ($Z^{1/6}$ ranges from 1 to about 2.2), and the sum over k in (229) converges rapidly, so that δ_k is needed only for the first few k's. When keeping this in mind, there is little danger in using (236) instead of (230).

The first partial E integration of (227) produced the leading λ oscillation of E_{osc}, given by Eqs. (229) and (230). A second partial E integration supplies a contribution to the next-to-leading λ oscillation. Another contribution comes from the next-to-leading term in $\left(N_{osc}(E)\right)_\lambda$ which supplements (227). These two contributions combined lead to a refinement of (229),

$$\left(\frac{-E_{osc}}{Z^{4/3}}\right)_\lambda = \sum_{k=1}^{\infty} \delta_k \sin(2\pi k\lambda_o) + Z^{-1/3}\sum_{k=1}^{\infty} c_k \cos(2\pi k\lambda_o) + \ldots \quad . \tag{5-237}$$

Table 5-3. Coefficients c_1, c_2, c_3, and c_4 for $Z^{1/3} = 1, 1.5, \ldots, 5; \gg 1$.

$Z^{1/3}$	c_1	c_2	c_3	c_4
1	-0.04584	0.00134	0.00098	-0.00059
1.5	-0.04843	0.00792	0.00121	-0.00176
2	-0.04973	0.01601	0.00138	-0.00311
2.5	-0.05047	0.02518	0.00151	-0.00459
3	-0.05092	0.03519	0.00161	-0.00616
3.5	-0.05122	0.04588	0.00168	-0.00780
4	-0.05142	0.05712	0.00173	-0.00948
4.5	-0.05157	0.06883	0.00178	-0.01120
5	-0.05168	0.08094	0.00181	-0.01294
$\gg 1$	-0.05218	4.06250	0.00200	-0.25342

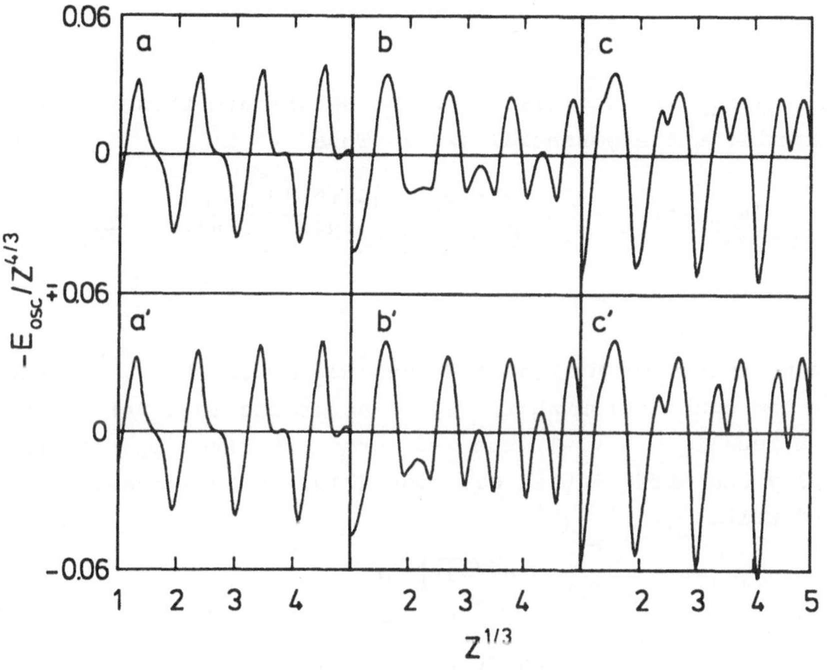

Fig. 5-17. Contributions to $-E_{osc}/Z^{4/3}$ from λ oscillations, as a function of $Z^{1/3}$; (a) leading λ oscillation with exact s_k from (230); (a') like (a), but using approximate s_k from (236); (b) next-to-leading λ oscillation with exact c_k; (b') like (b) with approximate c_k; (c) sum of (a) and (b); (c') sum of (a') and (b'). In (a), (a'), (b), and (b') six terms are summed.

The coefficients c_k are given by an expression analogous to (230) and can be approximated in the fashion of (236). We shall not exhibit these details, and refer the reader to the original publication.[18] Here we are content with the numbers listed in Table 3. We observe that these c_k's differ from their large-Z values as long as $Z^{1/3}$ is not very large, whereby the even-k ones are only a small fraction of their asymptotic value. Further, we note that the s_k's and c_k's are of roughly the same size. Consequently, the next-to-leading λ oscillation is not dominated by the leading one in the physical range of Z. This is also confirmed by Fig.17, where, as in Tables 2 and 3, we notice the growing importance of the even-k terms for increasing Z. This is a consequence of $\nu_0' \cong 2$. For example, when Z becomes very large, the ratio s_2/s_1 is 2.48, so that then the dominant period of the oscillations is halved. This effect is even more pronounced for the next-to-leading λ oscillation, as is visible in Fig.17 and is numerically expressed by $c_2/c_1 = -77.9$ for $Z \gg 1$. However, being suppressed by a factor of $Z^{1/3}$, it is nevertheless small compared to the leading λ oscillation for such enormous values of Z.

ν oscillations. The ν oscillations of $N_{qu}(E)$ are those terms in (208) which exhibit the exponential (217). Thus

$$\left(N_{osc}(E)\right)_\nu = 4\,\mathrm{Re}\sum_{j=1}^{\infty}\sum_{k=-\infty}^{\infty}\frac{(-1)^{k+j}}{i\pi j}\left[\frac{1}{i2\pi j\nu_E''}+\frac{\overline{\lambda}}{\sqrt{2j\nu_E''}}ih(\sqrt{2j\nu_E''}\,\overline{\lambda})\right] \times e^{i2\pi j\nu_E(0)} \quad . \tag{5-238}$$

The discontinuity of h(z) at z=0 does not matter here, because we encounter the combination zh(z). Since according to (194) $\overline{\lambda}=(k-j)/(j\nu_E'')$, where we recall that $'\nu_E=1$, the $\overline{\lambda}=0$ terms are given by k=j. For $k\neq j$, we can employ the differential equation (210) that is obeyed by h(z) for $z\neq 0$, and write

$$\begin{aligned}&\frac{1}{i2\pi j\nu_E''}+\frac{\overline{\lambda}}{\sqrt{2j\nu_E''}}\,i\,h(\sqrt{2j\nu_E''}\,\overline{\lambda})\\ &=\frac{1}{2\pi j\nu_E''}h'(\sqrt{2j\nu_E''}\,\overline{\lambda})=\frac{1}{2\pi j\nu_E''}h'\left(\sqrt{2/(j\nu_E'')}\,(k-j)\right)\quad .\end{aligned} \tag{5-239}$$

At this stage, we have

$$\left(N_{osc}(E)\right)_\nu = -\frac{2}{\nu_E''}\,\mathrm{Re}\sum_{j=1}^{\infty}\frac{(-1)^j}{(\pi j)^2}\,e^{i2\pi j\nu_E(0)}\ \times$$

$$\times \left[(-1)^j + i \sum_{k\neq j} (-1)^k \, h'\left(\sqrt{2/(j\nu_E'')}\,(k-j)\right)\right] \quad , \tag{5-240}$$

which is further simplified upon setting $m\equiv k-j$ and using $h'(-z)=h'(z)$:

$$\begin{aligned}\left(N_{osc}(E)\right)_\nu &= -\frac{2}{\nu_E''}\sum_{j=1}^{\infty}\frac{1}{(\pi j)^2}\cos\left(2\pi j\nu_E(0)\right) \\ &\quad -\frac{4}{\nu_E''}\,\mathrm{Re}\sum_{j=1}^{\infty}\frac{e^{i2\pi j\nu_E(0)}}{(\pi j)^2}\; i\sum_{m=1}^{\infty}(-1)^m h'\left(\sqrt{2/(j\nu_E'')}\;m\right) .\end{aligned} \tag{5-241}$$

In this form the first sum over j is immediately identified as the leading oscillation; since $h'(z) \sim z^{-2}$ for $z \gg 1$, the double sum is smaller by a factor of $\nu_E'' \sim Z^{-1/3}$ for large Z.

This leading ν oscillation is built like the leading ℓTF oscillation in (172). We can therefore quickly write down the leading ν oscillation of the binding energy:

$$(-E_{osc})_\nu = -\frac{1}{\nu_o''\dot{\nu}_o(0)}\sum_{j=1}^{\infty}\frac{1}{(\pi j)^3}\sin\left(2\pi j\nu_o(0)\right) + \dots \quad . \tag{5-242}$$

This is, of course, of the shape of the leading ℓTF oscillation in (174), plotted in Fig.14, with the period shortened by the fraction

$$\frac{\lambda_o}{\nu_o(0)} = \frac{\lambda_o}{\lambda_o\nu_o' - \frac{1}{2}\lambda_o^2\nu_o''} = \frac{2}{\nu_o' + {}'\nu_o} = \frac{1}{1.46884} \quad , \tag{5-243}$$

where (189) and (190) are used, and the amplitude reduced by the factor

$$\frac{1}{\nu_o''\dot{\nu}_o(0)}\,\frac{\lambda_o}{\lambda_o\nu_o'} = \frac{1}{16.0256} \quad . \tag{5-244}$$

Here one needs

$$\begin{aligned}\dot{\nu}_o(0) &= \frac{d}{dE}\left(\lambda_E\nu_E' - \frac{1}{2}\lambda_E^2\nu_E''\right)\Big|_{E=0} \\ &= \dot{\lambda}_o\nu_o' + \lambda_o\dot{\nu}_o' - \lambda_o\dot{\lambda}_o\nu_o'' - \frac{1}{2}\lambda_o^2\dot{\nu}_o'' \\ &= \frac{1}{2}[\nu_o'\lambda_o + (\nu_o'+1)\dot{\lambda}_o] = 32.9806\ Z^{-3/3} \quad ,\end{aligned} \tag{5-245}$$

which utilizes (198) and the numbers in (138) and (139). The amplitude of the leading TF oscillation is $0.015Z^{4/3}$, see Fig.(14) and Eq.(200), so that the leading ν oscillation has an amplitude of $0.001\ Z^{4/3}$. This is so small that we need not consider the next-to-leading ν oscillation.

Semiclassical prediction for E_{osc}. The time has come to put things together. We have identified various contributions to the binding-energy oscillations,

$$-E_{osc} = (-E_{osc})_{\lambda,\nu} + (-E_{osc})_{\lambda} + (-E_{osc})_{\nu} \quad . \tag{5-246}$$

The separate calculations of the three types of oscillations resulted in Eqs.(195) - (199) for the mixed λ,ν oscillations, in Eq.(237) with (230) for the λ oscillations (which contain also the ℓTF terms), and in Eq. (242) for the ν oscillations. Figure 17(c) tells us that the amplitude of the λ oscillations is about $0.05\ Z^{4/3}$, whereas the other two terms in (246) both have an amplitude of about $0.001\ Z^{4/3}$. As far as the mixed λ,ν oscillations are concerned, this statement is true for the small-Z range of physical interest. In contrast, when $Z \gg 1$, this oscillation is $\sim 0.001\ Z^{3/2}$ and the λ-oscillation is $\sim s_2 Z^{4/3} \cong 0.06\ Z^{4/3}$, so that one needs at least $Z^{1/6} \gtrsim 60$, or $Z \gtrsim 5\times10^{10}$, for the mixed λ,ν oscillations to be dominant. This is ridiculously far beyond the domain of physics. [In passing we remark that we have just delivered the justification of the statements following Eq.(150)].

Both the ν and the λ,ν oscillations are very small, and we shall neglect them completely. Inasmuch as the subsequent λ oscillations in (237) are expected to be of larger amplitude, this is thouroughly justified. We must also not forget that the approximation (189) with ν_E'' from (191) introduces an error, in view of which it is quite unnecessary to pay attention to the small corrections that the ν and the λ,ν oscillations represent. Consequently, our semiclassical prediction for E_{osc} is given by the two terms on the right-hand side of Eq.(237), the sum of the leading and the next-to-leading λ oscillation:

$$\left(\frac{-E_{osc}}{Z^{4/3}}\right)_{SC} = \sum_{k=1}^{\infty} s_k \sin(2\pi k\lambda_o) + Z^{-1/3}\sum_{k=1}^{\infty} c_k \cos(2\pi k\lambda_o) . \tag{5-247}$$

It is plotted in Fig.17(c). We compare it with the HF prediction of Fig.2 in Fig.18.

Both curves agree in a number of details. First, they have the same phase and period, which is given by λ_o, the maximum value of the

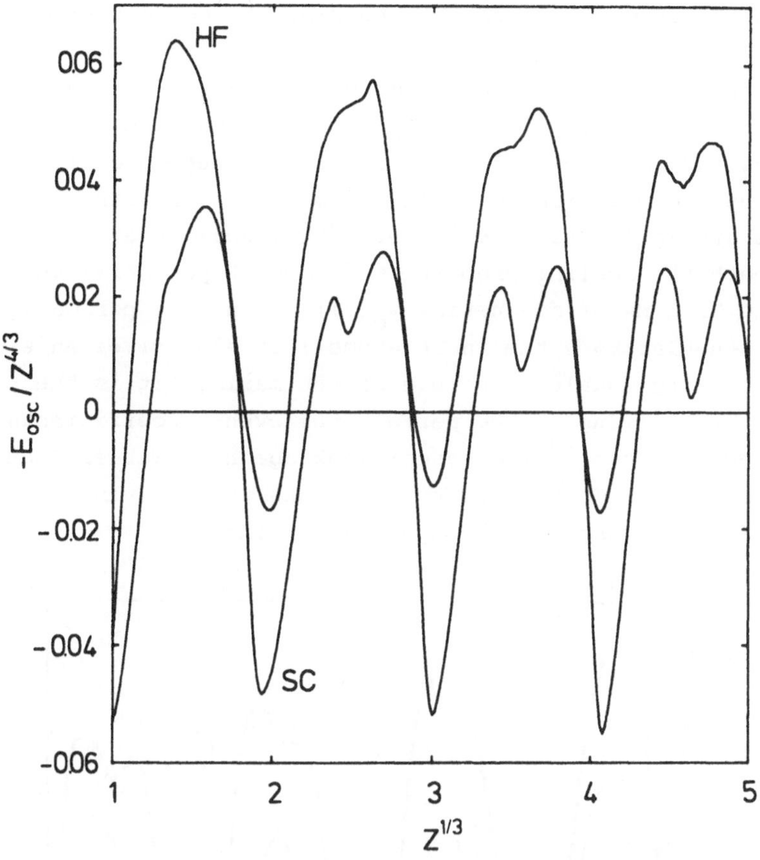

Fig.5-18. Comparison of our semiclassical prediction (SC) *for the binding-energy oscillations with the* HF *prediction. Ten terms are added in each sum of* (247).

angular quantum number in the TF limit. Then their amplitudes are about the same. Further, they both show rather sharp structureless minima, and maxima with an evolving double structure. The latter phenomenon is somewhat more pronounced in the semiclassical curve. But since we cannot really compare with experimental data, there is no way of judging which one is right.

The main difference between the HF prediction for E_{osc} and the semiclassical one is that the second curve is shifted down in Ref.18. A smooth term of order $Z^{4/3}$ is obviously missing in the binding-energy formula. Our calculation of E_{osc} concentrated on the oscillatory contributions and consistently disregarded all smooth contributions, so that this missing term could not be found. As we have remarked around Eq.(4), there are indications that the correlation energy (4-248) must be included into the description in order to be able to find the correct smooth term of order $Z^{4/3}$.

Although the HF method produces the curve of Fig.2 [strictly speaking, not even that, since E_{stat} of (1) is not a HF result], it provides no insight whatsoever for the origin of these oscillations. In contrast, the semiclassical calculation supplies us with an understanding of the over-all amplitude factor $Z^{4/3}$ [recall that this is a consequence of both the scaling properties of the TF potential and the bent shape of the TF line of degeneracy $\nu_o(\lambda)$] and of the period (given by the largest angular momentum in TF atoms); it also gives an explanation for the intriguing double structure of the maxima. It is the beginning of an effective halving of the period: the even-k coefficients in (247) change enormously as $Z^{1/3}$ increases; look again at Tables 2 and 3 as well as at Eq.(236). To see how this effects the evolving double-peak structure of the maxima, we decompose (247) into

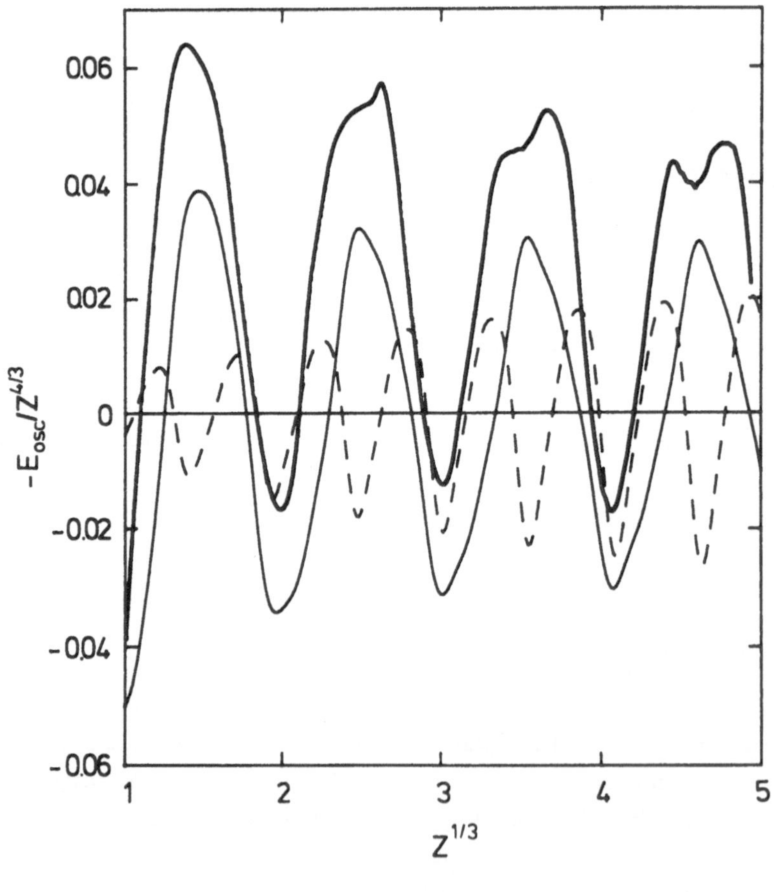

Fig.5-19. Thick curve: HF; *thin curve;* SC, *odd; dashed curve;* SC, *even.*

$$\left(\frac{-E_{osc}}{Z^{4/3}}\right)_{SC,odd} = \sum_{k=1,3,5...}\left(s_k \sin(2\pi k\lambda_o) + Z^{-1/3}c_k\cos(2\pi k\lambda_o)\right) \tag{5-248}$$

and

$$\left(\frac{-E_{osc}}{Z^{4/3}}\right)_{SC,even} = \sum_{k=2,4,6,...}\left(s_k \sin(2\pi k\lambda_o) + Z^{-1/3}c_k\cos(2\pi k\lambda_o)\right). \tag{5-249}$$

These are plotted, along with the HF oscillations, in Fig.19. We observe that the relative phase of (248) and (249) is such that the extrema of the first coincide with the minima of the second. The sum has, therefore, sharp structureless minima and broad doubly peaked maxima. And the increase in amplitude of (249) is responsible for the growing dip between the pairs of maxima.

Other manifestations of shell structure. In this section we shall briefly discuss other quantities than the binding energy which show shell effects.

In Chapter Three we found that the density of electrons at the site of the nucleus is given by

$$n_o/\frac{(2Z)^3}{4\pi} = 1.2021 - 1.7937\ Z^{-2/3} + O(1/Z) \tag{5-250}$$

for a neutral atom; see Eq.(3-166). The estimate of (3-167),

$$O(1/Z) \cong 1.82/Z - 0.82/Z^{5/3} \tag{5-251}$$

is compared to the HF prediction of $O(1/Z)$ in Fig.3-5, where we notice that n_o contains on oscillatory part. Since the amplitude of the oscillations around the smooth curve in this plot decreases with Z whereby the period gets longer, the natural surmise is that we have to supplement (251) by a term periodic in $Z^{1/3}$ with amplitude $Z^{-4/3}$. We gain some insight by considering Bohr atoms, for which

$$n_o/\frac{(2Z)^3}{4\pi} = 1.2021 - \frac{1}{2}\left(\frac{3}{2}N\right)^{-2/3} + \frac{5}{2}\left(\frac{3}{2}N\right)^{-4/3}\left(\langle y\rangle^2 - \frac{1}{6}\right) + O(N^{-5/3})\ , \tag{5-252}$$

as obtained by calculating the term of order $N^{-4/3}$ in Problem 1-5. Here y is the solution of Eq.(6). For neutral Bohr atoms, the leading oscillatory contribution to n_o is, therefore,

$$\left(n_o/\frac{(2Z)^3}{4\pi}\right)_{osc} \cong 1.46\ Z^{-4/3} \sum_{k=1}^{\infty} \frac{(-1)^k}{(\pi k)^2} \cos(2\pi k \times 1.145\ Z^{1/3}) \quad , \tag{5-253}$$

which has the expected structure. For real atoms, the oscillations in n_o are still waiting to be calculated.

In Chapter Four we found the ionization energy as predicted by the statistical model. The result is reported in Eq.(4-294) and compared to experimental data in Fig.4-8. Obviously, there are very pronounced shell effects, which we isolate by writing

$$I_{osc}(Z) = I(Z) - I_{stat}(Z) \quad . \tag{5-254}$$

One naturally presumes that $I_{osc}(Z)$ equals $Z^{-1/3}$ times a fluctuating function of $Z^{1/3}$. This is confirmed by the experimental data presented in Fig.20. A semiclassical prediction for I_{osc} has not been calculated as yet. For the evaluation of

$$I_{osc}(Z) = E_{osc}(Z,N=Z-1) - E_{osc}(Z,N=Z) \tag{5-255}$$

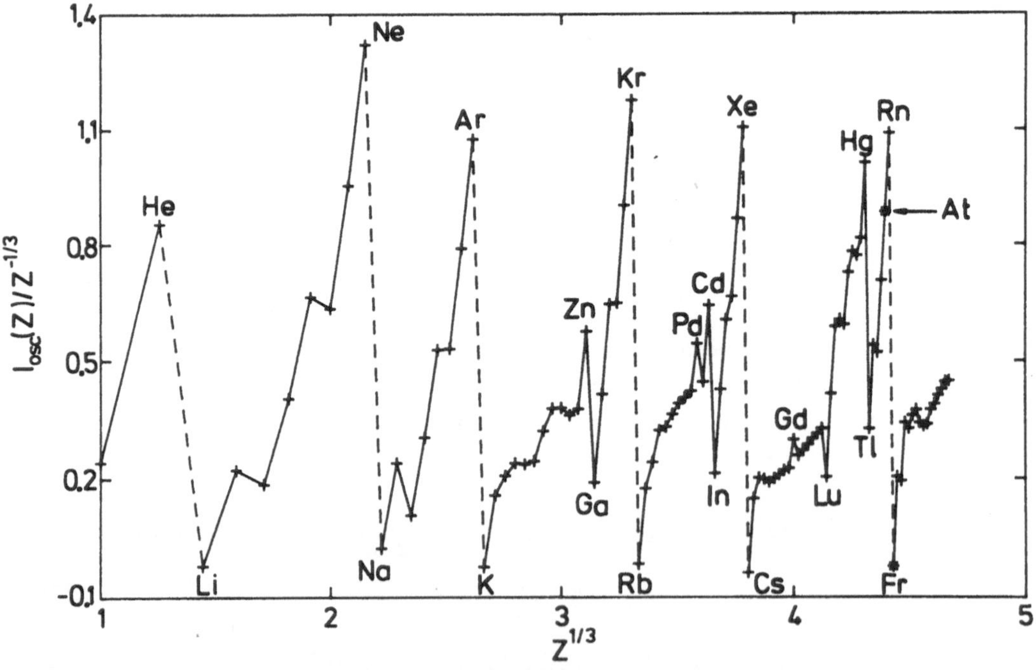

Fig.5-20. Experimental values of $I_{osc}(Z)/Z^{-1/3}$ as a function of $Z^{1/3}$. See also Fig.4-8.

one needs E_{osc} for weakly ionized atoms. This requires a study of the TF lines of degeneracy for such ions, which differ substantially from those of neutral TF atoms because of the long-range Coulomb part in the effective potential of an ion.

Also in Chapter Four we observed, in Table 4-1, that the statistical model prediction for the expectation value of 1/r, for neutral atoms,

$$\langle\frac{1}{r}\rangle_{stat} = 1.79374\ Z^{4/3} - Z + 0.44983\ Z^{2/3} \tag{5-256}$$

agrees well with the corresponding HF prediction. The numbers of Table 4-1 indicate that the difference is $Z^{2/3}$ times an oscillatory function of $Z^{1/3}$. For this function we write

$$\langle\frac{1}{r}\rangle_{osc} = \langle\frac{1}{r}\rangle - \langle\frac{1}{r}\rangle_{stat} \quad . \tag{5-257}$$

In Fig.21 the HF predictions for $\langle 1/r\rangle_{osc}$ are plotted. Since, see Eq.

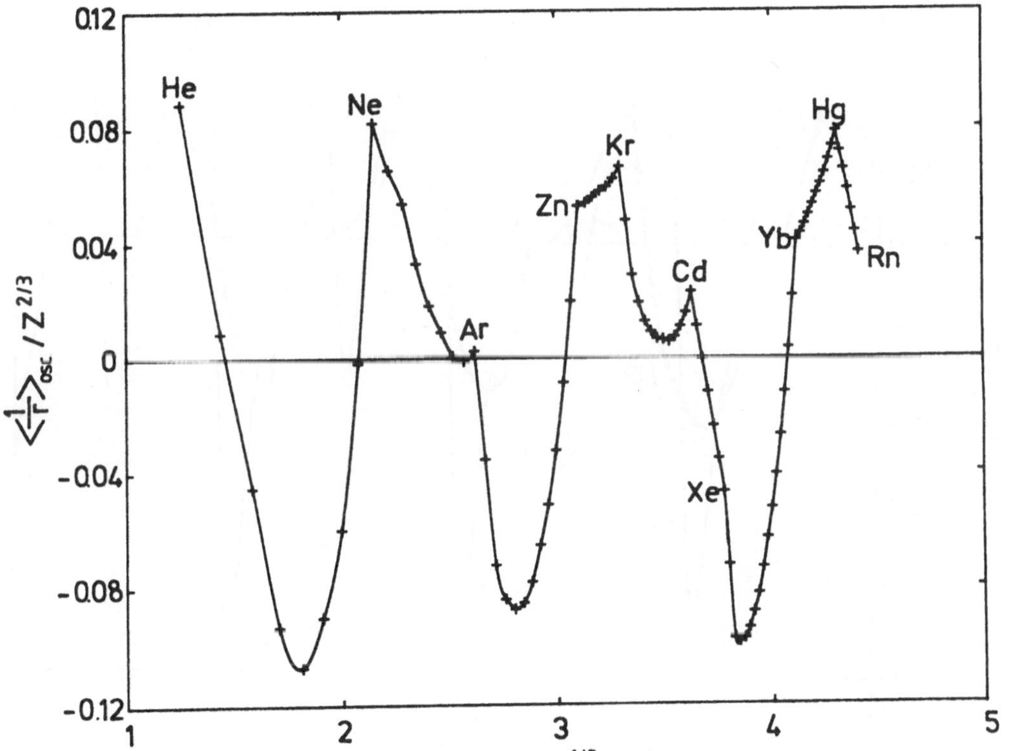

Fig.5-21. HF prediction for $\langle\frac{1}{r}\rangle_{osc}/Z^{2/3}$ *as a function of* $Z^{1/3}$.

(4-318),

$$\langle\frac{1}{r}\rangle = -\frac{\partial}{\partial Z}E(Z,N)\Big|_{N=Z}$$

$$= -\frac{d}{dZ}E(Z,Z) + \frac{\partial}{\partial N}E(Z,N)\Big|_{N=Z} \qquad (5\text{-}258)$$

$$= -\frac{d}{dZ}E(Z,Z) - \zeta(Z,Z)$$

and

$$\langle\frac{1}{r}\rangle_{stat} = -\frac{d}{dZ}E_{stat}(Z,Z) \qquad (5\text{-}259)$$

we find

$$\langle\frac{1}{r}\rangle_{osc} = \frac{d}{dZ}\left(-E_{osc}(Z,Z)\right) - \zeta(Z,Z) \quad . \qquad (5\text{-}260)$$

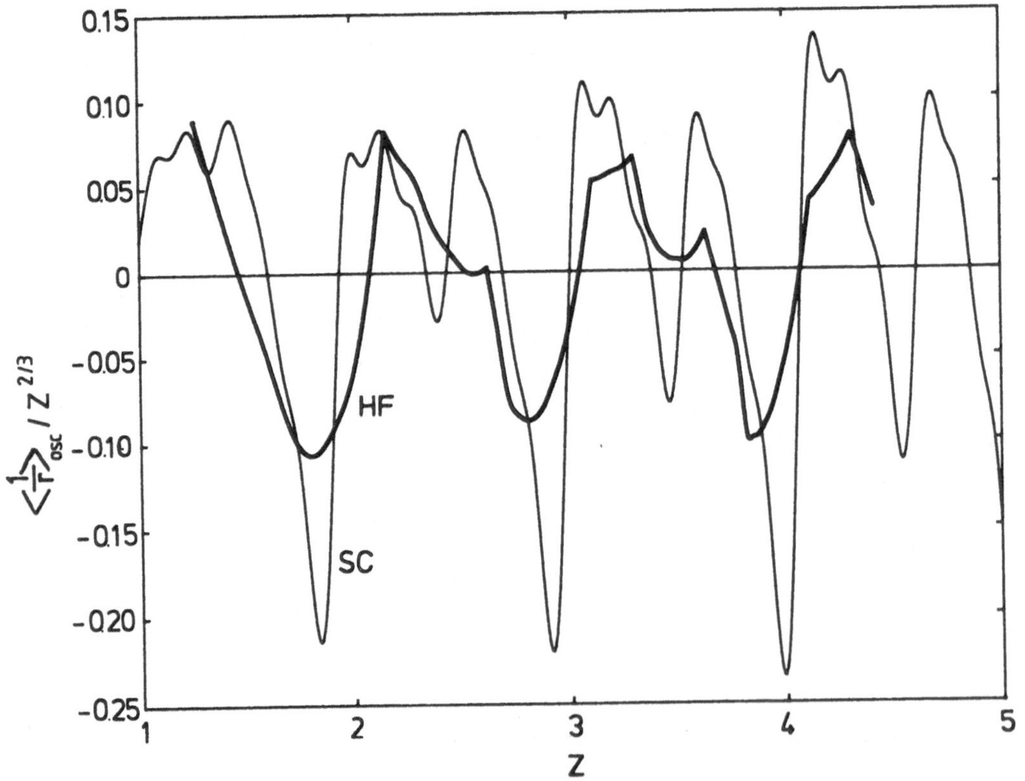

Fig.5-22: HF prediction for $\langle 1/r\rangle_{osc}$ compared to the semiclassical (SC) prediction for $\langle 1/r\rangle_{osc} + \zeta$.

Inserting the semiclassical prediction for E_{osc}, Eq.(247), supplies the leading terms

$$\left(\langle\frac{1}{r}\rangle_{osc} + \zeta\right)/Z^{2/3}$$

$$\cong \frac{2\pi}{3}\frac{\lambda_o}{Z^{1/3}}\sum_{k=1}^{\infty} k\,\delta_k \cos(2\pi k\lambda_o) \tag{5-261}$$

$$+\frac{1}{Z^{1/3}}\sum_{k=1}^{\infty}\left(\frac{4}{3}\delta_k - \frac{2\pi}{3}\frac{\lambda_o}{Z^{1/3}} k\, c_k\right)\sin(2\pi k\lambda_o) \quad .$$

This is compared to $(\langle\frac{1}{r}\rangle_{osc})_{HF}$ in Fig.5-22. The agreement is as good as one could expect, but not better. Certainly, an improvement upon the semiclassical prediction for $\langle 1/r\rangle_{osc}$ is asked for. Possibly the subsequent terms in (247) are significant, and presumably some knowledge of $\zeta(Z,Z)$ is required. Nothing has been done along these lines so far.

Problems

5-1. Show that, more precisely than Eq.(9), Eq.(6) is solved by

$$y = (\tfrac{3}{2}N)^{1/3} + \tfrac{1}{12}(\tfrac{3}{2}N)^{-1/3} - \tfrac{1}{3}(\tfrac{1}{12})^3(\tfrac{3}{2}N)^{-5/3} + \tfrac{1}{3}(\tfrac{1}{12})^4(\tfrac{3}{2}N)^{-7/3} + \ldots \quad .$$

observe, in particular, that there is no term proportional to $N^{-3/3}$.

5-2. Show that a potential $V(r)$, which approaches $-Z/r$ for $r\to 0$, must be of the form "$Z^{4/3}$ times a function of $Z^{1/3}r$", if Eq.(17) holds (with κ independent of Z, of course).

5-3. The Hamilton operator of a particle moving along the x-axis is

$$H_{1D} = \tfrac{1}{2}p_x^2 + \sqrt{2}\,|x| \quad .$$

Show that the eigenvalues of H_{1D} are related to the zeros of $F_{-1}(y)$, given in (4-158). Find approximations to these eigenvalues both by employing the TF quantization (37) and by utilizing (4-170). What do you notice? (Incidentally, this produced Fig.7.)

5-4. Show that, for $r_2 > r_1 > 0$,

$$\int_{r_1}^{r_2}\frac{dr}{r}\sqrt{(r-r_1)(r_2-r)} = \pi\left(\frac{r_1+r_2}{2} - \sqrt{r_1 r_2}\right) \ ,$$

$$\int_{r_1}^{r_2} \frac{dr}{r}/\sqrt{(r-r_1)(r_2-r)} = \pi/\sqrt{r_1 r_2} \quad ,$$

$$\int_{r_1}^{r_2} dr\; r/\sqrt{(r-r_1)(r_2-r)} = \pi\,\frac{r_1+r_2}{2} \quad .$$

Use these integrals to confirm Eqs.(56), (115), and (122).

5-5. Use (84) in (97), in conjunction with the known result for the Coulomb potential, to derive (90).

5-6. Show that for the Tietz potential (4-222) one has, for $E=0$,

$$r_o = R, \quad \lambda_o = \sqrt{ZR/2} \quad , \quad \omega_o = \frac{1}{2}\sqrt{Z/R} \; , \quad \nu_o' = {}'\nu_o = 2, \quad \nu_o'' = 0.$$

Can you confirm the suspicion that $\nu_o(\lambda)$ is the straight line $\nu_o(\lambda) = 2(\lambda_o-\lambda)$?

5-7. Evaluate (92) for the neutral-atom TF potential and confirm (140).

5-8. The Bernoulli polynomials $B_m(x)$ are generated by

$$\frac{t\,e^{xt}}{e^t-1} = \sum_{m=0}^{\infty} B_m(x)\,\frac{t^m}{m!} \quad .$$

Show that Eq.(171) is equivalent to

$$\left(N_{osc}(E)\right)_{\ell TF} = 4\sum_{m=1}^{\infty} \frac{(-1)^{m+1}}{(m+1)!}\, B_{m+1}\left(\frac{1}{2} + \langle\lambda_E\rangle\right)\left(\frac{\partial}{\partial\lambda}\right)^m \left(\lambda\nu_E(\lambda)\right)\Big|_{\lambda=\lambda_E} \quad .$$

(Footnote 4 to Chapter Three may prove useful.)

5-9. Find a corresponding expression for $(-E_{osc})_{\ell TF}$.

5-10. Suppose E is such that for some j the number $j\nu_E'$ is an integer. Then the right-hand side of (188) is discontinuous for this E. By which amount? As a consequence of the discontinuity of $h(z)$ at $z=0$, there is also, for this E, a discontinuity of the right-hand side of (218). Show that it exactly cancels the one of (188).

5-11. If $j\nu_o' - k = \epsilon$, $|\epsilon| \ll 1$, then the sine function in (230) is particularly small and the first $\tilde{f}(z)$ is very large. Conclude that, for such a k,

$$\delta_k \cong \left(\frac{\lambda_o/Z^{1/3}}{\dot{\lambda}_o/Z^{-3/3}}\right) \frac{(-1)^k}{\pi k} \frac{(-1)^j}{\pi j} \left[\frac{1}{\pi k} \pm \frac{1}{\sqrt{2j\nu_o''}}\right] \quad ,$$

where the sign agrees with that of ϵ. Insert numbers and compute δ_{31} and δ_{62}.

<u>5-12.</u> Use (232) to infer that, for $\dot{z} \gg 1$,

$$\delta_k/\delta_1 \cong \begin{cases} (-1)^{k/2} \dfrac{20}{k^3} & \text{for} \quad k \text{ even} \quad , \\ (-1)^{(k-1)/2} \dfrac{1}{k^2} & \text{for} \quad k \text{ odd} \quad . \end{cases}$$

Insert this into (229) and arrive at

$$(-E_{osc})_\lambda \cong \frac{\lambda}{\dot{\lambda}_o} \frac{1}{\pi}\left(\frac{1}{4} - \left|\langle\lambda_o - \tfrac{1}{4}\rangle\right|\right)$$

$$+ \frac{\lambda_o}{\dot{\lambda}_o} \frac{5\pi}{3} \langle 2\lambda_o\rangle \left(\langle 2\lambda_o\rangle^2 - \tfrac{1}{4}\right) \quad .$$

Find the amplitude of these oscillations, and their period.

<u>5-13.</u> For the exact $\nu_o(\lambda)$, Eqs. (71) and (72) imply

$$4 \int_0^{\lambda_o} d\lambda \, \lambda \, \nu_o(\lambda) = Z \quad .$$

Show that the approximate $\nu_o(\lambda)$ of (189) does not obey this equation. Why does this not discredit the approximation?

Chapter Six

MISCELLANEA

In this final Chapter we shall briefly discuss a few topics without presenting thorough treatments. Keeping with the general theme of these lectures, we shall stick to ground state properties of atoms. There are, of course, lots of other applications of semiclassical methods. Maybe one should draw the reader's attention to two conference proceedings,[1] where many aspects of semiclassical approximations are dealt with.

We shall also not concern ourselves with applications of the models developed in the preceding Chapters. Instead, we shall focus on additional refinements and point to possible future developments.

Relativistic corrections. In Fig.5-3 we observed that in larger atoms relativistic effects contribute more to the total binding energy than the shell effects of Chapter Five. An extension of the theory to include relativistic corrections is, therefore, called for. Now, if we simply replace the kinetic energy $\frac{1}{2}p^2$ in the independent-particle Hamilton operator (2-1) by the relativistic expression, so that ($\alpha = 1/137.036$ is the fine structure constant)

$$H = \frac{1}{\alpha^2}(\sqrt{1+\alpha^2 p^2} - 1) + V(\vec{r}) \quad , \tag{6-1}$$

then we find in the TF limit the relations

$$\begin{aligned} -\frac{1}{4\pi}\nabla^2(V+\frac{Z}{r}) &= n \\ &= \frac{1}{3\pi^2}[-2(V+\zeta)]^{3/2}\left(1+\frac{\alpha^2}{4}[-2(V+\zeta)]\right)^{3/2} \end{aligned} \tag{6-2}$$

In particular, at small distances when $V \cong -Z/r$, the density is

$$n \cong \frac{1}{3\pi^2}\left(\frac{Z\alpha}{r}\right)^3 \quad \text{for} \quad r \to 0 \quad . \tag{6-3}$$

This density does not integrate to a finite number of electrons. We have arrived at complete nonsense.

Equation (2) was first obtained by Vallarta and Rosen in 1932, and by Jensen in 1933.[2] Ever since people have argued that the divergent behavior (3) is overcome by recognizing that the electrostatic energy with the nucleus is different from $-Z/r$ for $r \to 0$, because of the finite size of the nuclear charge distribution. We know better than that: the breakdown of the semiclassical approximation at small r (small on the atomic scale, but still enormous on the nuclear scale) requires a special treatment of the strongly bound electrons. In other words: relativistic effects can only be included into the description if they are accompanied by the Scott correction.

To illustrate this remark we compute the leading relativistic correction to the binding energy of a neutral atom. That is: to order $(Z\alpha)^2$. The relevant additional terms in the Hamilton operator are

$$\Delta_{rel} H = -\frac{1}{8}\alpha^2 p^4 + \frac{1}{8}\alpha^2 \nabla^2 V \quad , \tag{6-4}$$

of which the first one is the correction in (1) to order α^2 and the second one is the well-known Darwin term. As a matter of fact there is also the term representing spin-orbit coupling, but this results in a level splitting with no effect on the total energy, so that we need not take it into account.

After splitting $E_1(\zeta)$ as in Eq.(3-1) we can look at the various contributions separately. We begin with the effect of the p^4 term on E_s, the energy of the strongly bound electrons. Since the expectation value of p^4 in the m-th Bohr shell is, averaged over the angular quantum number, (see Problem 1)

$$\langle p^4 \rangle_m = 5(Z/m)^4 \quad , \tag{6-5}$$

the energy of a full Bohr shell is changed according to[3]

$$\begin{aligned} 2m^2\left(-\frac{Z}{2m^2}\right) &= -Z^2 \\ \to 2m^2\left(-\frac{Z^2}{2m^2}\right)&\left(1+(Z\alpha)^2\frac{5}{4m^2}\right) \\ &= -Z^2\left(1+(Z\alpha)^2\frac{5}{4m^2}\right) \quad . \end{aligned} \tag{6-6}$$

Consequently, when n_s Bohr shell are filled, we have

$$E_s = -Z^2\sum_{m=1}^{n_s-}\left(1+(Z\alpha)^2\frac{5}{4m^2}\right) \qquad =$$

$$= -Z^2 n_s - Z^2 (Z\alpha)^2 \frac{5}{4}\left(\sum_{m=1}^{\infty} \frac{1}{m^2} - \sum_{m=n_s+1}^{\infty} \frac{1}{m^2}\right) , \tag{6-7}$$

or

$$E_s = -Z^2 n_s - Z^2 (Z\alpha)^2 \frac{5}{4}\left(\frac{\pi^2}{6} - \sum_{m=n_s+1}^{\infty} \frac{1}{m^2}\right) . \tag{6-8}$$

The connection between n_s and the separating binding energy ζ_s is

$$\frac{Z^2}{2n_s^2}\left(1+(Z\alpha)^2 \frac{5}{4n_s^2}\right) > \zeta_s > \frac{Z^2}{2(n_s+1)^2}\left(1+(Z\alpha)^2 \frac{5}{4(n_s+1)^2}\right) , \tag{6-9}$$

which is the analog of Eq. (3-6). Thus $n_s = [\nu_s]$ with

$$\begin{aligned} \frac{Z^2}{2\nu_s^2}\left(1+(Z\alpha)^2 \frac{5}{4\nu_s^2}\right) &= \zeta_s , \\ \nu_s &= \frac{Z}{\sqrt{2\zeta_s}}\left(1+(Z\alpha)^2 \frac{5}{8} \frac{2\zeta_s}{Z^2}\right) , \end{aligned} \tag{6-10}$$

where we consistently discard terms of order $(Z\alpha)^4$. Now we exhibit the nonoscillatory terms in E_s,

$$\begin{aligned} E_s &= -Z^2\left\{\nu_s - \frac{1}{2} + (Z\alpha)^2 \frac{5}{4}\left(\frac{\pi^2}{6} - \frac{1}{\nu_s}\right)\right\} \\ &= -Z^2\left\{\frac{Z}{\sqrt{2\zeta_s}} - \frac{1}{2} + (Z\alpha)^2 \left(\frac{5\pi^2}{24} - \frac{5}{8} \frac{\sqrt{2\zeta_s}}{Z}\right)\right\} , \end{aligned} \tag{6-11}$$

which identifies the relativistic correction to E_s (without the contribution from the Darwin term) as

$$\Delta_{\text{rel}}^{(1)} E_s = -Z^2 (Z\alpha)^2 \left(\frac{5\pi^2}{24} - \frac{5}{8} \frac{\sqrt{2\zeta_s}}{Z}\right) . \tag{6-12}$$

The corresponding correction to $E_{\zeta\zeta_s}$ is

$$\begin{aligned} \Delta_{\text{rel}}^{(1)} E_{\zeta\zeta_s} = -\frac{\alpha^2}{56\pi^2} \int (d\vec{r}) \{ &[-2(V+\zeta)]^{7/2} - [-2(V+\zeta_s)]^{7/2} \\ &- 7(\zeta_s - \zeta)[-2(V+\zeta_s)]^{5/2} \} , \end{aligned} \tag{6-13}$$

where one should not fail to notice the typical strong cancellations for small r. We insert the neutral-atom TF potential along with $\zeta = 0$, whereby $V \cong -Z/r$ suffices in the terms referring to the strongly bound electrons, and arrive at

$$\Delta_{\text{rel}}^{(1)} E_{\zeta\zeta_s} = -\frac{3}{14a^2} Z^{5/3}(Z\alpha)^2 \int_0^\infty \frac{dx}{x^{3/2}}\{[F(x)]^{7/2} - (1-x/x_s)^{7/2} - \frac{7}{2}\frac{x}{x_s}(1-x/x_s)^{5/2}\} \quad (6\text{-}14)$$

with $x_s = Z^{4/3}/(a\zeta_s)$. A partial integration turns this into

$$\Delta_{\text{rel}}^{(1)} E_{\zeta\zeta_s} = -\frac{3}{2a^2} Z^{5/3}(Z\alpha)^2 \int_0^\infty dx\{x^{-1/2}[F(x)]^{5/2}F'(x) + \frac{5}{2}\frac{x^{1/2}}{x_s^2}(1-x/x_s)^{3/2}\} , \quad (6\text{-}15)$$

where the two contributions can be integrated separately. The individual results are

$$\int_0^\infty dx \frac{x^{1/2}}{x_s^2}(1-x/x_s)^{3/2} = \frac{\pi}{16} x_s^{-1/2} = Z^{1/3}\frac{a^2}{6}\frac{\sqrt{2\zeta_s}}{Z} , \quad (6\text{-}16)$$

and after using the differential equation obeyed by $F(x)$ followed by another partial integration,

$$\begin{aligned}\int_0^\infty dx\, x^{-1/2}[F(x)]^{5/2} F'(x) &= \int_0^\infty dx\, F(x)F'(x)F''(x) \\ &= -\frac{1}{2}\left(B^2 - \int_0^\infty dx[-F'(x)]^3\right) \quad (6\text{-}17) \\ &= -\frac{1}{2} \times 2.16864 \quad ;\end{aligned}$$

the relevant numbers can be found in Eq. (2-203) and Problem 2-3. Thus,

$$\Delta_{\text{rel}}^{(1)} E_{\zeta\zeta_s} = -Z^2(Z\alpha)^2\{\frac{5}{8}\frac{\sqrt{2\zeta_s}}{Z} - Z^{-1/3}\frac{3}{4a^2}\left(B^2 - \int_0^\infty dx[-F'(x)]^3\right)\} , \quad (6\text{-}18)$$

which combines with (12) to the ζ_s independent result

$$\Delta_{\text{rel}}^{(1)} E = -Z^2(Z\alpha)^2\{\frac{5\pi^2}{24} - Z^{-1/3}\frac{3}{4a^2}\left(B^2 - \int_0^\infty dx[-F'(x)]^3\right)\} . \quad (6\text{-}19)$$

Next we consider the contribution from the Darwin term. Since

$$\nabla^2 V = -\nabla^2 \frac{Z}{r} + \nabla^2 (V + \frac{Z}{r}) \\ = 4\pi Z\, \delta(\vec{r}) - 4\pi\, n(\vec{r}) \quad , \tag{6-20}$$

the induced energy change is

$$\Delta_{rel}^{(2)} E = \frac{\alpha^2}{8} \{4\pi Z\, n(\vec{r}=0) - 4\pi \int (d\vec{r}) [n(\vec{r})]^2\} \quad . \tag{6-21}$$

Here we need $n_o = n(\vec{r}=0)$ which we have available in Eq.(3-166),

$$n_o = \frac{(2Z)^3}{4\pi} (\sum_{m=1}^{\infty} (\frac{1}{m})^3 - \frac{B}{a} Z^{-2/3}) \quad . \tag{6-22}$$

The main contribution to the integral of the squared density is supplied by the strongly bound electrons. In this small-r regime, the density is of the form "Z^3 times a function of Zr." Consequently, the integral in (21) is (to leading order) proportional to Z^3, so that its contribution to the energy is of relative order $Z^{-3/3}$ as compared to the n_o term. At the present level of accuracy this is to be neglected. Thus

$$\Delta_{rel}^{(2)} E = Z^2 (Z\alpha)^2 \{\sum_{m=1}^{\infty} (\frac{1}{m})^3 - \frac{B}{a} Z^{-2/3}\} \quad , \tag{6-23}$$

which we add to (19) to obtain the relativistic binding-energy correction to order $(Z\alpha)^2$:

$$-\Delta_{rel} E = Z^2 (Z\alpha)^2 \Big\{ (\frac{5\pi^2}{24} - \sum_{m=1}^{\infty} (\frac{1}{m})^3) \\ - Z^{-1/3} \frac{3}{4a^2} (B^2 - \int_0^{\infty} dx [-F'(x)]^3) \\ + Z^{-2/3} \frac{B}{a} \Big\} \quad . \tag{6-24}$$

The numerical version hereof is

$$\frac{-\Delta_{rel} E}{\frac{1}{8} Z^2 (Z\alpha)^2} = 6.833 - 16.600\, Z^{-1/3} + 14.350\, Z^{-2/3} \quad . \tag{6-25}$$

As the natural unit of this relativistic energy correction we chose its amount for the one-electron ion,

$$\frac{1}{\alpha^2}\left(1-\sqrt{1-(Z\alpha)^2}\right)-\frac{1}{2}Z^2$$

$$=\frac{1}{8}Z^2(Z\alpha)^2\left(\frac{2}{1+\sqrt{1-(Z\alpha)^2}}\right)^2 \tag{6-26}$$

$$=\frac{1}{8}Z^2(Z\alpha)^2\left[1+O\left((Z\alpha)^2\right)\right] \quad ,$$

which is quite analogous to measuring the nonrelativistic binding energy in multiples of $\frac{1}{2}Z^2$.

The leading and the next-to-leading terms in (24), or (25), were first derived by Schwinger in 1980.[4] The third term, of relative order $Z^{-2/3}$, was found by Dmitrieva and Plindov in 1982.[5] We compare the semiclassical prediction (25) with the corresponding HF numbers[6] in Fig. 1. Except for very small values of Z, the agreement is marvelous. The relative deviation is less than 10; 1.0; 0.25% for $Z \geq 13$; 40; 54, respectively. The larger discrepancy for $Z \lesssim 10$ is not unexpected; it is

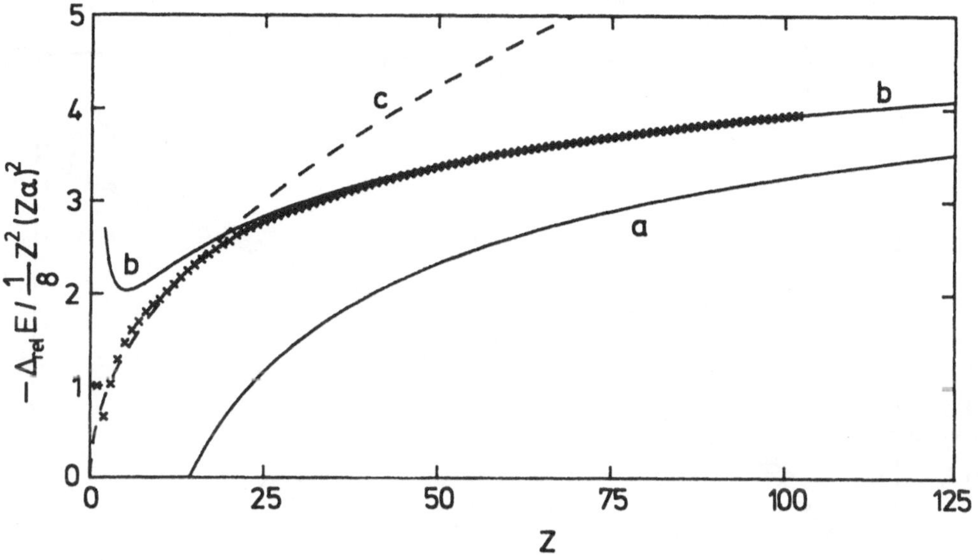

Fig.6-1. Comparison between Eq.(26) and corresponding HF data (crosses). Curve (a): leading plus next-to-leading term; curve (b): all three terms. The dashed curve (c) is Scott's estimate (27).

also nothing to worry about because for these small atoms the relativistic energy correction is practically negligible in the first place. In Fig.1 we have additionally displayed Scott's estimate of 1952,[7]

$$\frac{-\Delta_{rel}E}{\frac{1}{8}Z^2(Z\alpha)^2} \cong 0.6\, Z^{1/2} \quad , \tag{6-27}$$

which is surprisingly good for $Z \lesssim 20$. Unfortunately, Scott does not report how he arrived at (27).

The corrections of order $(Z\alpha)^4$ are much harder to come by and have not been calculated as yet. One of the difficulties lies in the iterated effect of the $(Z\alpha)^2$ corrections, inasmuch as the $(Z\alpha)^2$ corrections to the effective potential and the density produce $(Z\alpha)^4$ corrections to the energy when the expectation value of (4) is evaluated. In spite of these principal obstacles it is possible to estimate the $(Z\alpha)^4$ corrections to the binding energy by utilizing the physical insight that the lion's share is supplied by the strongly bound electrons, and their contribution does not differ much from the corresponding amount that one finds for noninteracting electrons. This way Dmitrieva and Plindov[5] obtained[8]

$$\frac{-\Delta_{rel}E}{\frac{1}{8}Z^2(Z\alpha)^2} \cong [\text{Eq.}(25)] + (Z\alpha)^2\left[2.248 - \frac{\pi^2}{3}\left(\frac{3}{2}Z\right)^{-2/3}\right]$$
$$= [\text{Eq.}(25)] + 1.197\times10^{-4}\, Z^2 - 1.337\times10^{-4}\, Z^{4/3} \ . \quad (6\text{-}28)$$

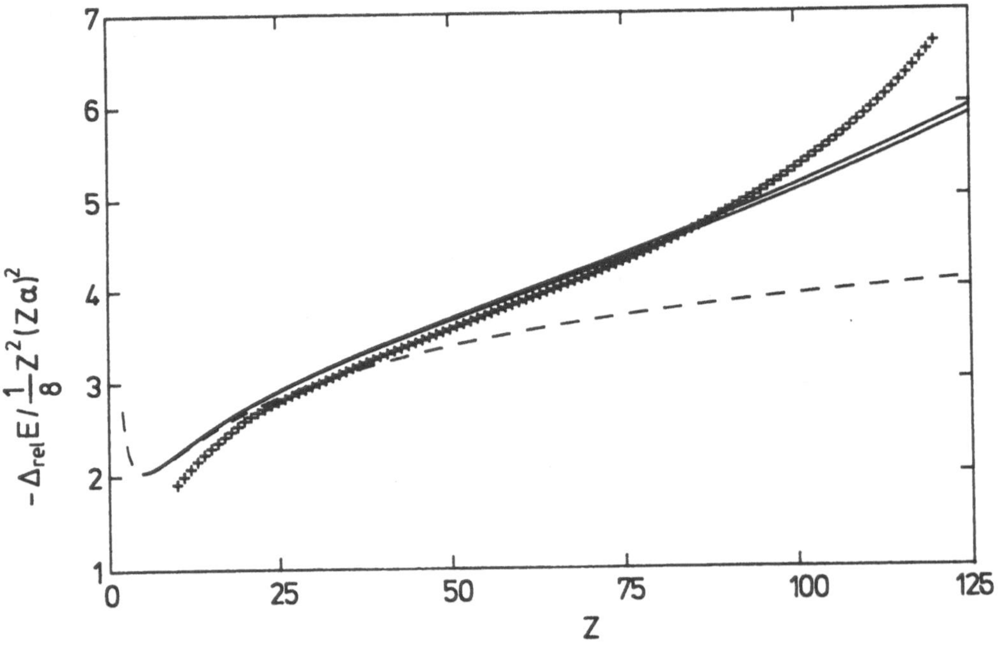

Fig.6-2. Comparison of the estimate (28) with corresponding HF data (crosses). The dashed curve represents (25), it is identical with curve (b) of Fig.1.

For the comparison with HF predictions we need HF data that include higher order corrections. It can be found in Ref.8 of Chapter Two. We see in Fig.2 that the estimate (28) is quite good for $25 \lesssim Z \lesssim 100$. For larger Z values the relativistic $(Z\alpha)^2$ expansion does not work and for small Z values the semiclassical $Z^{-1/3}$ series converges slowly. Further improvement can, in my opinion, only be achieved if no $(Z\alpha)^2$ expansion is performed at all. A possible starting point is the relativistic analog of (5-55), which - and this is essential for any relativistic treatment - has the correct energies of the strongly bound electrons built in from the beginning. This field remains to be tilled.

Kohn-Sham equations. The stationary property of the energy functional (2-434)

$$E(V,n,\zeta) = E_1(V+\zeta)-\int(d\vec{r}')(V-V_{ext})n+E_{ee}(n)-\zeta N \tag{6-29}$$

implies the set of coupled equations

$$\delta V: \quad n(\vec{r}') = \frac{\delta}{\delta V(\vec{r}')} E_1(V+\zeta) \quad , \tag{6-30a}$$

$$\delta n: \quad V(\vec{r}') = V_{ext}(\vec{r}') + \frac{\delta}{\delta n(\vec{r}')} E_{ee}(n) \quad , \tag{6-30b}$$

$$\delta\zeta: \quad N = \frac{\partial}{\partial\zeta} E_1(V+\zeta) \quad , \tag{6-30c}$$

of which the first and the third combine to

$$N = \int(d\vec{r}')n(\vec{r}') \quad , \tag{6-31}$$

expressing the normalization of the actual density. The simple structure of $E_1(V+\zeta)$,

$$E_1(V+\zeta) = \mathrm{tr}\left(\tfrac{1}{2}p^2+V+\zeta\right)\eta\left(-\tfrac{1}{2}p^2-V-\zeta\right) \quad , \tag{6-32}$$

has the consequence

$$n(\vec{r}') = 2 < \vec{r}' | \eta\left(-\tfrac{1}{2}p^2-V(\vec{r})-\zeta\right) | \vec{r}' > \quad , \tag{6-33}$$

which we have seen early in the game, in (2-20). Now one could think of evaluating (33) as in (2-21), that is: solve the effective Schrödinger equation

$$\left(\tfrac{1}{2}p^2+V(\vec{r})\right)|k\ ,\ \kappa > = |k\ ,\ \kappa > E_k \quad , \tag{6-34}$$

where κ stands for additional quantum numbers, and employ these eigenstates $|k,\kappa\rangle$ in writing

$$n(\vec{r}') = 2\sum_{k,\kappa} \langle\vec{r}'|k,\kappa\rangle\eta(-E_k-\zeta)\langle k,\kappa|\vec{r}'\rangle \tag{6-35}$$

$$= 2\sum_{k,\kappa} |\psi_{k,\kappa}(\vec{r}')|^2\eta(-E_k-\zeta) \quad ,$$

with the orbital wave functions

$$\psi_{k,\kappa}(\vec{r}') = \langle\vec{r}'|k,\kappa\rangle \quad . \tag{6-36}$$

In view of the normalization of these wave functions, Eq.(31) now appears as

$$N = 2\sum_{k,\kappa} \eta(-E_k-\zeta) \quad . \tag{6-37}$$

This determines ζ, once the E_k are known. Please note that because of the discreteness of the energy eigenvalues, no unique value is assigned to the minimal binding energy ζ. It is only determined within a (small) range, as we have stated already in the paragraph following Eq.(4-306).

The set of equations (30b), (34), (35), and (37) are the so-called Kohn-Sham (KS) equations.[9] In principle, they can be used to find both $n(\vec{r}')$ and $V(\vec{r}')$, and ζ. In practice, their usefulness is limited by our lack of knowledge about $E_{ee}(n)$.

In developing the HF method one has learnt how to solve Schrödinger equations like (34) numerically. It is, therefore, not difficult to find, with a high numerical precision, the density that corresponds via (33) to a given effective potential. But what is the point of this extreme accuracy, as long as the relation (30b) can only be explored approximately? In my opinion, the numerical precision achieved by the KS equations is both a luxury and a danger. For, it suggests an accuracy of the results obtained, which is only apparent because of the physical approximations that enter the $E_{ee}(n)$ functional used in (30b). In other words: the evaluation of $E_1(V+\zeta)$ need not, should not, and (I think) must not be more accurately than that of $E_{ee}(n)$. For example, if $E_{ee}(n)$ is approximated by the electrostatic energy only, the TFS version of $E_1(V+\zeta)$ suffices; if the Dirac-Jensen expression for the exchange energy is included into $E_{ee}(n)$, the consistent treatment of $E_1(V+\zeta)$ leads to the ES model.

Nevertheless, the KS equations are valuable, since they constitute an additional tool for studying $E_1(V+\zeta)$. Future investigations should, in my opinion, therefore focus on the analytical content of the KS equations. Maybe one can learn something about the unification of the quantum corrections of Chapter Four with the shell effects of Chapter Five, aiming at a refined model that contains both. Once obtained, this model will enable one to study, for instance, shell effects as they are manifested in the electron density.[10]

<u>Wigner's phase-space functions.</u> We have been mainly interested in the effective potential $V(\vec{r})$ and the spatial density $n(\vec{r}')$, which we used, for example, to compute diamagnetic susceptibilities. In other applications, however, knowledge of the spatial density does not suffice. Nonlocal quantities like the one-particle density matrix (2-422) appear, whenever the expectation value of a momentum dependent operator is asked for. The kinetic energy (2-42) is one example; another one is the Compton profile

$$\begin{aligned} J(Q) &= \left\langle \sum_{j=1}^{N} \delta(\vec{p}_j \cdot \vec{e}_z - Q) \right\rangle \\ &= \int \frac{(d\vec{r}')(d\vec{r}'')(d\vec{p}')}{(2\pi)^3}\, n^{(1)}(\vec{r}';\vec{r}'')\, e^{-i\vec{p}'\cdot(\vec{r}'-\vec{r}'')}\, \delta(p'_z - Q) \quad , \end{aligned} \tag{6-38}$$

which, in the situation of spherical symmetry, can be expressed equivalently by

$$J(Q) = \int \frac{(d\vec{r}')(d\vec{r}'')(d\vec{p}')}{(2\pi)^3}\, n^{(1)}(\vec{r}';\vec{r}'')\, e^{-i\vec{p}'\cdot(\vec{r}'-\vec{r}'')}\, \frac{1}{2p'}\eta(p'-Q) . \tag{6-39}$$

Similarly, the kinetic energy (2-421) equals

$$\begin{aligned} E_{kin} &= \left\langle \sum_{j=1}^{N} \frac{1}{2} p_j^2 \right\rangle \\ &= \int \frac{(d\vec{r}')(d\vec{r}'')(d\vec{p}')}{(2\pi)^3}\, n^{(1)}(\vec{r}';\vec{r}'')\, e^{-i\vec{p}'\cdot(\vec{r}'-\vec{r}'')}\, \frac{1}{2}p'^2 \end{aligned} \tag{6-40}$$

Quite obviously, we meet in (39) and (40) the density in momentum space

$$n(\vec{p}') = \int \frac{(d\vec{r}')(d\vec{r}'')}{(2\pi)^3}\, n^{(1)}(\vec{r}';\vec{r}'')\, e^{-i\vec{p}'\cdot(\vec{r}'-\vec{r}'')} \quad , \tag{6-41}$$

which we could have also obtained by starting with the momentum-space wave-function.

A change of integration variables turns (41) into

$$n(\vec{p}\,') = \frac{1}{(2\pi)^3}\int(d\vec{r}\,')\; n_W^{(1)}(\vec{r}\,',\vec{p}\,') \quad , \tag{6-42}$$

where

$$n_W^{(1)}(\vec{r}\,',\vec{p}\,') \equiv \int(d\vec{s})\,n^{(1)}(\vec{r}\,'+\tfrac{1}{2}\vec{s};\vec{r}\,'-\tfrac{1}{2}\vec{s})\; e^{-i\vec{p}\,'\cdot\vec{s}} \tag{6-43}$$

is the so-called Wigner function of the one-particle density.[11] The spatial density is obtained, in perfect analogy to (42), by

$$n(\vec{r}\,') = \frac{1}{(2\pi)^3}\int(d\vec{p}\,')\,n_W^{(1)}(\vec{r}\,',\vec{p}\,') \quad . \tag{6-44}$$

Given the Wigner function, one can evaluate the expectation value of any (one-particle) operator $F(\vec{r},\vec{p})$ by means of

$$\langle F(\vec{r},\vec{p})\rangle = \int\frac{(d\vec{r}\,')(d\vec{p}\,')}{(2\pi)^3}\; F_W(\vec{r}\,',\vec{p}\,')\,n_W^{(1)}(\vec{r}\,',\vec{p}\,') \;, \tag{6-45}$$

where $F_W(\vec{r}\,',\vec{p}\,')$ is the Wigner function of $F(\vec{r},\vec{p})$,

$$F_W(\vec{r}\,',\vec{p}\,') = \int(d\vec{s})\,\langle\vec{r}\,'+\tfrac{1}{2}\vec{s}\,|F(\vec{r},\vec{p})|\,\vec{r}\,'-\tfrac{1}{2}\vec{s}\rangle e^{-i\vec{p}\,'\cdot\vec{s}} \quad . \tag{6-46}$$

Equations (42) and (44) are special realizations of (45) with $F(\vec{r},\vec{p}) = \delta(\vec{p}-\vec{p}\,')$ in (42) and $F(\vec{r},\vec{p}) = \delta(\vec{r}-\vec{r}\,')$ in (44). Some properties of Wigner functions are the subject of Problems 3 to 7.

In Chapter Four, one of the central quantities was the time transformation function

$$\langle\vec{r}\,',t|\vec{r}\,'',0\rangle = \langle\vec{r}\,'|e^{-i(\frac{1}{2}p^2+V(\vec{r}))t}|\vec{r}\,''\rangle \quad , \tag{6-47}$$

for which we wrote $(\vec{r}\,'\to\vec{r}\,'+\frac{1}{2}\vec{s}\;,\;\vec{r}\,''\to\vec{r}\,'-\frac{1}{2}\vec{s})$

$$\langle\vec{r}\,'+\tfrac{1}{2}\vec{s},t|\vec{r}\,'-\tfrac{1}{2}\vec{s},0\rangle = \left(\frac{1}{2\pi iT}\right)^{3/2} e^{-i\Phi} \tag{6-48}$$

and found approximations for the phase Φ and the tyme T, both being functions of $\vec{r}\,',\vec{s}$, and t. Let us now find the corresponding approximation to the Wigner function of $\exp[-i(\frac{1}{2}p^2+V(\vec{r}))]$. In the TF limit we have

$$T \cong t, \quad \Phi \cong V(\vec{r}\,')t - \frac{s^2}{2t} \quad , \tag{6-49}$$

which result in

$$\left(e^{-i(\frac{1}{2}p^2+V(\vec{r}))t}\right)_W(\vec{r}\,',\vec{p}\,') \cong e^{-i(\frac{1}{2}p'^2+V(\vec{r}\,'))t} \quad ; \tag{6-50}$$

thus, the TF approximation simply replaces $\vec{r}$ and $\vec{p}$ by their eigenvalues. [To some extent, this was our starting point, see Eq.(1-43).] With the quantum corrections of Chapter Four we have, instead of (49), T and Φ given by Eqs.(4-45), (4-47), and (4-50), which produce

$$\left(e^{-i\left(\frac{1}{2}p^2+V(\vec{r})\right)t}\right)_W(\vec{r}\,',\vec{p}\,')$$

$$\cong \left(1+\frac{t^2}{12}\nabla'^2V(\vec{r}\,')\right)\left[\det\left(\overleftrightarrow{1}-\frac{t^2}{12}\vec{\nabla}'\vec{\nabla}'V(\vec{r}\,')\right)\right]^{-1/2}$$
$$\times\exp\left\{-i\left[\frac{1}{2}\vec{p}\,'\cdot\left(\overleftrightarrow{1}-\frac{t^2}{12}\vec{\nabla}'\vec{\nabla}'V(\vec{r}\,')\right)^{-1}\cdot\vec{p}\,' + V(\vec{r}\,')t+\frac{t^3}{24}\left(\vec{\nabla}'V(\vec{r}\,')\right)^2\right]\right\} \quad . \tag{6-51}$$

The machinery of Chapter Four (Airy averages, corrections for the strongly bound electrons, ...) can now be employed to derive the quantum corrected version of the TF approximation to $n_W^{(1)}(\vec{r}\,',\vec{p}\,')$, which is immediately available from (50):

$$n_W^{(1)}(\vec{r}\,',\vec{p}\,') \cong 2\,\eta\left(-\frac{1}{2}p'^2-V(\vec{r}\,')-\zeta\right) \quad . \tag{6-52}$$

Chapter Two deals with the implications of this equation. In contrast, the consequences of (51) are unexplored territory.

Problems

6-1. The expectation value of $1/r^2$ in a Bohr orbit with principal quantum number m and angular quantum number ℓ is

$$\left\langle\frac{1}{r^2}\right\rangle_{m,\ell} = \frac{Z^2}{m^3(\ell+1/2)} \quad .$$

Why? Use this, the Schrödinger equation, and the virial theorem to derive

$$\langle p^4\rangle_{m,\ell} = \frac{Z^4}{m^4}\left(\frac{4m}{\ell+1/2}-3\right) \quad .$$

Average over ℓ and arrive at Eq.(5). - For an alternative derivation employ the momentum space analog of (3-53), which is

$$|\psi_m|^2_{av}(p) = \frac{8}{\pi^2}\,\frac{(Z/m)^5}{[p^2+(Z/m)^2]^4} \quad .$$

6-2. The relativistic $(Z\alpha)^2$ expansion and the semiclassical $Z^{-1/3}$ expansion compete with each other, inasmuch as one should have $(Z\alpha)^2 \ll 1$ and $Z^{-1/3} \ll 1$ simultaneously. For which range of Z does one get a compromise? Compare with Fig.2.

6-3. Show that in one dimension, with position q and momentum p, the Wigner function of F(q,p) is given by

$$F_W(q',p') = \mathrm{tr}\, F(q,p)V(p-p';q-q')$$

$$= \mathrm{tr}\, F(q+q',p+p')V(p;q) \quad ,$$

where V(p;q) is the ordered exponential

$$V(p;q) = 2e^{2ip;q} = 2\sum_{k=0}^{\infty} \frac{(2i)^k}{k!}\, p^k q^k \quad .$$

What is the three dimensional analog?

6-4. Show that V(p;q) is hermitian, and that $[V(p;q)]^2 = 4$, that is: $\frac{1}{2}V(p;q)$ is unitary. Find the Wigner function of V(p;q) and interpret the result.

6-5. Show that $V(p;q) = V(p\cos\phi + q\sin\phi;\, q\cos\phi - p\sin\phi)$ for arbitrary (real) ϕ. Is there a corresponding property of the Wigner functions?

6-6. Show that for F(q) and G(p) one obtains $F_W(q') = F(q')$ and $G_W(p') = G(p')$. Find a F(q,p) such that $F_W(q',p') = q'p'$.

6-7. Find the Wigner function of the one dimensional Hamilton operator (5-54).

FOOTNOTES

Chapter One.

1 The outstanding textbook on TF theory is still P. Gombás, Die statistische Theorie des Atoms und ihre Anwendungen (Springer, Wien, 1949). Then there is the review article N.H.March, Adv.Phys.6,1 (1957). A more recent introductory text is N.H.March, Self-consistent Fields in Atoms (Pergamon, Oxford, 1975).

2 L.H.Thomas, Proc.Cambridge Phil.Soc. 23, 542 (1926); E.Fermi, Rend. Lincei 6, 602 (1927); D.R.Hartree, Proc.Cambridge Phil.Soc. 24, 89 (1928); V.Fock, Zschr.f.Phys. 61, 126 (1930).

3 In the form of a textbook this material has been presented by W. Thirring, Lehrbuch der Mathematischen Physik, Vol.4: Quantenmechanik großer Systeme (Springer, Wien-New York, 1980). A recent review is E.H.Lieb, Rev.Mod.Phys.53, 603 (1981). A treatment with emphasis on thermal properties is given by J.Messer, Temperature Dependent Thomas-Fermi Theory (Lecture Notes in Physics, Vol.147) (Springer, Berlin-Heidelberg-New York, 1981).

4 Models of screened Bohr atoms have been studied to some extent by R.Shakeshaft and L.Spruch, Phys.Rev. A 23, 2118 (1981). Their emphasis is on the oscillatory terms, about which we shall have to say something in Chapter Five.

5 This statement is frequently called the Hellmann-Feynman theorem. Both Hellmann (1933) and Feynman (1939), however, only rediscovered what had been known before. It is, indeed, difficult to imagine how quantum mechanics could have been developed without such a central tool. The theorem appears explicitly in Pauli's review of 1933, in Van Vleck's book of 1932, and in a paper by Güttinger in 1931. The latter contains, to my knowledge, the explicit statement for the first time. The various references are: P.Güttinger, Zschr.f.Phys. 73, 169(1931); J.H. Van Vleck, The Theory of Electric and Magnetic Susceptibilities (Oxford University Press, 1932);

W.Pauli, Handbuch der Physik, Vol.24,Part I, p.83, edited by H. Geiger and K.Scheel (Springer, Berlin, 1933); H. Hellmann, Zschr. f.Phys. 85, 180 (1933); R.P.Feynman, Phys.Rev. 56, 340 (1939).

Chapter Two.

1 This is written for the situation of an isolated atom. For applications to molecules and solids, the term $-Z/r$ has to be replaced by the respective external electrostatic potential. Or, if the atom is part of a gas, an additional term is needed to describe the pressure exerted by the other atoms. If the formalism is being applied to isolated self-binding systems of fermions, like a nucleus with its strong interactions, or a neutron star held together by gravity, the fermion-fermion interaction part is the entire effective potential.

2 The potential is not physically unique, because there is the freedom of adding a numerical constant. In writing Eq.(1), we have opted for the usual normalization: $V \rightarrow o$ for $r \rightarrow \infty$, which means that this constant is set equal to zero.

3 Since the potential is subject to the usual normalization $V(r \rightarrow \infty) = o$, δV has to vanish at infinity. This does, however, not affect the argument, because it suffices to set $\delta V(\vec{r}) = -\delta\zeta$ for those $\vec{r}$ for which the density is nonzero.

4 The vacuum is not essential; one could, with little additional complications, equally well consider a dielectric surrounding.

5 Many applications of this and other, related stationary principles can be found in J.Schwinger's (unfortunately still unpublished) lecture notes on Electromagnetic Theory (University of California, Los Angeles, 1975 ... 1984).

6 Strictly speaking, the first equality holds only for nonvanishing density, since $n = o$ implies no more than $V + \zeta \geq o$ in Eq.(51). This subtlety does not affect the argument, however.

7 In rewriting E_2, one has to make use of the identity (the spherical symmetry is essential here)

$$[\vec{\nabla}(V+\frac{Z}{r})]^2 = [\frac{d}{dr}(V+\frac{Z}{r})]^2$$

$$= \frac{1}{r^2}[\frac{d}{dr}(r(V+\frac{Z}{r}))]^2 - \frac{1}{r^2}\frac{d}{dr}[r(V+\frac{Z}{r})^2] \quad .$$

The latter term equals

$$-\vec{\nabla}\cdot[\frac{\vec{r}}{r^2}(V+\frac{Z}{r})^2] \quad ;$$

it is a divergence that integrates to a null result. Consequently,

$$E_2 = -\frac{1}{8\pi}\int(d\vec{r})\,[\vec{\nabla}(V+\frac{Z}{r})]^2 = -\frac{1}{2}\int_0^\infty dr[(rV)]^2 \quad ,$$

which then leads to the second integral in (108).

8 J.P.Desclaux, At.Data Nucl.Data Tables 12, 311 (1973).

9 E.Baker, Phys.Rev.36, 630 (1930).

10 E. Hille, J. d'analyse Math. 23, 147 (1970) showed rigourously that the series (164) converges for small values of $\sqrt{x}$, certainly for $\sqrt{x} < (108/3125)^{1/4}/\sqrt{B+2} = 0.227$, possibly for somewhat larger values.

11 A. Sommerfeld, Zschr. f. Phys. 78, 283 (1932).

12 Some people call this the Coulson-March expansion [C.A.Coulson and N.H.March, Proc.Phys.Soc.London A 63, 367 (1950)].

13 Hille (see Footnote 10) proved that the series (193) converges for sufficiently large values of x.

14 Such a computer program was realized the first time by S.Kobayashi, T.Matsukuma, S.Nagai, and K.Umeda, J.Phys.Soc.Japan 10, 759 (1955). They truncated the expansion (193) after the k=17 term. It is funny to ovserve that their coefficients $c_1 \dots c_{10}$ are correct whereas $c_{11} \dots c_{17}$ are wrong (with increasing error). This did, however, not affect their results as far as the values of B and β, given in the Abstract, are concerned. Also the decimales in their table of the TF function are correct.

[15] The usefulness of this change of variables was first noticed by G.I.Plindov and I.K.Dmitrieva, Dokl.Akad.Nauk BSSR 19, 788 (1975).

[16] A little bit more detail is reported in B.-G. Englert, Phys.Rev. A 33, 2146 (1986).

[17] E.Fermi, Mem.Accad.d'Italia 1, 149 (1930). Fermi did not go beyond the first order. A systematic study of the expansion (316) was first presented by G.I.Plindov and I.K.Dmitrieva, Dokl.Akad.Nauk BSSR 21, 209 (1977).

[18] G.I.Plindov and I.K.Dmitrieva, J.Phys. (Paris) 38, 1061 (1977).

[19] B.-G.Englert, Zschr.f.Naturforschung 42a, 825 (1987).

[20] This (physically rather obvious) statement has become known as the Lieb-Simon theorem after a formal proof was given by E.H.Lieb and B.Simon, Phys.Rev.Lett.31, 681 (1973). For more detail see the review by Lieb cited in Footnote 3 of Chapter One.

[21] Cited in Footnote 2 of Chapter One.

[22] Cited in Footnote 1 of Chapter One.

[23] P. Hohenberg and W. Kohn, Phys.Rev. 136, B 864 (1964).

[24] M. Levy and J.P. Perdew, Phys.Rev.A 32, 2010 (1985).

[25] Cited in Footnote 2 of Chapter One.

[26] Actually, (502), is a variant of Hartree's equations, inasmuch as the self energy is included and no averaging of $V(\vec{r}\,')$ over its angular dependence is performed, which is a reasonable procedure in the situation of a spherically symmetric V_{ext}.

[27] E.H. Lieb, Rev.Mod.Phys. 48, 553 (1976).

[28] The corresponding $\tilde{V}(r) = -(Z/r)\tilde{F}(x)$ is known as the Tietz potential [T.Tietz, Acta Phys.Hung.9, 73 (1958)].

Chapter Three.

1 J.M.S. Scott, Philos. Mag. 43, 859(1952).

2 J. Schwinger, Phys. Rev. A 22, 1827 (1980).

3 The additive constant that should, in principle, be included into the Coulomb potential, $V(r) \cong -Z/r + \text{const.}$, can be regarded as being part of ζ_s and ζ, respectively. Nothing is gained by displaying this constant explicitly, but the algebra is more transparent if one is not forced to keep track of this term, which for the present discussion is irrelevant anyhow.

4 These (and all corresponding) polynomials in <y> are closely related to the Bernoulli polynomials (Jakob Bernoulli, 1689). For example,

$$B_1(x+\tfrac{1}{2}) = x \quad ,$$

$$B_2(x+\tfrac{1}{2}) = x^2 - \tfrac{1}{12} \quad ,$$

and

$$B_3(x+\tfrac{1}{2}) = x^3 - \tfrac{1}{4}x \quad .$$

I owe this remark to Prof. G. Süßmann.

5 Cited in Footnote 1 of Chapter One.

6 In Lieb's review of 1981 (cited in Footnote 3 of Chapter One) there is the statement that "the Scott correction (...) is very plausible," but "has not yet been proved." This article was a contribution to a conference at Erice in June, 1980, which is a couple of months before Schwinger's paper appeared in print (November, 1980).

7 B.-G. Englert and J. Schwinger, Phys. Rev. A 29, 2331 (1984). This paper has been the victim of absent-minded proofreading. Misprints that I am aware of are:
(1) in Eq. (24), V^2 should read: ∇^2 ;
(2) the left-hand side of Eq. (40) should read: $-\frac{1}{4\pi}\nabla^2 V$;
(3) in the first paragraph of the section "Scaling" read "scaling property" instead of "rescaling property;"
(4) in the second sentence of the same paragraph read "do two things for us" instead of "do the two things for us;"

(5) Eq.(63) should read: $B = 1.5880710\ldots$;

(6) after Eq.(73) read "as Z scales to $\lambda^{\beta-1}$ Z" instead of "as r scales to $\lambda^{\beta-1}$ r;"

(7) in the sum over j in Eq.(87) replace ζ_i by ζ_j ;

(8) in Eq.(88) replace "$\frac{b}{a} + Z^{4/3}$" by "$\frac{b}{a}\, Z^{4/3}$;"

(9) the right-hand side of Eq.(102) should read: $\frac{1}{3}\, Z^{-4/3}$;

(10) the last number in the first line of Table I should read: (-0.04).

[8] In Ref.7, there is the (wrong) statement that this [i.e., here Eq.(89), there Eq.(101)] is only approximately true — a misunderstanding caused by confusing the different meanings of Z in Eqs. (92) and (101).

[9] If a_1 and a_2 are, indeed, constants, the closed expression is

$$E_{TFS}(Z,N) = E_{TF}(Z,N-\frac{a_2}{6Z}) + \frac{1}{2}\, Z^2 - \frac{1}{4}\, a_1 Z \quad ,$$

which can be regarded as evidence in favor of the notion $a_2 \geq 0$.

[10] This result has also been found, independently and almost simultaneously, by Bander whose argument is reminiscent of Scott's way of reasoning, and by Dmitrieva and Plindov who make an educated guess. The references are M. Bander, Ann. Phys. (NY) 144, 1 (1982); I.K. Dmitrieva and G.I. Plindov, J. Phys. (Paris) 43, 1599 (1982).

[11] The HF predictions have been compiled on the basis of the orbital parameters given by S. Fraga, J. Karwowski, and K.M.S. Saxena, Handbook of Atomic Data (Elsevier, Amsterdam, 1976).

[12] This is the main obstacle. The second-order TFS model has not been formulated as yet.

[13] For more detail, consult Ref.7.

Chapter Four.

1 Consult, for instance, R.Finkelstein, Nonrelativistic Mechanics (Benjamin, Reading, Massachusetts, 1973), Chapter 3.

2 Of course, some such "wrong answers" are still better than others. For example, taking into account all contributions to Φ and T that are linear in V, improves the approximation significantly over the TF result of Eq.(19). In particular, the density at r=o turns out to be finite - but it does not have the correct value. For detail see R.K.Bhaduri, M.Brack, H.Gräf, and P.Schuck, J.Phys.Lett. (Paris) 41, L 347 (1980).

3 The derivation of (50) and (70) by R.K.Bhaduri, Phys.Rev.Lett. 39, 329 (1977) is, indeed, an expansion in powers of t (in $\beta=it$ to be precise, but no matter). Bhaduri does not consider the $\vec{s}$ dependence of Φ.

4 M.Durand, M.Brack, and P.Schuck, Zschr.f.Phys. A286, 381 (1978) arrive at Eqs.(45), (47), and (50) by "expanding in powers of $\hbar$" and in powers of $\vec{s}$.

5 E.Wigner, Phys.Rev. 40, 749 (1932); J.G.Kirkwood, Phys.Rev. 44, 31 (1933).

6 A standard reference is H.A.Antosiewicz, in Handbook of Mathematical Functions, edited by M.Abramowitz and I.Stegun (Dover, New York, 1972).

7 J.Schwinger, Phys.Rev. A 24, 2353 (1981).

8 B.-G.Englert and J.Schwinger, Phys.Rev. A 29, 2339 (1984).

9 C.F.von Weizsäcker, Zschr.f.Phys. 96, 431 (1935).

10 This statement is not entirely true, since a deficiency of von Weizsäcker's approach caused his result to be too large by a factor of nine. This, unfortunately, has induced people to consider that numerical factor as an adjustable parameter. (The "optimal" coefficient is then believed to be about 1/40 instead of 1/72.) I do not see the slightest justification for such a point of view. - To my knowledge the correct numerical multiple of $(\vec{\nabla}n)^2/n$ was first found

by A.Kompaneets and E.Pavlovskii, Zh.Eksp.Teor.Fiz. 31, 427 (1956) [Sov.Phys. - JETP 4, 328 (1957)], whose method is very different from the one used in the text. A procedure more closely related to ours is the one employed by D.Kirzhnits, Zh.Eksp.Teor.Fiz. 32, 115 (1957) [Sov.Phys. - JETP 5, 64 (1957)].

11 G.I.Plindov and I.K.Dmitrieva, Phys.Lett. 64A, 348 (1978), obtain a finite answer by introducing properly chosen cut-offs at small and large distances. They succeed in deriving (103) at the price of an electron cloud reminiscent of "an apricot without a stone," and of a wrong numerical coefficient for the Scott correction. Their procedure is therefore hardly convincing, although the correct $Z^{5/3}$ term emerges.

12 Unfortunately, some (independent) earlier attempts of developing a quantum corrected description got either stuck or misled because the investigators did not succeed in evaluating the Airy integrals explicitly. See, in particular, R.Baltin, Zschr.f.Naturforschung A 27, 1176 (1972), and Ref.4.

13 Of some historical significance is the so-called "Amaldi correction" which aims at improving the original TF model by a rough guess of the electrostatic self-energy [E. Fermi and E. Amaldi, Mem. Acc. Ital. 6, 117 (1934)]. The arguments in favor of the Amaldi correction are not very strong in the first place (since certainly the Scott correction is more important), and they collapse totally as soon as exchange is included. It is, therefore, depressing to see people still handle models which contain both the exchange energy and the Amaldi correction.

14 Cited in Footnote 2 of Chapter One.

15 E. A. Milne, Proc. Cambridge Philos. Soc. 23, 794 (1927).

16 F. Rasetti, in: E. Fermi, Collected Papers (E. Amaldi et al., eds., Univ. of Chicago Press, 1962), p.277.

17 Cited in Footnote 9 of Chapter Two.

18 V. Bush and S.H. Caldwell, Phys. Rev. 38, 1898 (1931).

[19] Cited in Footnote 14 of Chapter Two.

[20] Cited in Footnote 1 of Chapter Three.

[21] Cited in Footnote 1 of Chapter One.

[22] Cited in Footnote 2 of Chapter Three.

[23] See Footnote 7 of Chapter Three.

[24] P.A.M. Dirac, Proc. Cambridge Philos. Soc. 26, 376 (1930).

[25] Ref. 16, pp. 291-304.

[26] H. Jensen, Zschr. f. Phys. 89, 713 (1934).

[27] I.K. Dmitrieva and G.I. Plindov, cited in Footnote 10 of Chapter Three.

[28] Cited in Footnote 11 of Chapter Three.

[29] See Ref. 19 of Chapter Two.

[30] Experimental ionization energies are tabulated by A.A. Radzig and B.M. Smirnov, Reference Data on Atoms, Molecules, and Ions (Springer, Berlin-Heidelberg, 1985) (Springer Series in Chemical Physics, Vol.31), for Z=1 to Z=102, except for Z=85 (astatine) and Z=87 (francium) for which no spectroscopic data is reported. The stars for these elements in Fig.8 are the predictions of Ref. 19 of Chapter Two.

[31] An illustration of this remark are the utterly wrong coefficients, corresponding to our Eq.(289), which are reported by S.H. Hill, P.J. Grout, and N.H. March, J. Phys. B 20, 11 (1987), in the appendix.

[32] See Fig.20 in Gombás' textbook, cited in Footnote 1 of Chapter One.

[33] I.K. Dmitrieva and G.I. Plindov, J. Phys. (Paris) 45, 85 (1984) give Padé approximants that interpolate between (302) and (303), and agree with these two limiting forms under the respective cir-

cumstances.

34 R.G. Parr, R.A. Donelly, M. Levy, and W.E. Palke, J. Chem. Phys. 68, 3801 (1978).

35 W.E. Lamb, Phys. Rev. 60, 817 (1941).

36 The subscript $_m$ is non-standard; it is introduced only to exclude any confusion with $\sigma = (7 + \sqrt{73})/2$, defined in Eq. (2-309).

37 For $Z \leq 86$ the HF numbers were compiled from the data given by C. Froese-Fischer, The Hartree-Fock Method for Atoms (Wiley, New York, 1977). For $Z > 86$, the (less precise) numbers of Ref. 11 of Chapter Three were used.

38 B.-G. Englert and J. Schwinger, Phys. Rev. A 26, 2322 (1982).

39 The potential $\overline{V}$ in Eq.(13) of Ref. 38 equals $U - U_{ex} + \zeta$; it is thus an electrostatic pseudo-potential. With this identification, (13) of Ref.38 is equivalent to Eq.(342) in the text.

40 The external potential $-Z/r$ has to be replaced by a sum over the Coulomb potentials of all the nuclei in the molecule; also the Scott term $\frac{1}{2}Z^2$ becomes a sum over the contributions from individual nuclei. The level of sophistication in performing the CSBE in Eq. (331b) depends on the particular application. Further, if one is interested in the dependence of the energy on the parameters that specify the configuration of the nuclei, the electrostatic energy due to the Coulomb repulsion between the nuclei must be included. Effects of the finite nuclear masses are very small and certainly irrelevant at this level of accuracy.

41 As a matter of fact, what is discussed in the text is not the "usual argument" but a variant of it. The discussion is both simpler and more transparent this way.

42 Handbook of Chemistry and Physics (Chemical Rubber Co., Cleveland, Ohio, 1979/80).

43 F. Hoare and G. Brindley, Proc. Roy. Soc. London 159A, 395 (1937).

[44] Sommerfeld's approximation to F(x), that is $F(x) \cong [1+(12^{-2/3}x)^{\gamma}]^{-3/\gamma}$, gives 8.71 for this integral. When inserted after undoing the twofold partial integration, it produces 8.67. Both numbers are in satisfactory agreement with the actual value.

[45] From Ref.11 of Chapter Three.

[46] In Ref.24 of Chapter Two, Levy and Perdew try to put the blame for the discrepancy between experiment and the HF predictions entirely onto the experiments (or the experimentalists). This does not seem plausible to me. Also, relativistic corrections cannot account for differences this large.

[47] B.-G. Englert and J. Schwinger, Phys. Rev. A29, 2353 (1984).

[48] Another application is reported by A. Mañanes and E. Santos, Phys. Rev. B 34, 5874 (1986). Unfortunately, these authors confuse the potentials U and U_{es} (our denotation) with, luckily, no consequences as far as the conclusions of the paper are concerned. The appendix, however, and all related remarks in the text are erroneous.

[49] This modified TF density (or its one-dimensional analog) has been derived prior to the publication of Ref.8 for the special situation of a linear potential, when (394) is the whole answer. I am aware of the following three papers: W. Kohn and L.J. Sham, Phys. Rev.137, A1697 (1965); S.F. Timashev, Elektrokhimia 40, 730 (1979); H. Gräf, Nucl. Phys. A349, 349 (1980). All these authors squared the wave function in a linear potential [this is essentially an Airy function, see Eq.(143)] to arrive, finally, at (394) for that special potential. Our derivation is more general in not making such assumptions about V.

[50] The subscript $_p$ is non-standard; it is introduced only to exclude any confusion with $\alpha = 1.04018...$ of Eq.(2-313).

[51] See Radzig and Smirnov, cited in Footnote 30.

[52] The evaluation of the integral (140) and its y-derivative for x=0 is a simple exercise in performing complex contour integrals. At worst, the results can be looked up in Ref.6.

[53] The HF density is compiled from D.R. Hartree, Proc.Roy.Soc. London,

Ser. A 151, 96 (1935).

[54] L.L. DeRaad and J. Schwinger, Phys. Rev. A 25, 2399 (1982).

Chapter Five.

[1] The first plot of this kind is contained in the paper by Dmitrieva and Plindov, cited in Footnote 10 to Chapter Three.

[2] To avoid a possible misunderstanding: it is not the numerical accuracy of HF numbers that is questioned here, but the physical reliability of the HF approximation.

[3] C.E. Moore, Atomic Energy Levels, Natl. Bur. Stand. Ref. Data Ser., Natl. Bur. Stand. (U.S.) Circ. No. 35 (U.S. GPO, Washington, D.C., 1970).

[4] J.P. Desclaux, cited in Footnote 8 to Chapter Two.

[5] J.M.S. Scott, cited in Footnote 1 to Chapter Three.

[6] H.A. Kramers, Zschr. f. Phys. 39, 836 (1926).

[7] R.E. Langer, Phys. Rev. 51, 669 (1937).

[8] N.H. March and J.S. Plaskett, Proc. R. Soc. London, Ser. A235, 419 (1956) already noticed that the semiclassical sum contains the TF approximation in the continuum limit. However, their method is very different from the one discussed in this text, and they did not develop a systematic way of analyzing these quantum corrections.

[9] B.-G. Englert and J. Schwinger, Phys. Rev. A32, 26 (1985).

[10] The circular orbit is also the one which, for given angular momentum, has least energy.

[11] Presumably due to too crude an approximation for F(x), Fermi reported 3.2 (instead of 3.916) for this integral in his classical paper on the systematics of the Periodic Table. The reference is:

E. Fermi, Rend. Lincei 7, 342 (1928).

12 B.-G. Englert and J. Schwinger, Phys. Rev. A 32, 36 (1985)

13 H. Hellmann, Acta Physicochim. URSS 4, 225 (1936), was the first to consider the ℓTF model. Since he used the original sum over ℓ, not the Fourier formulation, Hellmann failed to recognize that one can split the energy into E_{TF} plus the quantum correction E_{qu}.

14 A simple counting of the respective points on a j,k lattice shows that this fraction equals $[\arctan(\nu_o') - \pi/4]/\pi = 0.098$, a little bit less than ten percent.

15 The factor k-j is missing in Eqs.(107) and (108) of the original publication [B.-G. Englert and J. Schwinger, Phys. Rev. A32, 47 (1986)]. Fortunately, this inadvertence did not cause any harm.

16 A standard reference is M. Abramowitz in the Handbook of Mathematical Functions, cited in Footnote 6 of Chapter Four.

17 Another way of plotting the Fresnel functions, and a particularly charming one, is presented in Fig.16 of F. Lösch, Tafeln höherer Funktionen/Tables of Higher Functions (7-th edition, Teubner, Stuttgart, 1966).

18 See the paper cited in Footnote 15.

Chapter Six.

1 Density Functional Methods in Physics (Alcabideche/Portugal, 1983), edited by R.M. Dreizler and J. da Providência (Plenum Press, New York, 1985); Semiclassical Methods in Nuclear Physics (Grenoble/France, 1984), edited by R.W. Hasse, R. Arvieu, and P. Schuck, J. de Physique 45, Coll. C 6.

2 M.S. Vallarta and N. Rosen, Phys. Rev. 41, 708 (1932); H. Jensen, Zschr. f. Phys. 82, 794 (1933).

3 Footnote 3 of Chapter Three applies here as well.

[4] J. Schwinger, cited in Footnote 2 of Chapter Three.

[5] I.K. Dmitrieva and G.I. Plindov, cited in Footnote 10 of Chapter Three.

[6] From Ref.11 of Chapter Three.

[7] J.M.S. Scott, cited in Footnote 1 of Chapter Three.

[8] The number appearing here is

$$2.248 = 4\sum_{n=1}^{\infty}\frac{1}{n^5} - 19\sum_{n=1}^{\infty}\frac{1}{n^4} + 4\sum_{n=1}^{\infty}\sum_{m=1}^{n}\frac{1}{n^3}\frac{1}{m^2} + 12\sum_{n=1}^{\infty}\sum_{m=1}^{n}\frac{1}{n^4}\frac{1}{m}$$

$$= 4\times 1.036928 - 19\times\frac{\pi^4}{90} + 4\times 1.265738 + 12\times 1.133479 \quad .$$

[9] W. Kohn and L.J. Sham, Phys. Rev. 140, A1133 (1965)

[10] D.A. Kirzhnits and G.V. Shpatakovskaya, Zh. Eksp. Teor. Fiz. 62, 2082 (1972) [Sov. Phys. - JETP 35, 1088 (1972)] used methods related of those of Chapter Five to study shell effects in atomic densities. Their results are encouraging, but hardly satisfactory.

[11] E.P. Wigner, cited in Footnote 5 of Chapter Four. We do not include the factors of 2π into the definition of the Wigner function. - A recent review of this subject is M. Hillary, R.F. O'Connell, M.O. Scully, and E.P. Wigner, Phys. Rep. 106, 121 (1984).

INDEX

A superscript at the page number signifies that the respective name or subject is not mentioned in the main text itself, but in a footnote to the indicated page.

Names

Abramowitz, M., 348[16]
Airy, G.B., 190
Amaldi, E., 225[13]
Antosiewicz, H.A., 190[6]
Arvieu, R., 370[1]
Avogadro, A., 257

Baker, E., 57,231
Baltin, R., 199[12]
Bander, M., 165[10]
Bernoulli, J., 136[4]
Bhaduri, R.K., 187[2,3]
Bohr, N., 3,231
Brack, M., 187[2],188[4],199[12]
Brindley, G., 257[43]
Bush, V., 231

Caldwell, S.H., 231
Clausius, R.J.E., 263
Compton, A.H., 379
de Coulomb, C.-A., 1
Coulson, C.A., 62[12]

Darwin, C.G., 371
Desclaux, J.P., 56[8],298[4]
Dirac, P.A.M., 173,175,231f,245
Dmitrieva, I.K., 74[15],83[17],98[18], 165[10],196[11],231,234[27],240[33], 297[1],375f
Donelly, R.A., 241[34]
Dreizler, R.M., 370[1]
Durand, M., 188[4],199[12]

Emden, R., 230
Euler, L., 26,50

Fermi, E., 2,83[17],104,225[13], 230f,324[11]
Feynman, R.P., 23[5]
Finkelstein, R., 181[1]
Fock, V., 2
Fourier, J., 24,312
Fraga, S., 166[11],235[28],243[37], 259[45],264[45],375[6]
Fresnel, A.J., 348
Froese-Fischer, C., 243[37]

Gauß, C.F., 36
Gombás, P., 1[1],104,229,239[32]
Gräf, H., 187[2],261[49]
Grout, P.J., 239[31]
Güttinger, P., 23[5]

Hartree, D.R., 2,119,289[53]
Hasse, R.W., 370[1]
Heaviside, O., 7
Hellmann, H., 23[5],338[13]
Hill, S.H., 239[31]
Hillary, M., 380[11]
Hille, E., 60[10],64[13]
Hoare, F., 257[43]
Hohenberg, P., 104[23]

Jensen, H., 231f,245,371

Karwowski, J., 166[11],235[28],243[37], 259[45],264[45],375[6]
Kirkwood, J.G., 189[5]
Kirzhnits, D., 196[10],379[10]
Kobayashi, S., 65[14],231
Kohn, W., 104[23],261[49],378
Kompaneets, A., 196[10]
Kramers, H.A., 231,311
Kronecker, L., 58

Lamb, W.E., 241
Langer, R.E., 311
Levy, M., 112,115,241[34],259[46]
Lieb, E.H., 4[3],101[20],124,139[6]
Lösch, F., 349[17]

Mañanes, A., 260[48]
March, N.H., 1[1],62[12],139,231,239[31], 315[8]
Matsukuma, T., 65[14]
Messer, J., 4[3]
Milne, E.A., 230
Moore, C.E., 298[3]
Mosotti, O.F., 263

Nagai, S., 65[14]

O'Connell, R.F., 380[11]

Palke, W.E., 241[34]
Parr, R.G., 241[34]
Pauli, W., 23[5]

Pavlovskii, E., 196[10]
Perdew, J.P., 112,115,259[46]
Plaskett, J.S., 315[8]
Plindov, G.I., 74[15],83[17],89[18], 165[10],196[11],231,234[17],240[33], 297[1],375f
Poisson, S.D., 27,173
da Providência, J., 370[1]

DeRaad, L.L., 294[54]
Radzig, A.A., 238[30],264[51]
Rasetti, F. 231[17]
Riemann, B., 165
Rosen, N., 371

Santos, E., 260[48]
Saxena, K.M.S., 166[11],235[28], 243[37],259[45],264[45],375[6]
Schrödinger, E., 105
Schuck, P., 187[2],188[4],199[12],370[1]
Schwinger, J., 41[5],130,139,143[7], 193,193[8],196,217[8],231,249[38], 252[8],254[38],256[38],260[47],278[47], 281[47,8],289[47],294[54],317[9], 324[9],335[12],346[15],358[18],375
Scott, J.M.S., 130f,138f,173,231,298, 375f
Scully, M.O., 380[11]
Shakeshaft, R., 12[4]
Sham, L.J., 261[49],378
Shpatakovskaya, G.V., 379[10]
Simon, B., 101[20]
Slater, J.C., 223
Smirnov, B.M., 238[30],264[51]
Sommerfeld, A., 61[11],124,259[44]
Spruch, L., 12[4]
Süßmann, G., 136[4]

Thirring, W., 4[3]
Thomas, L.H., 2,104,230f
Tietz, T., 126[28]
Timashev, S.F., 261[49]

Umeda, K., 65[14]

Vallarta, M.S., 371
Van Vleck, J.H., 23[5]

von Weizsäcker, C.F., 175,193f,196, 232
Wigner, E., 189[5],380

Subjects

a, 36
actinides, 329
Ai(x), 190f,197,201,203ff,244, 260,270,281
Airy average(s), 191,199ff,311, 381
Airy's function *see* Ai(x)
alkaline metals, 238
Amaldi correction, 225[13]
argon, 260
astatine, 238[30]
Avogadro's number, 257

B, 38,49f,53f,65
Baker's constant *see* B
Baker's series, 57,60,60[10]
Bernoulli polynomials, 136[4],368
binding energy(-ies)
  of Bohr atoms, 6,300
  of Bohr atoms with shielding, 13,21
  of neutral HF atoms, 56
  of neutral TF atoms, 38,56,66
  of HF ions, 234
  of TF ions, 78,81,83,96
  of neutral TFS atoms, 138
  of neutral SM atoms, 230
  of SM ions, 232ff
  experimental, 297f
  relativistic correction, 374,376
binding-energy oscillations, 297ff
  double-peak structure, 297,341, 361ff
Bohr atoms, 4ff,132ff,300ff,335
  with shielding, 11ff
  size, 21ff,26
Bohr radius, 3,256
Bohr shell(s), 4ff,132,138ff,169, 220,287,304,371
Bohr-shell artifacts, 136,143,336, 210
de Broglie wavelength, 11,100,182,187

c_k, 63,88f,92,235
c_k, 356ff
centrifugal barrier, 311
cesium, 288f
chemical potential, 239ff
Clausius-Mosotti formula, 263
closed-shell atoms, 256,286f,299
Compton profile, 379
correlation energy, 225,298,361
Coulson-March expansion, 62[12],64,64[13]
count of states *see* N(E)
CSBE, 245

d_k, 89f,93f
Darwin term, 371ff
degeneracy, 299f
  in the TF potential, 323ff
  linear $\sim$, 335f
  lines of $\sim$, 313,315ff,335,338
density (of electrons), 27,29ff,105, 121
  momentum space $\sim$, 379f

at the nucleus, 147,164ff,363f
pseudo ~, 246
quantum corrected ~, 216
radial ~, 101,220f,288ff
shells effects in ~, 379
TF ~, 35,102ff
TFS ~, 147

density functional(s), 104ff,172, 196
of the energy (general), 106
of the exchange energy, 227f
of the interaction energy, 106,221ff,245
of the kinetic energy, 106, 129,194
minimum property, 46,106,112
of the TF energy, 44

density matrix
one-particle ~, 107,222ff,379
two-particle ~, 222ff

density operator, 108,222
diamagnetic susceptibilities, 225f,281,379
dielectric constant, 263f
dipole moment, induced electric ~, 263
double counting of pairs, 17,31

$\tilde{E}$, 228,232
E_{osc}, 296,301f,304f,334,338
ℓTF prediction, 340ff,345ff
λ,ν contribution to ~, 346f
λ contribution to ~, 354,356
semiclassical prediction for ~, 360ff

E_s, 131ff,211ff,371ff
$E_{\zeta\zeta_j}$, $E_{\zeta\zeta_s}$, 131,134,144,157,177f, 192f,198,210ff,336,372f
E_1, 28ff,108ff
discrete sums for ~, 306
quantum corrections to ~, 193,210,335
TF version, 34
TFS version, 130,141

e(q), 71,78,80,86,97,98
$\tilde{e}$(q), 232ff
edge of the atom, 36,251f,260ff, 268,270,274f,280,283,291f
electric polarizability, 263f, 266ff,275,294
electronegativity, 241
ES model, 249ff,270,292,378
exchange energy, 16,162,167,175f, 193,221ff,245,262,270,338
Dirac-Jensen approximation, 228,261,276,378
self ~, 16,225

exchange interaction, 16,32,119, 338
expectation value of 1/r, 22f, 241ff,365ff

F(x), 38,57,63ff,126,259,323ff
F_m(y), 199ff,206f,282,293f
$F_m(V,|\vec{\nabla}V|)$, 211ff,278f,282
f(z),$\tilde{f}$(z), 349f,354,356
f_k(x), 83,88
f_q(x), 37,83,85,90,232f
Fermi momentum, 34
Fermi statistics, 3
field, electrostatic, 33,36,263ff
fine structure constant, 124,242,370
francium, 238
Fresnel integral, 348ff

G(y), 61ff
g(z), $\tilde{g}$(z), 349f,354
Gauss's law, 36
Gaussian notation, 5
gradient corrections,176

h_q, 90ff,235
$\hbar$, "expanding in powers of ~," 4,188
Hamilton operator
effective one-dimensional ~, 309
independent-particle ~, 27,241, 370
many-particle ~, 2,104,119
single-particle ~, 7,123

Hartree energy, 122
Hartree equations, 124
Hartree-Fock ~ *see* HF ~
Hartree's method, relation to TF approximation, 119ff
Hellmann-Feynman theorem, 23[5]
HF, 1,2,119
HF predictions
binding energy, 55f, 234f
density at the nucleus, 166
mean square radius, 259
polarization radius, 264
relativistic energy correction, 298,375
shielding of nuclear magnetic moment, 244

Hohenberg-Kohn theorem, 104,106, 107,222
hydrogen, 297

independent-particle energy, 16ff,28
inert gases *see* closed-shell atoms
inhomogeneity corrections, 176,194
interaction energy, 31,104,106,111ff, 120,221ff,245
ionization energy, 162,236ff,298,364

J, 144,168ff,214ff

kinetic energy, 44,104,106,111ff, 117,119,129,222,379
Kohn-Sham equations, 377ff
krypton, 221

Lagrangian multiplier, 44,107,121
lanthanides, 329
Lieb-Simon theorem, 101[20]

local oscillator approximation, 183
ℓTF model, 338ff

MES model, 272ff
mercury, 164,170ff,243
MIT Differential Analyzer, 231

N, 2
N(ζ), N(E), 28,35,312ff,329ff
$N_{osc}(E)$, 339,344,351ff,358f
$N_{qu}(E)$, 314,329,332ff,337ff, 343ff,348,350
N_s, 133,211
$n(\vec{r})$ *see* density
ñ, 147,161,215ff
n_{IME}, 147,214,281
n_s, 132,211,372
n_o, 164ff,282f
NIE, 4ff,11,379
noninteracting electrons *see* NIE
nuclear magnetic moment, shielding of ∿, 241ff,(365ff)

orbits, classical ∿, 320ff
ordered operator, 10f,382

Periodic Table, 2,298f,327ff
phase, 180,183ff,226,380
Poisson equation, 27,31,35,36, 168,193,265,273,275,278
Poisson's identity, 173,313
polarization radius, 264
potassium, 288f
potential(s)
  atomic ∿, 317f
  effective ∿, 14,16,27f,31,104, 108,122,149,280f,285f,306
  effective vector ∿, 124
  electrostatic ∿, 27,31,337
  electrostatic pseudo ∿, 250, 268ff
  exchange ∿, 245,263,275ff, 286,291f,337
  external ∿, 104ff,116,149, 153,265
  interaction ∿, 31
  pseudo ∿, 247
  pseudo exchange ∿, 247,273
potential functional(s), 104ff, 196,337
  in electrostatics, 41
  of the ES energy, 253
  maximum property, 40
  of the TF energy, 34
  of the TFS energy, 145
potential-density functional(s), 110
  of the ES model, 248
  of the MES model, 272
  of the TF model, 110
  of the TFD model, 249

Q_j,146,168ff,213ff,281
q, 37
quantum corrections, 11,175ff,209, 314,335
  to the count of states, 314
  to exchange, 275
  to the time transformation function, 177ff
  to traces, 191f
quantum number(s), 30,299,378
  angular ∿, 25,140,299,306,312, 327ff,371,381
  magnetic ∿, 140,306
  principal ∿, 5,25,132,140,169, 381

r_E,r_o, 315ff,319ff,325ff
radium, 328
radius, mean-square ∿, 256ff
relativistic corrections, 259[46],298, 370ff
rubidium, 288ff
Rydberg energy, 3

$\vec{s}$, 181,183ff
s_k, 57ff
$\mathcal{s}_k$, 354ff,369
scale
  atomic ∿, 3,371
  nuclear ∿, 371
  TF ∿, 131,153
scaling properties
  of the external potential, 116
  of the interaction-energy density functional, 111ff
  of the kinetic-energy density functional, 111f
  of the E_1 potential functional, 115ff
  of the TF potential functional, 66ff
  of the TFS potential functional, 155ff
Schrödinger equation, 105,112,122, 181,306,308,377
Scott correction, 131,136,139,143, 145,153,155,158,162,163,172,186, 209,225[13],240,277,335f
screening, inner-shell ∿, 12,13
self energy, 14ff,20,225
shell(s)
  closed ∿, 299
  closed ∿ of NIE, 7,305
  filling of ∿, 7,299,302,329
  partly filled ∿, 8,304
shell effects, 4,7,135,167,238,258, 295fff,379
size, atomic ∿, 21ff,26
Slater determinant, 223,225
SM *see* statistical model

spectrum
  semiclassical ~, 189
  quantum corrected ~, 190f
statistical model, 228
statistical model predictions, 287ff
  binding energy, 230,234
  density at the nucleus, 283
  exchange potential, 291
  ionization energy, 237
  mean square radius, 260
  radial densities, 288ff
  shielding of nuclear magnetic moment, 244
strongly bound electrons, 124, 130-172,186f,194,196,278, 292,299,311,316,326,335f, 371ff,376f,381

$T(\vec{r}',\vec{s},t)$ *see* tyme
TF, 1,2,7
TF
  degeneracy, 323ff
  density, 35,102ff,193
  equation, 35,37
  function *see* $F(x)$ and $f_q(x)$
  model, 33ff,99ff,207ff
  quantization, 308,311,322
  variables, 46
TF predictions
  binding energy, 55,66,78,96ff
  mean square radius, 259
TFD model, 175,249,251f,255,292
TFS, 130
TFS
  density, 147
  model, 130ff
TFS predictions
  binding energy, 138
  density at the nucleus, 165
Thomas-Fermi ~ *see* TF ~
Thomas-Fermi-Dirac ~ *see* TFD ~
Thomas-Fermi-Scott ~ *see* TFS ~
Thomas-Fermi-Scott-Weizsäcker-Dirac model, 228
Thomas-Fermi-von Weizsäcker model, 175
Tietz potential, 126[28],220,368
time transformation function, 177ff,225,293,380
traces
  and phase-space integrals, 10f
  quantum corrected ~, 191,198
  semiclassical evaluation of ~, 9,11,178
  variation of ~, 29
tyme, 180,183ff,226,380

units, atomic, 2,3,4,9,124
uranium, 299

$V(\vec{r})$ *see* potential(s), effective ~
virial theorem(s), 68,70,112,140, 163

w_j, 144,168ff,214ff
von Weizsäcker term, 194ff,196[10]
Wigner functions, 379ff
Wigner-Kirkwood expansion, 189
WKB quantization *see* TF quantization

$x_o(q)$, 37,78,86,98

y, y_s, y_j, 199ff,210ff,282,284,292
$y(x)$, 62

Z, 2
z, z_s, 197ff

α, 82

β, 62,65

γ, 61,62

ζ, 28,239ff,378
ζ_j, ζ_s, 131,142,168ff,210ff,281, 335f,372f

$\eta(x)$, 7,304

Λ, 80ff,127ff
λ, λ_E, λ_O, 303,312,314ff,323,325ff, 329,340
$\lambda(q)$, 74f,78,87
λ oscillations, 350ff
λ,ν oscillations, 344,351

$\nu,\nu_E(\lambda),\nu_O(\lambda)$, 307,312f,316,318ff, 322ff,329,332,344
ν_E', ν_O', ν_E'', ν_O'', 317,320,325f,343, 345
$'\nu_E$, $'\nu_O$, 318f,324,343,345
ν_s, 133ff,372
ν oscillations, 351,358ff

ρ_s, 140,150,156,211

σ, 82

$\Phi(t)$, 81f,89,236
$\Phi(\vec{r}',\vec{s},t)$ *see* phase
$\phi_k(t)$, 77,127,233
$\phi_\lambda(t)$, 74ff,232f

$\psi_m(t)$, 90,236

$|\psi_n|^2_{av}$, 142,148,150,156,214,381

ω_E, ω_O, 316f,319f,325ff

[y], <y>, 5

$\langle f(x)\rangle^\circ$, 191,197ff